Scientific Societies
in the
United States

Scientific Societies in the United States

Ralph S. Bates

THIRD EDITION

The M.I.T. Press
Massachusetts Institute of Technology
Cambridge, Massachusetts

Library of Congress Catalog Card Number: 57-10143
Manufactured in the United States of America

PREFACE TO THE THIRD EDITION

This book deals with the role of scientific societies in American life from the days of witchcraft to the age of spacecraft. Seven years have elapsed since the publication of the second edition of *Scientific Societies in the United States*. When that edition appeared, the world had just entered the space age. Now, cheap atomic power is rapidly becoming a reality, rockets speed to the planets, men circle the globe and race for the moon.

In this new era, the lone scientist has given place to the research team. Government, industry, and university are now frequently found as a troika, yoked together for the accomplishment of science projects. It is here that the scientific society plays a most useful role as coordinator. Multinational undertakings of global scope often involve a variety of international scientific unions and national academies of science. Thus, both the march of time and a new approach have made it necessary to add a new chapter to the lengthening history of American scientific societies.

While the author has been concerned primarily with updating the book, he has also made minor revisions and corrections in the earlier chapters and has added many items to the bibliography.

The inclusion of *Scientific Societies in the United States* in the science section of the White House Library selected for the late President John F. Kennedy would seem indicative of the need for continuing research into the various American agencies of scientific advancement, including the scientific societies, as the "endless frontier" of science plays its part in "The New Frontier" and "The Great Society."

A new feature of this edition is the inclusion of "A Chronology of Science in the United States."

My heartfelt thanks are hereby acknowledged to the M.I.T. Press for making possible this new edition.

RALPH S. BATES

MASSACHUSETTS STATE COLLEGE AT
  BRIDGEWATER
July 1, 1965

PREFACE TO THE SECOND EDITION

Twelve years have elapsed since the publication of the first edition of *Scientific Societies in the United States,* and the book has been out of print now for several years. Continued calls for the book and the need to bring the story up to date have led to the preparation of a new edition. Accordingly, the text of the first edition has been supplemented by a new chapter (Chapter VI) in order to note the many changes that have occurred since the close of World War II in scientific societies. At the same time the bibliography has been revised completely and enlarged by the addition of many recent items.

Concerning the material of the new chapter (Chapter VI) a few words may be said. The past decade has served more than ever to bring into focus the basic need for international organizations—political, economic, social, intellectual, and scientific. Much of the new leadership in coordinating the activities of existing national and international scientific organizations has centered in the United Nations, especially in that agency of it known as UNESCO (United Nations Educational, Scientific and Cultural Organization).

Then, too, science and scientific organizations within the United States have new rallying points—the National Science Foundation, the Department of Health, Education, and Welfare, the National Health Institutes, the Atomic Energy Commission, the Office of Defense Mobilization, the agencies promoting the International Geophysical Year, and the Earth Satellite program—to mention only a few examples.

The author wishes to express his indebtedness to Mr. F. G. Fassett, Jr., of the Technology Press of the Massachusetts Institute of Technology, and to Mr. Henry H. Wiggins, Manager of the Publications Department of Columbia University Press, for making publication of this new edition possible; to Mr. William F. Bernhardt, of the editorial staff of Columbia University Press; and also to the staffs of many libraries for assistance—especially to those of the Library of Congress, the Widener Library of Harvard University, the library of the Massachusetts Institute of Technology, and the library of the Massachusetts State Teachers College at Bridgewater.

MASSACHUSETTS STATE TEACHERS   RALPH S. BATES
 COLLEGE AT BRIDGEWATER
October 15, 1957

PREFACE TO THE FIRST EDITION

The history and the influence of the scientific societies of the United States are the themes dealt with in this book. More than two centuries have elapsed since 1727, when Benjamin Franklin's Junto laid the foundations of the American Philosophical Society, the oldest scientific society now in existence in America. Since that time, the American Academy of Arts and Sciences, the Smithsonian Institution, the American Association for the Advancement of Science, the National Academy of Sciences, the National Research Council, and the American Council of Learned Societies have all arisen as new generations of American scholars have marched forward under the banner of science. Then, too, academies of science have come into existence in almost every state in the union; and, now, municipal scientific societies are flourishing in practically all our principal cities. National and local associations have been formed for the pursuit of the many specialized branches of science, from anthropology to zoology. As hitherto no extensive account of the history and work of American scientific societies has appeared, perhaps this book on the subject will help to fill a gap in the literature dealing with the intellectual history of our country.

There is a second and a more immediate need for a book at this time dealing with scientific societies. The past few years have taught the nation even more than did World War I the value of such societies in mustering out a roll of American men of science for war. This phase of national mobilization has necessitated up-to-the-minute revision of many of the later pages of this book in order to keep it abreast of the changes occurring in our national scientific set-up as World War II has progressed. It is hoped that this book looks toward the future, as well as back to the past and at the present, for it now appears likely that at least a skeleton roster of the nation's scientific personnel will be maintained into the post-war period, and that it will function in part through the scientific societies. It should be borne in mind also that international scientific organization, which partly collapsed both in the last war and then again under the rise of dictatorships and the eclipse of the League of Nations, will doubtless experience a revival, particularly in the United Nations, at the close of World War II.

In the preparation of this book, constant use was made of such primary materials as the proceedings, transactions, and other publications of some hundreds of national, state, and local societies. Numerous secondary sources of information were also consulted, including handbooks, histories of the societies, biographies, memoirs, and scientific articles. It has been necessary for me to be in frequent consultation with scientists in many of the specialized fields of learning, without whose assistance a book of this kind would not be possible.

Although my material has been gathered from almost every part of the United States, I am under particular indebtedness to the resources of the following libraries: the Widener Memorial Library at Harvard University, the Boston Public Library, the New York Public Library, the Library of Congress, the John Crearer Library in Chicago, Cornell University Library, the University of Michigan Library, and the Massachusetts Institute of Technology Library.

Since, unfortunately, there is considerable discrepancy in the literature as to which men are to be regarded as the founders of some of the societies, my policy has been to sift the evidence and to give principal credit to those who I believe were most actively engaged in bringing the societies into actual existence. An even more serious problem is the rather wide variation in the published dates for the founding of many of the societies. Some sources give the dates of the first informal organization, others of the formal organization or the adoption of a constitution; still others give the date of incorporation or chartering. My policy has generally been to consider as the date of founding the first effective organization of the particular body in question as determined by a search of the minutes and transactions of the societies, or, in the case of recently formed organizations, to use the dates furnished by correspondence with the secretaries of the societies. Where neither method has proved practicable, I have relied on the dates given in *Bulletin 106* of the National Research Council, *Handbook of Scientific and Technical Societies and Institutions of the United States and Canada* (Washington, 1942).

I must here express a grateful word of acknowledgment to Professor Arthur Meier Schlesinger of Harvard University, my doctoral mentor of a decade ago, who first mentioned to me the possibility of venturing into this unexplored field, and who has made many fruitful suggestions at every stage of the investigation. My heartfelt thanks also go to Mr. J. R. Killian, Jr., and Mr. Frederick G. Fassett, Jr., of the

Technology Press, Massachusetts Institute of Technology, for their assistance in bringing the book into print. I must also thank the American Council of Learned Societies and the Technology Press for financial assistance which has made possible the publication of this book.

R. S. B.

MASSACHUSETTS INSTITUTE OF TECHNOLOGY
January, 1945

CONTENTS

Scientific Societies
in the
United States

Chapter I

SCIENTIFIC SOCIETIES IN EIGHTEENTH CENTURY AMERICA

Scientific societies are among the outstanding agencies for increasing and diffusing the world's store of knowledge. Originating in Renaissance Italy in the sixteenth century, spreading across Europe in the seventeenth, and over the American continents in the eighteenth and nineteenth centuries, and now, in the twentieth century, rapidly invading the cities of Asia, these organizations of scientific men have circled the globe. Especially have scientific societies multiplied rapidly in the United States; hence, the story of the rise of the societies to prominence and their stimulation of scientific advancement in America should be of particular interest to the student of cultural history.

It has long been traditional to trace the lineage of all subsequent learned and scientific societies back to Plato's famous Academy. The distinction of having been the first scientific society in the modern world, however, is generally accorded to the Academia Secretorum Naturae, established at Naples in 1560. As the movement got under way in the seventeenth century, at a time when the foundations of British America were being laid, many notable academies of science were being formed in Europe, including the Accademia dei Lincei at Rome in 1603, the Academia Naturae Curiosum at Leipzig in 1651, the Accademia del Cimento at Florence in 1657, the Royal Society of London in 1662, the Académie des Sciences at Paris in 1666, the Accademia delle Scienze at Bologna in 1690, and the Societas Regia Scientiarum at Berlin in 1700.[1] In the course of the eighteenth century similar academies were established at Edinburgh, Dublin, Brussels, Bordeaux, Lyons, Orléans, Marseilles, Lisbon, Prague, Munich, Göttingen, Danzig, and St. Petersburg, and in countless other European cities.

---

[1] Seventeenth century scientific organizations are especially treated in Martha Ornstein, *The Rôle of Scientific Societies in the Seventeenth Century* (3d ed., Chicago, 1938); in Harcourt Brown, *Scientific Organizations in Seventeenth Century France* (History of Science Society Publications, n. s., V, Baltimore, 1934); and in Abraham Wolf, edit., *A History of Science, Technology, and Philosophy in the 16th and 17th Centuries* (New York, 1935), ch. IV, "Scientific Academies."

In England, Francis Bacon, in his fable, *New Atlantis,* advocated a "Solomon's House" whose learned Fellows would make inquiries into every branch of human knowledge. The idea took root in contemporary English literary and scientific circles, and finally bore fruit in a club which grew into the Royal Society of London. The Royal Society originated according to Wallis about 1645, when "divers worthy persons, inquisitive into natural philosophy, and other parts of human learning; and particularly of what hath been called the New Philosophy or Experimental Philosophy" met weekly in London to "discourse and consider of philosophical enquiries, and such as related thereunto; as Physics, Anatomy, Geometry, Astronomy, Navigation, Staticks, Magneticks, Chymicks, Mechanics, and Natural Experiments."[2]

The Royal Society, chartered in 1662, was the principal scientific society to serve the American colonists down to the Revolutionary War. It is interesting to note that the embryo Royal Society nearly migrated to Connecticut at one time, since it is recorded that John Winthrop

> . . . was one of those, who first formed the Plan of the *Royal Society;* and had not the Civil Wars happily ended as they did, Mr. Boyle[3] and Dr. Wilkins with several other learned men, would have left England, and, out of esteem for the most excellent and valuable Governor, John Winthrop the Younger, would have retir'd to his new-born Colony, and there have established that *Society for Promoting Natural Knowledge,* which these gentlemen had formed, as it were, in embryo among themselves; but which afterwards receiving the protection of King Charles II obtained the Style of Royal. . . .[4]

The proposed migration did not take place; but " . . . by a letter from Mr. Oldenburg, written by order of the Royal Society, Mr. Winthrop was in a particular manner invited to take upon him the charge of being the chief correspondent of the Royal Society in the West as Sir Philiberto Vernatti was in the East Indies."[5] He furnished the society with numerous scientific papers. Elected to membership in the ranks of the Royal Society were the following colonists: Cotton Mather,[6] the

---

[2] "Dr. Wallis's Account of Some Passages of His Own Life," Publisher's Appendix to Preface of *Peter Langtoft's Chronicle,* Thomas Hearne, edit., cited in C. R. Weld, *History of the Royal Society* (2 vols., London, 1848) , I, 30-31.

[3] Letters from Mr. Boyle, Dr. Wilkins, Sir K. Digby, etc., to Mr. Winthrop.

[4] Mortimer Cromwell, Secretary to the Royal Society, "to the Honorable John Winthrop, Esq." Dedication of vol. XL (1737-1738) of the *Philosophical Transactions* (London, 1741), b_1b_2.

[5] *Philosophical Transactions* (London, 1741) , b_2.

[6] For an account of his mooted election, see G. L. Kittredge, "Cotton Mather's Election into the Royal Society," Colonial Society of Massachusetts, *Publications,* XIV (1913) , 81-114; and "Further Notes on Cotton Mather and the Royal Society," *ibid.,* 291-292.

three John Winthrops, James Bowdoin, Paul Dudley, Roger Williams, and Zabdiel Boylston of New England; Benjamin Franklin, David Rittenhouse, and John Morgan of Pennsylvania; William Byrd II, John Mitchell, and John Tennent of Virginia; and Alexander Garden of South Carolina. Many Americans, including notably Benjamin Franklin, contributed to the *Philosophical Transactions,* the journal which the Royal Society began to publish in 1756.[7]

The first learned and scientific society to be organized in the American colonies was probably the Boston Philosophical Society. Its existence has been recorded by Increase Mather, the leader of the group of New Englanders who undertook its establishment. He writes, "In 1683 I promoted a design for a private philosophical society in Boston, which I hope may have laid the foundation for that which will be for future edification."[8] It was later described by Cotton Mather as "A Philosophical Society of Agreeable Gentlemen, who met once a Fortnight for a Conference upon Improvements in Philosophy and Additions to the stores of Natural History."[9] He tells that the famous Dutch scholar Wolferdus Senguerdius made use of materials collected by the society and deposited in its proceedings.[10] Moreover, Cotton Mather, referring apparently to himself, says: "One that had a share in that combination . . . now a fellow of the Royal Society in London, afterwards transmitted communications thither."[11] These communications were based on observations made by members of the society.

The Boston Philosophical Society seems soon to have expired. Boston had become involved in the political and theological disputations,

---

[7] The scientific work and communications of the American colonists are ably discussed at length in F. E. Brasch, "The Royal Society of London and Its Influence upon Scientific Thought in the American Colonies," *Scientific Monthly,* XXXIII (1931), 337-355, 448-449. For the contributions of Cotton Mather, see especially G. L. Kittredge, "Cotton Mather's Scientific Communications to the Royal Society," American Antiquarian Society, *Proceedings,* n. s., XXVI (1916), pt. 1, 18-57. See also S. E. Morison, *The Puritan Pronaos; Studies in the Intellectual Life of New England in the Seventeenth Century* (New York, 1936); Michael Kraus, "Scientific Relations between Europe and America in the Eighteenth Century," *Scientific Monthly,* LV (1942), 259-272; and Bernard Fäy, "Learned Societies in Europe and America in the Eighteenth Century," *American Historical Review,* XXXVII (1932), 255-266.

[8] Increase Mather, *Autobiography,* MS; in the possession of the American Antiquarian Society.

[9] Cotton Mather, *Parentator, Memoirs of Remarkables in the Life and Death of the Ever-Memorable Dr. Increase Mather* (Boston, 1724), 86.

[10] "From which the learned Wolferdus Senguerdius, a professor at Leyden had some of the materials, wherewith his Philosophia Naturalis was Enriched," *ibid.*

[11] *Ibid.*

spoken of in those days as "the calamity of the times";[12] and the study of science was but a side line in the busy life of Increase Mather, the society's prime mover, who, moreover, presently became absorbed in clerical controversies. It is not remarkable, therefore, that the society perished. It is more remarkable rather that such an organization appeared in America at such an early date[13] and that it was capable of doing valuable work for a time. Its existence may be taken as evidence of the "keenness of interest in science in the colonies"[14] in seventeenth century America.

As the eighteenth century came on, the conditions grew somewhat more favorable for the study of science in America. There were, first of all, more wealth and more leisure, and there were more sizable urban communities where those of like interest in scientific pursuits congregated. Then, too, the Newtonian philosophy gradually won acceptance as copies of the *Principia* found their way to American colleges in increasing numbers.[15] By the middle of the century the writings of the French rationalists were inseminating the minds of thoughtful Americans with new ways of viewing the phenomena of nature. There was one city in America whose citizens were especially prepared to receive the new leavening influences, and that was Philadelphia. From the time of its founding, the "City of Brotherly Love" had been a haven to all of liberal tendencies and bold inquiring turn of mind. Its cosmopolitan population comprised all manner of intellects, and it kept them supplied by the outpourings of its thriving printeries. Philadelphia was, moreover, an important medical center. It was quite natural, therefore, that Philadelphia should be the scene of the next American activities in the direction of establishing scientific societies.

Thither from Boston had come the youthful, but ardent-spirited, Benjamin Franklin. Seeking to affiliate himself with a scientific society at a time when none existed in Philadelphia or elsewhere in all the American colonies, he decided to start such an organization himself. The outcome was the Junto, formed in 1727, a secret literary and scientific society which he later described in his inimitable *Autobiography*.

---

12 *Ibid.*

13 As late as 1760, so competent a judge as Jonathan Mayhew was to express the belief that Boston was not yet ready for such an organization. See MS of the *Hollis Papers*, vol. V, in the possession of the Massachusetts Historical Society.

14 Preserved Smith, *A History of Modern Culture* (2 vols., New York, 1930-1934), I, 172.

15 See, in this connection, F. E. Brasch, *The Newtonian Epoch in the American Colonies (1680-1783)*, (Worcester, Mass., 1940). Reprinted from the *Proceedings* of the American Antiquarian Society.

I should have mentioned before, that . . . I had form'd most of my in-genious acquaintance into a club of mutual improvement, which we called the Junto; we met on Friday evenings. The rules that I drew up required that every member, in his turn, should produce one or more queries on any point of Morals, Politics, or Natural Philosophy, to be discussed by the com-pany; and once in three months produce and read an essay of his own writ-ing, on any subject he pleased. Our debates were to be under the direction of a president, and were to be conducted in the sincere spirit of inquiry after truth, without fondness for dispute or desire of victory; and to prevent warmth, all expressions of positiveness in opinions, or direct contradiction, were after some time made contraband, and prohibited under small pecuni-ary penalties.[16]

The activities and government of the Junto were determined in ac-cordance with a constitution which he drafted and entitled, "Rules for a Club for Mutual Improvement."[17] The history of the Junto has passed into oblivion, although it apparently functioned successfully. About 1736, we hear of a demand to extend its membership which, however, did not meet with Franklin's approval. He advanced, instead, the ingenious idea of having each member of the Junto start a sub-sidiary club along similar lines, and five or six were actually formed.[18]

Seeing the need for a society of larger scope and influence than the Junto, Franklin suggested the establishment of the American Philo-sophical Society in a circular letter, issued May 13, 1743, and entitled "A Proposal for Promoting Useful Knowledge among the British Plantations in America." This read in part:

The first drudgery of settling new colonies, which confines the attention of people to mere necessaries, is now pretty well over; and there are many in every province in circumstances that set them at ease, and afford leisure to cultivate the finer arts and improve the common stock of knowledge.

But as from the extent of the country such persons are widely separated, and seldom can see and converse or be acquainted with each other, so that many useful particulars remain uncommunicated, die with the discoverers, and are lost to mankind; it is to remedy this inconvenience for the future, proposed,

That one society be formed of *virtuosi* or ingenious men, residing in the several colonies, to be called *The American Philosophical Society*, who are to maintain a constant correspondence.

That these members meet once a month, or oftener, . . . to communicate to each other their observations and experiments, to receive, read, and consider such letters, communications or queries as shall be sent from dis-

16 Benjamin Franklin, *Writings* (A. H. Smyth, edit., 10 vols., New York, 1907) , I, 298-299.
17 *Ibid.*, II, 88-90.
18 *Ibid.*, I, 298-299.

tant members, to direct the dispersing of copies of such communications as are valuable, to other distant members, in order to procure their sentiments thereupon.[19]

Perhaps it ought to be mentioned that the word "philosophical" had a somewhat different connotation in Franklin's day from what it has in ours. As used then, it referred to the systematic study of practically any category of phenomena. Thus, we find that the members of the American Philosophical Society discussed questions of "natural philosophy, moral science, history, politics,"[20] and that they carried on "investigations in botany; in medicine; in mineralogy and mining; in mathematics; in chemistry; in mechanics; in arts, trades, and manufactures; in geography and topography; in agriculture."[21] In short, its activities embraced practically every field of learning, but the classics and theology absorbed far less of the society's interest than they did in the customary college curricula of the times.

Franklin proposed that the society meet once a month to discuss observations and experiments and to consider correspondence from distant members on:

> All new-discovered plants, herbs, trees, roots, their virtues, uses, etc.; methods of propagating them, and making such as are useful, but particular to some plantations, more general; improvements of vegetable juices, as ciders, wines, etc.; new methods of curing or preventing diseases; all new-discovered fossils in different countries, as mines, minerals, and quarries; new and useful improvements in any branch of mathematics; new discoveries in chemistry, such as improvements in distillation, brewing and assaying of ores; new mechanical inventions for saving labour, as mills and carriages, and for raising and conveying of water, draining of meadows, etc.; all new arts, trades, and manufactures, that may be proposed or thought of; surveys, maps. and charts of particular parts of the seacoasts or inland countries; course and junction of rivers and great roads, situation of lakes and mountains, nature of the soil and productions; new methods of improving the breed of useful animals; introducing other sorts from foreign countries; new improvements in planting, gardening and clearing land; *and all philosophical·experiments that let light into the nature of things, tend to increase the power of man over matter, and multiply the conveniences or pleasures of life.*[22]

The society advocated by Franklin came into actual existence either in 1743 or in the following year. It had its headquarters in Philadelphia, which had the advantage of being centrally located in the colonies.

19 *Ibid.*, II, 228-229.
20 American Philosophical Society, *Proceedings*, III (1843), 9.
21 *Ibid.*
22 Benjamin Franklin, *Writings*, II, 229-230; italics mine.

On April 5, 1744, we find Franklin writing to Cadwallader Colden of New York telling him that the society had been formed and giving its membership as being comprised of Dr. Thomas Bond, physician, John Bartram, botanist, Thomas Godfrey, mathematician, Samuel Rhodes, mechanician, William Parsons, geographer, and Dr. Phineas Bond, general natural philosopher, with Thomas Hopkinson, president, William Coleman, treasurer, and Benjamin Franklin, secretary.[23] After a time the organization languished, but it was revived in 1767.

Meanwhile the original Junto had become dormant, though, by 1750 we hear of a Junto of mostly different membership which was either a revival of the earlier club or possibly a new organization modeled upon it. This later Junto began to issue proceedings in 1758, became inactive in 1760, was revived in 1761; but no records of it have been found for the period of October, 1762, to April, 1766. It met on September 23, 1766, and took as its name the American Society for Promoting and Propagating Useful Knowledge. Two years later, on September 23, 1768, it reorganized, becoming this time the American Society Held at Philadelphia for Promoting Useful Knowledge. The Philadelphia Medical Society, formed in 1765, became incorporated with it on November 4, 1768. On the same day, Benjamin Franklin, absent in Europe at the time, was elected president.

Carl Bridenbaugh has recently used the absence of Franklin and other facts to show that the rôle played by Franklin in the 1768–1769 resuscitation of the American Philosophical Society may have been somewhat less important than is usually credited to him and that of John Bartram may have been considerably greater.[23a] The rôle played by Bartram is even more stressed by Carl Van Doren in his still more recent analysis of the origins of the American Philosophical Society. Bartram was thinking of such a society as early as 1739.[23b]

On January 2, 1769, the American Society Held at Philadelphia for Promoting Useful Knowledge united with the American Philosophical Society on terms of equality. The body effected by this union took a name compounded from those of the constituents in the merger, becoming known as the American Philosophical Society Held at Philadelphia for Promoting Useful Knowledge.[24] Although this still remains

[23] *Ibid.*, 276-277.

[23a] Carl Bridenbaugh and J. H. Bridenbaugh, *Rebels and Gentlemen; Philadelphia in the Age of Franklin* (New York, 1942).

[23b] Carl Van Doren, "The Beginnings of the American Philosophical Society," *Proceedings of the American Philosophical Society*, LXXXVII, No. 3 (1943), 278.

[24] See *American Philosophical Society; an Historical Account of the Origin and Formation of the American Philosophical Society, . . . with the Communication of*

the official name of the society, it is ordinarily referred to by the shorter name, American Philosophical Society. Benjamin Franklin became its first president, was annually re-elected until his death in 1790, and added luster and prestige to the society for many years by his splendid achievements and voluminous scientific writings.

The first major undertaking of the American Philosophical Society was the observation of a transit of Venus on June 3, 1769, a rare phenomenon not destined to occur again for more than a century. An observatory, partly financed by a grant from the Assembly of Pennsylvania, was erected in the State House Yard, and apparatus was procured from England. It was here and in other near-by observatories that David Rittenhouse and his fellow astronomers won international distinction by securing the first accurate data for measuring the distance of the earth from the sun, that is, for determining the astronomical unit as it is now called. The tall clock which Rittenhouse devised for timing his observations still runs and is among the treasured possessions of the society in its quarters in Philadelphia.[25]

J. Francis Fisher, Esq., and the Report of the Committee to Which These Papers Were Referred, Read October 15, 1841, . . . and the Report of the Committee on the Date of the Foundation of the Society Accepted May 1, 1914 (Philadelphia, 1914) for an exhaustive inquiry into the origin of the American Philosophical Society. The report of the 1914 committee endorses the view that the Junto existing in 1750 was founded in 1727. The committee of 1841, perhaps with an eye to staging a centennial celebration in 1843, had held that there were two separate and distinct Juntos.

E. P. Cheyney, issuing a minority report in 1914, expressed his opinion that the Junto of 1750 was a branch or offshoot of that of 1727, and said that "they were both Juntos, an older and a younger" (*ibid.*, 193). He added: "The Philosophical Society is derived from the ancient Junto, although through a younger branch" (*ibid.*, 193).

He, therefore, was able to concur with the main conclusion reached by the committee, namely, accepting 1727 as the date of founding. Thus, he says, summarizing his own slightly different reasoning: "If the 'date of founding,' as it has been formulated for the use of colleges by the Carnegie Foundation for the Advancement of Teaching, means 'the year in which the institution was established out of which the present college or university (institution) has developed,' the Junto of 1750 certainly developed out of the Junto of 1727, and the Society may claim its descent through a younger line just as fairly as by primogeniture, so that the date of the origin of the Society should be considered that of the formation of Franklin's Junto in 1727."

[25] For the astronomical activities of the American Philosophical Society, see S. A. Mitchell, "Astronomy during the Early Years of the American Philosophical Society, *Science*, n. s., XCV (1942), 489-495; and S. A. Mitchell, "Early American Astronomers," *Journal of the Royal Astronomical Society of Canada*," XXX (1942), 352; also A. E. Lownes, "The 1769 Transit of Venus," *Sky and Telescope*, II, no. 6 (1943), 3-5. See also especially Harry Woolf, *The Transits of Venus: A Study of Eighteenth Century Science* (Princeton, N.J., 1959).

In 1773 Benjamin Rush presented an interesting paper before the society entitled "Inquiry into Dreams and Sleep." Benjamin Franklin also contributed important papers and made notable gifts to the library of the society. *Transactions* were first published in 1771. Celebrated foreign scientists such as Lavoisier, Linnaeus, Buffon, and Condorcet were honored with membership, and for the remainder of the century the American Philosophical Society received stimulus and encouragement from eminent French thinkers. The invasion of Philadelphia by the British during the Revolution caused the society hastily to cease its activities for a time. A charter granted to the American Philosophical Society on March 15, 1780, by the General Assembly of Pennsylvania declared: "The experience of the ages shows that improvements of a public nature are best carried on by societies of liberal and ingenious men uniting their labors without regard to nation, sect, or party, in one grand pursuit."[26] It, moreover, authorized the society

at all times in peace or war, to correspond with learned societies, as well as individual learned men of any nation and country upon matters merely belonging to the business of the said Society, such as mutual communication of their discoveries and proceedings in philosophy and science, the procuring of books, apparatus, natural curiosities, and such other articles and intelligence as are usually exchanged between learned bodies for furthering their common pursuits.[27]

The Commonwealth of Pennsylvania granted a lot to the society in State House Square upon which Philosophical Hall was erected, being first occupied in 1789.

The influence of the American Philosophical Society upon science in America in the eighteenth century can hardly be overestimated.[28] It also exerted some influence on science in Europe as well.

The scientific activities centering in Philadelphia around the American Philosophical Society did not pass unnoticed in Boston, which had its own reputation for erudition to uphold. The exciting controversy with England delayed immediate action by the Bostonians, but the closing days of the Revolutionary War brought the matter into the foreground of attention. John Adams proved to be the prime mover in organizing the rival society, the American Academy of Arts and Sciences.

[26] American Philosophical Society, *Charter*, 4.   [27] *Ibid.*, 11.

[28] For the history and work of the society, see especially: American Philosophical Society, *Proceedings*, LVI (1927), *Record of the Celebration of the Two Hundredth Anniversary of the Founding of the American Philosophical Society; Mankind Advancing* (American Philosophical Society, Philadelphia, 1929); and American Philosophical Society, *Proceedings*, LXXXVI, no. 1 (1942), "The Early History of Science and Learning in America." Numerous other references are in the Bibliography at the end of this book.

It is likely that Adams had his interest in such a project aroused long before he went to France by a conversation on the subject with the then Hollis Professor of Mathematics and Natural Philosophy at Harvard, John Winthrop. Adams' account of the circumstances leading to the formation of the American Academy of Arts and Sciences reveals the part that local pride played in the event.

> In France, among the academicians and other men of science and letters, I was frequently entertained with inquiries, concerning the Philosophical Society of Philadelphia, and with eulogiums on some publications in their transactions. These conversations suggested to me the idea of such an establishment at Boston, where I knew there was as much love of science, and as many gentlemen who were capable of pursuing it, as in any other city of its size. In 1779, I returned to Boston in the *Frigate La Sensible,* with the Chevalier de la Luzerne and M. Marbois. The Corporation of Harvard College gave a public dinner in honor of the French Ambassador and his suite, and did me the honor of an invitation to dine with them. At the table, in the Philosophy Chamber, I chanced to sit next to Dr. Cooper. [Dr. Samuel Cooper was pastor of the Brattle Square Church in Boston, an active champion of the Revolution, a member of the Corporation of Harvard College, and was destined to become the first vice-president of the American Academy of Arts and Sciences.] I entertained him during the whole time we were together, with an account of Arnold's collections, the collections I had seen in Europe, the compliments I had heard in France upon the Philosophical Society at Philadelphia, and concluded with proposing that the future legislature of Massachusetts should institute an Academy of Arts and Sciences. . . .The Doctor accordingly did diffuse and project so judiciously and effectually that the first legislature under the new constitution adopted and established it by law.[29]

The charter of incorporation, granted May 4, 1780, named the new body the American Academy of Arts and Sciences and set forth its objects as being:

> To promote and encourage the knowledge of the antiquities of America and of the natural history of the country, and to determine the uses to which the various natural productions of the country may be applied; to promote and encourage medical discoveries, mathematical disquisitions, philosophical inquiries and experiments; astronomical, meteorological and geographical observations, and improvements in agriculture, arts, manufactures and commerce, and in fine, to cultivate every art and science which may tend to advance the interest, honor, dignity and happiness of a free, independent, and virtuous people.[30]

[29] John Adams, *Works* (C. F. Adams, edit., Boston, 1851-1856) , IV, 302, in footnote.

[30] American Academy of Arts and Sciences, *Memoirs,* n. s., XI (1888) , 78.

Governor James Bowdoin became the first president of the society, and was succeeded by John Adams in 1791. Samuel Adams and John Hancock were among the charter members. George Washington and Benjamin Franklin were chosen fellows at the first election held in 1781. David Rittenhouse and Thomas Jefferson were soon taken into the fold. Among esteemed foreign honorary members of the early years were Lalande, Buffon, Euler, Priestley, Sir William Herschel, and the erstwhile American, Benjamin Thompson.

The first research project of the Academy was a joint expedition with Harvard to what is now Islesboro, Maine, to observe the total solar eclipse of October 27, 1780. The expedition, headed by Professor Williams, had to get permission to pass through a British naval blockade, was hampered by bad weather, and finally missed seeing totality through an error in latitude. This very mistake, however, resulted in the important discovery of the phenomenon known today as Baily's Beads.[31] The Academy published an account of this expedition in the first volume of its *Memoirs,* which appeared in 1785.[32]

French influence was not limited to inspiring the creation of the American Academy of Arts and Sciences and leavening the American Philosophical Society. Under French guidance an ambitious attempt was made in 1788 to launch at Richmond, Virginia, an institution similar to the l'Académie Française. It was planned to open branch academies in Baltimore, Philadelphia, and New York, and to maintain correspondence and exchange with the great learned and scientific societies of the Old World.

The principal proponent of this somewhat grandiose enterprise was Chevalier Quesnay de Beaurepaire, a grandson of Quesnay, the well-known economic theorist of the physiocratic school. Like his countryman, Lafayette, the Chevalier Quesnay served in the American Revolution, but he was forced by illness to relinquish a military career. The idea of founding an academy in America was suggested to him by John Page, lieutenant governor of Virginia, who urged him to bring European professors to America and promised to help him to secure the presidency of the academy. Needing no urging, the eager Frenchman diligently campaigned for the project in Virginia and northward through the colonies. Sixty thousand francs are said to have been raised

[31] R. S. Bates, "Baily's Beads or Williams' Beads," *Telescope,* VIII (March-April, 1941), 36-38.

[32] The early history of the American Academy of Arts and Sciences is traced in more detail in *The American Academy of Arts and Sciences* (Boston, 1941), and R. S. Bates, "The American Academy of Arts and Sciences," *Scientific Monthly,* LIV, No. 3 (March, 1942), 265-268.

by subscription, and there is still in existence a list of nearly a hundred of the original subscribers in Virginia, containing a number of prominent names.[33]

L'Académie des États-Unis de l'Amérique, as the new enterprise came to be called, was to be provided with an elaborate organization composed of a president, a vice-president, six counselors, a treasurer general, a secretary, French professors, masters, and artists, and twenty-five resident and one hundred and seventy-five non-resident associates representing the best scholarship of the Old World and the New. The academy was to circulate its memoirs widely, communicate a knowledge of the natural resources of America to European scientists, and furnish them with specimens of the unknown fauna and flora of the American hinterland. Doubtless French mineralogists and engineers were going to pave the way for exploitation of natural resources by syndicates of European capitalists. French experts were to be imported for the purpose of giving instruction to the youth of America, and to serve the government and private corporations in a consultative capacity for fees shared with the academy. Among those whom Quesnay was able to approach with his project were James Madison, president of William and Mary College, George Clinton, governor of New York, Baron von Steuben, Benjamin Franklin,[34] George Washington, and Thomas Jefferson.[35] Six counselors and a treasurer were appointed to serve with President Quesnay. A building constructed to be the home of the academy in Richmond was dedicated with impressive Masonic ceremonies in 1786. It later became the place where the Virginia convention ratified the Constitution and was, still later, to serve as Richmond's first theatre.

Quesnay himself returned to Paris in quest of friends for his new scheme to bind France and America by cultural ties. Scion of an influential family and a member of the idolized group of volunteers in the

---

[33] The entire list is given in H. B. Adams, "L'Académie des États-Unis de l'Amérique," *Academy*, II (1887) , 404-405.

[34] A letter to him from Mrs. Bache [Franklin's sister] has been preserved asking that he give Quesnay "every aid and assistance that may lie in your power." Printed in H. B. Adams, *loc. cit.*, 406-407.

[35] H. B. Adams, *loc. cit.*, 409, maintains: "He looked upon the project with favor, otherwise he would not have allowed his name to be so prominently used in connection with Quesnay's scheme, . . ." A recent writer says, however, ". . . Jefferson, while not actively opposing it, claimed that America was too poor to support such undertakings." K. L. Forsyth, "Quesnay, Alexandre-Marie," *Dictionary of American Biography* (edited by Allen Johnson and Dumas Malone, 20 vols., New York, 1928-1936) , XV, 301.

American Revolution against Britain, he gained access to the best circles of French society. His project was presented to the king and the queen in a memoir.[36] A report of the Royal Academy of Sciences approving his project was certified by Condorcet. Another report similarly favorable was signed by Vernet and other artists of the Royal Academy of Painting and Sculpture.

Tentative arrangements were made for instituting schools in Virginia to give advanced instruction in "Foreign languages; Mathematics; Design; Architecture, civil and military; Paintings; Sculpture; Engraving; Experimental Physics; Astronomy; Geography; Chemistry; Mineralogy; Botany; Anatomy, human and veterinary; and Natural History." [37] As it turned out, Dr. Jean Rouelle was the only professor ever actually appointed and it is doubtful whether he actually came to America, though it was arranged that he should sail in October of 1788, having been appointed to serve for ten years as mineralogist-in-chief of the academy and professor of natural history, chemistry, and botany.

The whole scheme for establishing l'Académie des États-Unis de l'Amérique was destined for speedy collapse. Quesnay himself was detained in Europe by family affairs. Aid from France was withdrawn at the outbreak of the French Revolution, and the scattered population of rural Virginia had not been sufficiently aroused to support the undertaking financially. Vain though speculation may be, one wonders what the consequences might have been had France been in a position to go through with the enterprise. In one scholar's opinion,

> . . . if circumstances had favored Quesnay's project, it is probable that the University of Virginia would never have been founded. There would have been no need of it. The Academy of the United States of America, established at Richmond, would have become the center of higher education, not only for Virginia, but for the whole South, and possibly for a large part of the North, if the Academy had been extended as proposed, to the cities of Baltimore, Philadelphia and New York. Supported by French capital, . . . strengthened by French prestige, by literary, scientific and artistic associations with Paris, then the intellectual capital of the world, the Academy at Richmond might have become an educational stronghold, comparable in some degree to the Jesuit in Canada, which has proved more

---

36 Alexandre-Marie Quesnay, Chevalier de Beaurepaire, *Mémoire, status et prospectus concernant l'Académie des Sciences et Beaux-Arts des États-Unis de l'Amérique, établie à Richemond, Capitale de la Virginie: présentés à Leurs Majestés, et à la Famille Royale, par le Chevalier Quesnay de Beaurepaire* (Paris, 1788).

37 Adams, *loc. cit.*, 411. "Quesnay's project was clearly something higher than an American College. He had in mind the highest special training of American students in the arts and sciences." *Ibid.*, 406.

lasting than French dominion, more impregnable than the fortress of Quebec.[38]

Destiny and circumstances had willed that things should be otherwise. "Quesnay's brilliant project attracted brief admiration and then sank into oblivion."[39] Although the Virginians were not to receive the writings of the French *philosophes* through the ministration of l'Académie, the scheme was not wholly chimerical, and the broad scope of its aims and its plan for graduate instruction were well known to Thomas Jefferson, principal founder of the University of Virginia.[40]

The oldest historical society in America, the Massachusetts Historical Society, formed in 1791 under the inspiration of Jeremy Belknap and his associates of Boston and Cambridge,[41] in 1792-1793 in the first volume of its collections began to publish papers on natural history. These were followed by other scientific papers appearing at intervals for several decades, although they became less frequent after about 1815.

The oldest of the state academies dates from the very close of the eighteenth century, when the Connecticut Academy of Arts and Sciences came into existence in 1799, only after severe birth pains. As early as 1765, Ezra Stiles considered the possibility of founding an American Academy of Arts and Sciences, and for years he continued to toy with various projects for a learned society. That the example set by the actual formation of the American Academy of Arts and Sciences at Boston had some influence is indicated by the fact that soon after its establishment, Benjamin Guild, a tutor at Harvard, sent a copy of the charter of the Academy to President Stiles of Yale who drafted a similar one for a proposed Connecticut Academy. The proposal was introduced in the General Assembly at its May session in 1781, but after some fruitless discussion and talk of giving the assembly the right of visitation and inspection, the matter was dropped.[42] A fresh effort on the part of Dr. Stiles in 1783 fared no better. The relations between Yale and the General Assembly were at the time strained and there was opposition to creating any new body which would gravitate to and be more or less affiliated with Yale.

[38] *Ibid.*, 410.

[39] *Ibid.*, 411.

[40] H. M. Jones, *America and French Culture, 1750-1848* (Chapel Hill, 1927) , 478.

[41] J. S. Bassett, *The Middle Group of Historians* (New York, 1917) , 35-37.

[42] S. E. Baldwin, "The First Century of the Connecticut Academy of Arts and Sciences," *Transactions of the Connecticut Academy of Arts and Sciences*, XI, Centennial Volume (1901-1903) , XV.

Dr. Stiles was next instrumental in the formation of a voluntary association at Hartford in 1786, during the meeting of the Assembly there, known as the Connecticut Society of Arts and Sciences.[43] This body, which was unincorporated, published only one paper, a rather notable one by Jonathan Edwards, "On the Language of the Muhhekaneew Indians," which he communicated to the Society in October of 1787. With the country presently in the throes of adoption of the federal constitution and absorbed in topics of conversation much more exciting than Dr. Edwards' analogies between the Hebrew and Mohican tongues, the society was soon moribund.

In 1799 a new organization was quietly formed under the name originally selected by Dr. Stiles, Connecticut Academy of Arts and Sciences. At first this was a voluntary association, but a few months later it was incorporated. The preamble of its charter affirmed that "literary Societies have been found to promote, diffuse and preserve the knowledge of those Arts and Sciences, which are the support of Agriculture, Manufactures, and Commerce, and to advance the dignity, virtue and happiness of a people." Under the presidency of Timothy Dwight, also president of Yale, the Connecticut Academy was launched upon an energetic career. *Transactions* were commenced in 1810, but after 1818, when Benjamin Silliman began his *American Journal of Science,* the Academy allowed many of its papers to be published in *Silliman's Journal,* as the magazine was popularly called, in order to give them wider publicity.

Meanwhile, in Maryland, the Academy Society had been formed in 1797, a rather ephemeral body, which, however, helped pave the way for the present Maryland Academy of Sciences, established twenty-five years later.

Not only were general scientific societies being founded in eighteenth century America, but also, and this fact is much less widely known, specialized societies had begun to come into existence embracing the fields of medicine, agriculture, chemistry, and mineralogy, together with organizations for the purpose of advancing manufactures and such technology as then existed.

The various societies discussed in the preceding pages embraced the field of science as a whole, and their "natural philosophers" carried out studies along a wide front. The unity rather than the diversity of science was emphasized. Yet already the concept of specialization was beginning to eat away at this fabric of unity.

[43] Ezra Stiles, *The Literary Diary of Ezra Stiles* (edited by F. B. Dexter, 3 vols., New York, 1901) , II, 263.

Medicine was a highly practical art, and more men in the colonies were devoting their time to the study and practice of healing than to any other field of scientific endeavor. Since the days of antiquity, physicians had banded together for the advancement of their knowledge and to protect and further the interest of their profession. Colonial Americans found their way in increasing numbers to Edinburgh and other centers of medical instruction noted for their schools and medical societies and, returning to America, resolved to band together in warfare on the quackery inevitably rampant in a frontier country inadequately supplied with properly trained physicians. The result was the formation of a number of medical societies which are entitled to be reckoned as the earliest specialized scientific societies in the United States.[44] Indeed, prior to the establishment of the national government, at least a dozen medical societies had been founded in America, four of which have continued in active existence down to the present time.

Boston was one of the principal medical centers in Colonial America. It was there in 1721 that Zabdiel Boylston performed the first inoculations for smallpox in America. It was there, too, that the medical profession in America began to organize. The first medical society in the American colonies was one mentioned in a letter by Dr. William Douglass to Cadwallader Colden, dated February 17, 1735, and stating: "We have lately in Boston formed a medical society; . . . . We designed from time to time to publish some short pieces, ..."[45] The society is known to have lasted for at least half a dozen years,[46] for an operation is reported to have been performed before it by Silvester Gardiner in 1741.[47]

In 1765, Cotton Tufts of Weymouth attempted to form a medical society in Massachusetts and issued a call for physicians to convene at Gardner's Tavern on Boston Neck on the third Monday in March, but the society seems not to have materialized. An Anatomical Society,

---

[44] For general treatments of early medical societies in America, see F. R. Packard, "Medical Societies in This Country Founded Prior to the Year 1787," *Philadelphia Medical Journal*, V (1900), 229-231; and, by the same author, *History of Medicine in the United States* (Philadelphia and London, 1901) , ch. VIII, "History of Medical Societies Founded before the Year 1800."

[45] Massachusetts Historical Society, *Collections*, 4th ser., II (1854) , 188. The original letter is among the Colden papers in the possession of the New York Historical Society.

[46] A short summary of what is known of the activities of the society is to be found in H. R. Viets, *A Brief History of Medicine in Massachusetts* (Boston and New York, 1930) , 65-68.

[47] *Boston News Letter*, November 13, 1741.

composed of Harvard students, was formed in 1771. The Revolutionary War resulted in setting apart the medical man in a distinct profession and paved the way for the Boston Medical Society, established in 1780. John Warren, although under thirty years of age, was already Boston's leading physician and was soon to be professor of anatomy and surgery at Harvard's Medical School, founded two years later.[48] He was especially active in the newly formed Boston Medical Society and in the Massachusetts Humane Society, which was also organized in 1780 and held its first meeting at his house.

In 1781, the Massachusetts Medical Society was incorporated in response to the petition of an active group of young physicians, including, of course, John Warren. Nearly one hundred fellows were admitted to the society during the years prior to 1800, about one third of whom were really active members. The society began a library in 1782 which it carried on independently for nearly a century and which now is merged with the Boston Medical Library.[49] The society began to publish "Medical Communications" in 1790.[50] Presently, district medical societies began to spring up under the aegis of the Massachusetts Medical Society.[51]

The earliest attempt to form a medical society in Connecticut was made in 1763 by physicians in New London County. The physicians of Litchfield County organized a society in 1766, and those of New Haven County in 1784.[52] After a discouraging series of attempts to secure a charter for a state medical society from a recalcitrant legislature, fearful lest it create a monopoly, success came at last with the incorporation of the Medical Society of Connecticut in 1792.[53] The New Hampshire Medical Society dates from 1791.

A medical body, known as a Weekly Society of Gentlemen, came into ephemeral existence in New York in 1749, and twenty years later another transitory medical group came into existence in the same city.

[48] T. F. Harrington, *The Harvard Medical School* (Chicago, 1905).

[49] J. N. Farlow, *The History of the Boston Medical Library* (Norwood, Massachusetts, 1918).

[50] For the early history of this society, see S. A. Green, *A Centennial Address Delivered in the Sanders Theatre at Cambridge, June 7, 1881, before the Massachusetts Medical Society* (Groton, 1881).

[51] W. L. Burrage, *A History of the Massachusetts Medical Society, 1781-1922* (Norwood, Massachusetts, 1923), ch. X, "The District Medical Societies," 323-349.

[52] G. W. Russell, *Early Medicine and Early Medical Men in Connecticut* (Hartford, 1892), 148.

[53] Henry Bronson, "Historical Account of the Origin of the Connecticut Medical Society," Connecticut Medical Society, *Proceedings*, 2d ser., IV (1873), 192-201.

Still another was formed in 1794, which, in 1806, merged into the Medical Society of the County of New York.[54]

The oldest of the province-wide medical societies, the Medical Society of New Jersey, formed at New Brunswick in 1766 under the name New Jersey Medical Society, is still in existence. The constitution which this body adopted in 1766 reflects the fact that its members not only were striving to promote their professional interest, but they were also quite consciously interested in the promotion of medical science;[55] and its wording became a prototype for the constitutions of many subsequent American medical societies.

In 1790 a second organization was effected under the name the Medical Society of the Eastern District of New Jersey. For a short time it was a rival of the older society, and then lapsed into oblivion.[56] Indeed, Paul Micheau, its founder and a member of the older society, was censured by that body for starting a second and competing organization.[57]

It was but natural that Philadelphia, the leading medical center of America, should have several medical organizations. The Philadelphia Medical Society was instituted in 1765 under the inspiration of John Morgan and continued in operation until November, 1868, when it united with the American Society for Promoting Useful Knowledge. Thus it is to be reckoned as a minor root of the present American Philosophical Society. Benjamin Rush, Thomas Cadwallader, and John Morgan were among its members.[58] A body calling itself the American Medical Society came into existence in Philadelphia in 1773, when

> a number of students, who had assembled in the city of Philadelphia, from different parts of the Continent, to hear the Lectures of the Medical Professors, thought that they might derive some advantage from associating themselves, in order to discuss various questions in the healing art, and to communicate to each other their observations on different subjects.

[54] Stephen Wickes, *History of Medicine in New Jersey, and of Its Medical Men, from the Settlement of the Province to A. D. 1800* (Newark, N. J., 1879), 52-53.

[55] "Instruments of Association and Constitution of the New Jersey Medical Society," reprinted in F. R. Packard, *History of Medicine in the United States* (Philadelphia and London, 1901), 379. The constitutions of many later medical societies contain similar language.

[56] Wickes, *op. cit.*, 49-50.

[57] *Ibid.*, 332.

[58] G. W. Norris, *The Early History of Medicine in Philadelphia* (Philadelphia, 1886), "Medical Societies," 115-120. See also F. P. Norris, *Standard History of the Medical Professor of Philadelphia* (Chicago, 1897).

Such associations had been found highly beneficial to the students of medicine in Europe, and it was thought might be still more so in a country, the diseases and remedies of which had not been fully explored. These ideas gave rise to the American Medical Society, which now ranks amongst its members many of the most respectable medical characters on the continent.[59]

This society lasted until 1792, and as important a medical figure as Dr. William Shippen is known to have served as its president in 1790.

The important medical society which bears the name College of Physicians of Philadelphia dates from January 2, 1787, when it held its first meeting and adopted its first constitution.[60] John Redman was its first president and a senior fellow; other senior fellows included John Morgan, William Shippen, Jr., Benjamin Rush, and Samuel Duffield.[61] In 1789, another Philadelphia Medical Society was formed which lasted for half a century, finally amalgamating with the College of Physicians of Philadelphia. An Academy of Medicine of Philadelphia was formed in 1797.

The Philadelphia Humane Society was formed in 1780, the same year as the humane society in Massachusetts, and was in reality a first-aid society to minister to victims of drowning, asphyxiation, sunstroke, and other catastrophes.[62]

In Baltimore, Charles Frederick Wiesenthal is said to have organized a medical society in 1788 of which he was the president.[63] Still another Maryland medical organization was in existence a decade later; it was

[59] "An Account of the American Medical Society," *The Columbian Magazine or Monthly Miscellany*, IV (1790), 206.

[60] "The objects of this College are, to advance the Science of Medicine, and thereby to lessen Human Misery, by investigating the diseases which are peculiar to our Country, by observing the effects of different seasons, climates, and situations upon the Human body, by recording the changes that are produced in diseases by the progress of Agriculture, Arts, Population, and Manners, by searching for Medicines in our Woods, Waters, and the bowels of the Earth, by enlarging our avenues to knowledge; from the discoveries and publications of foreign Countries; by appointing stated times for Literary intercourse and communications, and by cultivating order and uniformity in the practice of Physick.

". . . The College shall consist of twelve Senior Fellows and Associates." College of Physicians of Philadelphia, "Constitution of the College of Physicians of Philadelphia, January 2nd, 1787." Reprinted in F. R. Packard, *History of Medicine in the United States* (New York, Paul B. Hoeber, Inc., 1931). 2 vols.

[61] N. S. N. Ruschenberger, *An Account of the Institution and Progress of the College of Physicians of Philadelphia, during a Hundred Years from January, 1787* (Philadelphia, 1887).

[62] F. R. Packard, *op. cit.*, 371. (Philadelphia, J. B. Lippincott Co., 1901.)

[63] F. R. Packard, *op. cit.* (2 vols., New York, 1931), II, 958.

known as the Harford Medical Society.[64] More important was the Medical and Chirurgical Faculty of the State of Maryland, formed in 1798, and still in existence.

The Medical Society of Delaware was organized in 1789.[65] A Medical Society of South Carolina was formed in 1789.

The Boston Marine Society (1742), the earliest marine society in America, is still in active existence. The primary purpose of Captain William Starkey and his associates in founding the institution was, as stated in its charter, "to relieve one another in poverty or other adverse accidents," and they had other philanthropic aims such as aiding widows and orphans of mariners. The society also had as one of its objects the collection of scientific data, as is shown when the society voted in 1753

> . . . that every member of this Society upon his arrival from Sea give in to this Society to be recorded, his observations on the Variations of the needle, the Soundings, Courses, Distances, and all other things remarkable upon this Coast, and that it be inserted in the Petition of this Society to the General Assembly, and be read at every meeting of this Society.[66]

The Salem Marine Society dates from 1766.[67] Its records of early voyages of navigation, by sea captains who made it a point to record in their journals accurate information about the waters and coasts which they visited, are deposited with the Essex Institute of Salem and, together with those of the East India Marine Society, constitute a priceless legacy of American fishing and sailing vessels for a century and a half.

Much more widely known is the East India Marine Society, begun in the year 1799 by a group of sea captains who immediately started a museum or "cabinet," wherein Captain Cairnes and others began to deposit their curios.[68] Membership was limited to those who had

[64] E. F. Cordell, "Transactions of the Harford Medical Society, 1797-1798," *Johns Hopkins Hospital Bulletin*, 11, July-August (1900); 13, August-September (1902).

[65] I. P. Busey, "The Delaware State Medical Society and Its Founders in the Eighteenth Century," paper presented at the Annual Meeting of the American Academy of Medicine, 1885.

[66] *Constitution and By-Laws of the Boston Marine Society . . . together with a Brief History of the Society* (Boston, 1884).

[67] R. D. Paine, *The Ships and Sailors of Old Salem* (Chicago, 1912), 11-12, gives the aims of this society as set forth in the Act of Incorporation, dated 1772, together with a few facts of its history.

[68] William Bentley, *The Diary of William Bentley, D. D.* (4 vols., Salem, Mass., The Essex Institute, 1905-1941), II, 321, entry of October 22, 1799; *ibid.*, 322, November 22, 1799.

actually navigated the seas near the Cape of Good Hope or Cape Horn, and by 1867, when its collections were turned over to the Peabody Academy of Science in Salem, more than three hundred and fifty masters and supercargoes of Salem had qualified for membership.[69] The logs, sketches of landfall, pictures of ships, and journals which these hardy mariners made on their voyages to the South Seas or China present an unusual record of the heyday of the American clipper.

Specialized societies for the promotion of agriculture next emerged. In the latter half of the seventeenth century an agricultural revolution "was in progress involving changes in the methods of planting, stock-breeding, and the invention of farm machinery." Then, too, "the organizing habit which was bred in the American mind" by the American Revolution "gave an impetus to the much needed formation of agricultural societies.[70]

Although the American Philosophical Society published many articles on agricultural subjects during its early years, a feeling gradually grew that there should be societies in America devoted specifically to agriculture, similar to those which had begun to appear in the British Isles and in continental Europe during the latter half of the eighteenth century. This feeling led to the formal organization of the Philadelphia Society for Promoting Agriculture, in March, 1785, by twenty-three citizens of that city, acting under the initiative of Judge J. B. Bordley, a Maryland planter. Its object was to promote "a greater increase of the products of land within the American states,"[71] to foster which the society offered prizes, held meetings, and printed agricultural essays. Samuel Powell, twice mayor of Philadelphia, was the first president of the society. Benjamin Franklin and Timothy Pickering were among the early resident members of the society; George Washington and Robert R. Livingston were among the honorary members, who by 1789 included men from all thirteen states. Washington was simultaneously working without success for the formation of a similar society in Virginia.[72] In 1794, the society endeavored unsuccessfully to have the

[69] R. D. Paine, *op. cit.*, 10.

[70] J. F. Jameson, *The American Revolution Considered as a Social Movement* (Princeton, 1926), 80.

[71] Philadelphia Society for the Promotion of Agriculture, *Minutes of the Philadelphia Society for the Promotion of Agriculture, from its Institution in February, 1785 to March, 1810* (Philadelphia, 1854).

[72] Letter to Alexander Spotswood, dated February 13, 1788, *Writings* (Jared Sparks, edit., Boston, 1858), IX, 326.

Pennsylvania Legislature incorporate a state society for the promotion of agriculture.[73]

About 1740, a group of South Carolina planters interested in the indigo industry formed a club known as the Winyaw Indigo Society which held meetings, largely social, at which the cultivation of indigo was discussed.[74]

The South Carolina Society for Promoting and Improving Agriculture and Other Rural Concerns was organized in Charleston on August 24, 1785, and ten years later was incorporated under its present name, the Agricultural Society of South Carolina.[75] An agricultural society was formed at Hallowell, Maine (then in Massachusetts), in 1787.[76]

An organization called the New Jersey Society for Promoting Agriculture, Commerce and Arts came into existence in 1781, but not much is known about its work.[77] The Burlington Society for the Promotion of Agriculture and Domestic Manufacturers was permanently organized February 6, 1790, and it remained active for at least ten years, publishing numerous essays on agricultural subjects in newspapers and giving prizes in contests in agricultural production. A Morris County Society for Promoting Agriculture and Domestic Manufactures, formed in 1792, was principally concerned with establishing a library.

The New York Society for the Promotion of Agriculture, Arts and Manufactures, organized on February 26, 1791, was incorporated on March 12, 1792.[78] Robert R. Livingston, its first president, frequently reported to the society on his agricultural experiments as did also Samuel L. Mitchill, professor of natural history, chemistry, and agriculture in Columbia College. The rules and regulations of the society which were formulated by Robert R. Livingston, Simeon DeWitt, and Samuel L. Mitchill declared that "the objects of investigation for the

[73] Philadelphia Society for the Promotion of Agriculture, *Memoirs*, I-IV (1811-1818).

[74] Winyaw Indigo Society, *Rules of the Winyaw Indigo Society*, with a Short History of the Society and Lists of Living and Deceased Members (Charleston, S. C., 1874).

[75] C. I. Walker, *History of the South Carolina Agricultural Society* (Charleston, 1919).

[76] Maine Board of Agriculture, *Annual Report for 1865* (Augusta, 1865).

[77] C. R. Woodward, "The Development of Agriculture of New Jersey, 1640-1880," *N. J. Agricultural Experiment Station Bulletin* 451 (New Brunswick, 1927).

[78] Society Instituted in the State of New York, for the Promotion of Agriculture, Arts, and Manufactures, *Transactions*, I, pts. 1-4 (Albany, 1801).

society shall be Agriculture, Manufactures, and Arts, with such subjects of enquiry, as may tend to explain, or elucidate their principles."[79] In 1804, the name of the society was changed to the Society for the promotion of Useful Arts and, in 1824, it was united with the Albany Lyceum of Natural History to form the Albany Institute.

The Massachusetts Society for Promoting Agriculture was incorporated on May 31, 1792. In pursuance of its object of promoting "useful improvements in agriculture" it raised money for premiums, and during the first eight years of its existence it published many articles in newspapers and pamphlets.[80]

The Society for Promoting Agriculture in the State of Connecticut, organized on August 12, 1794, at Wallingford by "a number of citizens from different towns in the state," adopted a constitution stating that "the object of investigations for the society shall be agriculture, with such subjects of inquiry as may tend to explain its principles."[81] The transactions of the society record experiments with soils, fertilizers, grains, vegetables, fruit trees, and dairy products.[82]

In the closing years of the eighteenth century an attempt was made to have Congress sponsor the establishment of a national agricultural society. Washington had advocated the creation of a national board of agriculture in his final message, delivered December 7, 1796. The matter was referred to a committee of the House of Representatives which reported January 11, 1797.

> The only method which a government can with propriety adopt, to promote agricultural improvement, is to furnish the cultivators of the soil with the easiest means of acquiring the best information respecting the culture and management of their farms, and to excite a general spirit of inquiry, industry and experiment. This object can be best attained by the institution of societies for the encouragement of agriculture and internal improvement; a practice which has been already sanctioned by the experience of other countries.
>
> Societies have been established in many parts of the United States, but are on too limited a scale to answer the great national purpose of agricultural improvement throughout the United States; it is, therefore, necessary that a society should be established, under the patronage of the

[79] Society Instituted in the State of New York for the Promotion of Agriculture, Arts, and Manufactures, *Transactions*, I (1792), iv.

[80] Massachusetts Society for Promoting Agriculture, Centennial Year 1792-1892 (Salem, 1892).

[81] Society for Promoting Agriculture in the State of Connecticut, *Transactions* (New Haven, 1802).

[82] For the work of this society, see E. H. Jenkins, "Connecticut Agriculture," *History of Connecticut* (New Haven, 1926), 289-425.

General Government, which should extend its influence through the whole country. . . .[83]

The committee recommended the incorporation of a body to be known as the American Society of Agriculture, with headquarters at the national capital and a paid secretary. Senators, representatives, judges of the Supreme Court, the secretaries of state, treasury, and war, and the attorney general were to be members, and "such other persons as should choose to become members agreeably to the rules prescribed." The society was to hold annual meetings at which a board of agriculture was to be elected.

The bill also included a proposition for establishing a military academy, a thing which Jefferson presently opposed openly on the ground that Congress was not authorized by the Constitution to establish one. Despite Jefferson's fundamental interest in agriculture, he could not bring himself to support a proposition for a federal agricultural agency, as is revealed in a letter to Livingston in February, 1801, when he wrote, "I am against that because I think Congress cannot find in all the enumerated powers any one which authorizes the act, much less the giving of public money to that use." Although the bill was brought up in the House for consideration, it never came to a vote, its friends apparently fearing to press the matter that session. Washington, writing to Sinclair, thought it "highly probable that next session will bring the matter to maturity," but nothing more came of it. Thus, the strict constructionists blocked the attempt to create a national agricultural society under governmental patronage, and when, in the nineteenth century, national agricultural societies came into existence, they were voluntary and private organizations.

The credit for establishing the first chemical society[84] in the world is sometimes given to James Woodhouse,[85] principal founder of the Chemical Society of Philadelphia, formed in 1792 and continuing in existence for some seventeen years, being then succeeded, in 1811, by the Columbian Chemical Society. The Philadelphia Chemical Society early became the means for popularizing in this country the views of Lavoisier on combustion as opposed to the earlier phlogiston theory.[86]

[83] 1797 *Report on the Promotion of Agriculture*, by Z. Swift, U. S. Congress, 4th, 2d Session, House Executive Document.

[84] H. C. Bolton, "Early American Chemical Societies," *Journal of the American Chemical Society*, xix (1897), 718.

[85] E. F. Smith, *James Woodhouse; a Pioneer in Chemistry, 1770-1809* (Philadelphia, 1918), 39.

[86] E. F. Smith, *Chemistry in America* (New York and London, 1914), 12; and by the same author, *Chemistry in Old Philadelphia* (1919), 14-15.

Thus, in 1798, the society listened to an annual oration delivered by Thomas P. Smith entitled "A Sketch of the Revolutions in Chemistry," in which the newer views were ardently championed.[87] Woodhouse and other members of the society carried on a long controversy with the venerable Priestley, still a doughty defender of the phlogiston theory, after his arrival in America in 1794.

The society held weekly meetings. In pursuance of its principal purpose of acquiring information relative to the minerals of the United States, it issued advertisements soliciting minerals to be examined by a committee without charge.[88] Adam Seybert was among those on the committee, and it was he who conducted the laboratory of the society. On October 24, 1801, a committee of the society, including Robert Hare in its membership, was appointed to study how a greater concentration of heat might be obtained for chemical purposes. On December 10, Hare presented his famous "Memoir of the Supply and Application of the Blow-Pipe," setting forth the principles of his newly discovered oxyhydrogen blow-pipe capable of producing intense heat. Hare's memoir was published the following year by order of the society[89] and won for him instant recognition as an outstanding scientist.[90] Ultimately it brought him in 1839 the Rumford Medal of the Academy of Arts and Sciences, granted then for the first time. Benjamin Silliman and others presently took up the device invented by Hare and by new technical arrangements further enhanced its usefulness.[91]

The first geological society in the United States appears to have been the short-lived American Mineralogical Society which came into existence in 1798 in New York City. Its purposes were declared to be the "investigation of the mineral and fossil bodies which compose the

---

[87] T. P. Smith, *Annual Oration Delivered before the Chemical Society of Philadelphia, April 11, 1798. A Sketch of the Revolutions in Chemistry* (Philadelphia, 1798).

[88] *The Weekly Magazine*, Sat., Feb. 3, 1898; *Medical Repository*, II (1799), 120; *ibid.*, III (1800), 68.

[89] Robert Hare, Jr., *Memoir of the Supply and Application of the Blow-Pipe . . . Published by Order of the Chemical Society of Philadelphia* (Philadelphia, 1802); reprinted entire in E. F. Smith, *Chemistry in America*, 153-179; also in *Tilloch's Philosophical Magazine*, xiv (1802), 238-245, 298-308.

[90] E. F. Smith, *The Life of Robert Hare; an American Chemist, 1781-1858* (Philadelphia and London, 1917), 5 ff.

[91] The succeeding developments are fully discussed in Benjamin Silliman, "American Contributions to Chemistry," *Proceedings at the Centennial of Chemistry Held August 1, 1874, at Northumberland, Pa. Reprinted from the August-September and December Numbers (1874) of the American Chemist* (Philadelphia, 1875), 77-78.

fabric of the globe; and more especially for the natural and chemical history of the minerals and fossils of the United States."[92] Dr. Samuel L. Mitchill zealously proclaimed that it was the object of the organization "to arm every hand with a hammer, and every eye with a microscope."[93]

Several organizations, somewhat ephemeral in character, were instrumental in stimulating interest in various branches of applied science. The Society for the Encouragement of Arts, Manufactures and Commerce, of New York, dated from 1754, sponsored various projects to aid commerce and industry in the colonies. It served as a model for a Society for the Encouragement of Manufactures in New York in 1765, and the New York Society for Promoting Arts formed in the following year, about the same time that the American Society Held at Philadelphia for Promoting Useful Knowledge was coming into existence.

Among the more prominent of the utilitarian organizations formed toward the close of the eighteenth century were the Pennsylvania Society for the Encouragement of Manufactures and Useful Arts (1787), the Rumsean Society of about the same date, of which Franklin was a member, to aid the inventor Rumsey with his projected steamboat, the New York Society for the Encouragement of American Manufactures (1788), the Society for the Promotion of Agriculture, Arts and Manufactures (New York, 1791), the Society for Promoting the Improvement of Inland Navigation (Philadelphia, 1791), and the Society for the Promotion of Useful Arts (Albany, 1792). The Society for Establishing Useful Manufactures (1791), the so-called S.U.M., planned and organized by Alexander Hamilton as a part of his program for making the United States industrially independent, proved to be the origin of Paterson, N. J., at the falls of the Passaic River, where the society attempted to establish manufactures. Although the society soon retired from direct manufacturing operations, it is still in existence and has been a vital influence in the industrial development of New Jersey. Numerous similar groups were destined to be characteristic of the early nineteenth century. The Boston Mechanick Association (1795) was one of the earliest forerunners of many similar later groups of men about halfway between being tradesmen and technicians, who engaged in discussions of mechanical problems and scientific instruments, and occasionally even published transactions.

By way of summarizing the early American developments described

[92] *Medical Repository*, I (1800), 105.

[93] L. C. Beck, "History of the Progress of Mineralogy in the State of New York," *Natural History Survey of New York*, div. 3, *Mineralogy* (Albany, 1843), preface, 9-10.

in this chapter, it may be said that numerous scientific societies were springing up in Europe during the time that the British colonies in America were being settled. The Royal Society of London was the first scientific society to serve the needs of the American colonists, a goodly number of whom became enrolled among its fellows or contributed to its transactions. While the energies of several generations of pioneers were primarily consumed in subduing the primeval wilderness and in eking out a precarious livelihood, there is clear evidence that there were those, either of European birth or in close contact with the intellectual currents of the continent, who desired to turn their attention to affairs of the intellect. The Boston Philosophical Society, inaugurated under the leadership of Increase Mather in 1683, comprised such a group. Later, two notable societies that are still in existence, the American Philosophical Society and the American Academy of Arts and Sciences, were launched upon their long and distinguished careers. An even more ambitious project, the Académie des États-Unis de l'Amérique, attempted in Virginia during the troublous times under the Confederation, proved abortive for lack of financial support and collapsed altogether when European aid had to be withdrawn when the storm of revolution burst upon France in 1789. Under more favorable circumstances it might conceivably have become a genuine national academy of arts and sciences. Even prior to the establishment of our federal government, the beginnings of specialization within the field of science are clearly marked. Evidence of this fact may be seen in the scattered societies of the medical profession, the first to organize. Societies devoted to agriculture, geology, chemistry, and technological pursuits, too, were springing up by the close of the century in America.

Chapter II

NATIONAL GROWTH, 1800–1865

With the establishment of a stronger federal government in 1789, the United States came to depend less on Europe and to develop more and more along its own national lines. In the sphere of international relations, a self-reliant attitude manifested itself in a policy of isolation and in a desire to resist the territorial ambitions of European powers toward the American continents, which culminated in the enunciation of the Monroe Doctrine in 1823. Self-reliance in the domestic political sphere found expression in the triumphant democracy of the Jeffersonian and Jacksonian eras. In economic affairs there was an attempt to resist the influence of foreign capitalists on the banking system of the nation and to achieve financial independence. Protective tariffs in 1816, 1824, 1828, and other years were designed to foster the growth of native manufactures. In the intellectual realm the *literati* became restive of the shackles and conventions of tastes imported from abroad. Poets and prose writers began creating a distinctively American literature; language became more distinctively American at the hands of the lexicographer Noah Webster.

While science of necessity remained somewhat more international in outlook, Americans no longer relied upon stimulus from Europe in the development of their scientific societies, and the American societies came to differ rather sharply from the older academies of Europe. The new American organizations, born in the crucible of a democratic society, came to open their membership to all who were interested in science rather than to restrict their membership to the élite intellectually, as had the European academies. Confronted by grave problems of a political and economic nature for about half a century after 1789, Americans failed to establish additional permanent scientific societies of national scope, but they succeeded in creating many local ones. Toward the close of the eighteen thirties, the need for societies of nationwide membership, in addition to the American Philosophical Society and the American Academy of Arts and Sciences, became increasingly apparent. Not only were general scientific societies of national proportions sought, but an insistent demand arose for national specialized societies. The success attained in meeting these needs marked an important forward step in the history of the organization of American scientific activity.

The American scientists who lived during the first half of the nineteenth century were confronted with exceptionally challenging problems. A vast amount of what was both figuratively and literally the "spade work" of science was waiting to be performed to systematize knowledge in botany, zoology, and geology.[1] Newly discovered flora, fauna, and strata presented endless difficulties in the way of description and classification. Moreover, the conditions of work at the outset were discouraging. The facilities for scientific research were ludicrously insufficient. For example, we are told:

> About the time when Mr. Silliman was appointed a professor (1804), the entire mineralogical and geological collection of Yale College was transported to Philadelphia in one small box, that the specimens might be named by Dr. Adam Seybert, then fresh from Werner's School at Freiberg, the only man in this country who could be regarded as a mineralogist scientifically trained.[2]

Thomas Jefferson, even while struggling to keep American commerce free of the toils of European strife, found time to assemble a collection of fossil bones in a room of the White House. As is well known, his endeavors in paleontology were not wholly appreciated by many of his less scientific contemporaries. Among his critics was a boy poet, thirteen years of age, William Cullen Bryant, later to become a champion of Jeffersonian democracy and a writer of nature verse. In 1808, however, he savagely ridiculed Jefferson's scientific investigations in a pamphlet, *The Embargo, or Sketches of the Times:*

[1] "When Amos Eaton, Parker Cleveland, Robert Hare, Benjamin Silliman, Edward Hitchcock, and Chester Dewey began their labors, the natural sciences, as they are now understood, had hardly an existence. . . . These sciences, themselves, being in a formative state, were not differentiated, nor their limits marked out. These men of necessity studied nature in the mass, meeting often the unclassified subject-matter of several sciences in a single investigation. They constantly encountered the difficulties resulting from faulty and incoherent terminology. Their memories were burdened in keeping abreast with the changes which rapid scientific progress made necessary. Classes as well as names were in a state of constant flux." M. B. Anderson, "Sketch of the Life of Prof. Chester Dewey, D.D., LLD., printed as pp. 121-132 of *Proceedings of the Fifth Anniversary of the Convocation of the State of New York* (Albany, 1869), 125, 127.

[2] "Benjamin Silliman," an unsigned obituary in *American Journal of Science and Arts,* 2d ser., XXXIX (1865), 4.

Go, wretch, resign the Presidential chair,
Disclose thy secret measures, foul or fair.
Go search with curious eye for horrid frogs
Mid the wild wastes of Louisianian bogs;
Or where Ohio rolls his turbid stream
Dig for huge bones, thy glory and thy theme.[3]

Chairs of science, apart from professorships of medicine, were with few exceptions new in American colleges. Most American scientists in the first half of the nineteenth century did not find their careers as instructors in science in the colleges, but rather they either were amateur investigators or were employed in the service of the government. The men who did succeed in securing academic appointments were burdened with a variety of cares. Heavy teaching schedules were the rule and involved giving courses in several different branches of learning. Administrative duties were also often exacted of the college professors. Many professors were clergymen and performed, at least occasionally, ministerial functions. Learned men were expected to take a prominent part in the public and social life of their communities. Only a small portion of their time, therefore, was left for research. They were greatly handicapped for want of books, and scientific periodicals were scarce. Nevertheless, an able group of men, including Benjamin Silliman, Robert Hare, and John J. Audubon, set to work with astonishing energy to place American science upon firm foundations.

The rapid accumulation of scientific knowledge abroad and its cordial reception in America are important in explaining the accelerated multiplication of scientific societies in the United States from the close of the eighteenth century onward.

During the latter half of the eighteenth century, Euler, Clairaut, d'Alembert, Lagrange, and Laplace had been among those pre-eminent in solving a variety of mathematical problems bequeathed by the great Sir Isaac Newton, who died in 1727, the very year in which Franklin founded the Junto. Sir William Herschel had discovered the planet Uranus in 1781. Lavoisier had laid the foundations for quantitative chemistry before he suffered an untimely death at the hands of the French revolutionists in 1794. Linnaeus, Buffon, and Cuvier had performed notable work in classifying animal and plant life. Toward the close of the eighteenth century, Benjamin Thompson, an American by birth, who subsequently became Count Rumford of Bavaria, enunciated the vibratory theory of heat. Sir Humphry Davy and others followed in his footsteps.

[3] *The Embargo, or Sketches of the Times* (Boston, 1808).

In the opening decades of the nineteenth century, new and important advances were made on both sides of the Atlantic. In the field of astronomy, Sir William Herschel and, later, his son Sir John Herschel gave a new and grander picture of the stellar universe. Laplace continued to bring new methods of analysis to bear upon the problems of gravitational attraction. Piazzi discovered the first asteroid on the opening night of the nineteenth century. An able American seaman, Nathaniel Bowditch, published his *New American Practical Navigator* in 1802, followed by many subsequent editions. Between the years 1829 and 1839 he brought out a translation of Laplace's *Méchanique Céleste*. Physical theories underwent a revolution when Young, Fresnel, and Arago emerged the victors in a bitter fight that won recognition for the undulatory theory of light. Wollaston in 1802 and Fraunhofer in 1815 noticed the dark lines in the solar spectrum and thereby paved the way for the study of spectroscopy later in the century. Ever since Benjamin Franklin had demonstrated the identical nature of lightning and the electrical discharge from a Leyden jar in 1752, Americans had shown a particular interest in experimentation with electricity. Galvani and Volta in Europe were contemporaries of Franklin whose very names have passed into the nomenclature of electrical science. Oersted, Ampère, and Ohm in Europe, and Joseph Henry in America, continued the study of electrical currents in the opening decades of the nineteenth century. Davy won distinction by isolating the elements sodium and potassium. His great popular renown, however, was the result of his useful invention of the mine safety lamp. Dalton arrived at his atomic theory in the first decade of the century. Proust, Berzelius, Gay-Lussac, and Avogadro further investigated the properties of matter and the laws of its behavior under varying conditions. In America, Benjamin Silliman and Robert Hare, the latter the inventor of the oxyhydrogen blowpipe, became expounders of the new ideas. By 1823, Liebig in Germany was well launched upon a long career fraught with discoveries in inorganic chemistry and its application to agriculture.

As the century progressed, astronomers continued to be among the most active of the scientists. The first man to measure the distance to a star was Friedrich Bessel, who achieved the difficult feat in 1838. Doppler's principle, enunciated in 1842, predicting a change in observed frequency of vibrations from a moving source, proved, somewhat later, to be of great value to the astrophysicist by affording a method for measuring stellar motions. The planet Neptune was discovered in 1846 as a result of the calculations of Adams of England and Leverrier of France. In 1849 the American, Daniel Kirkwood, aroused great interest in the problem of planetary rotation by his writings in the *Proceedings* of the American Association for the Advancement of Science. Professor Olmsted of Yale, plotting the paths of meteors of the shower of 1833, gave a convincing demonstration that they were interplanetary bodies and not of terrestrial origin. His work was continued by another Yale professor, Hubert A. Newton. An observatory was founded at Harvard in 1840, the Naval Observatory at Washington in 1844, one at the University of Michigan in 1853, and the Dudley Observatory at Albany in 1856. As early as 1826, Daguerre in France was making successful experiments with photography. Celestial photography was attempted in 1839 by Arago in France and John W. Draper in the United States. William C. Bond, who took up the work at Harvard College Observatory a few years later, took the first photograph of a star in 1850, and in subsequent years went on to develop the intricate processes of faint light photography.

Toward the close of the eighteen twenties, Joseph Henry in America and Michael Faraday in England, working independently, made the great discovery of electromagnetic induction. Both men continued to invent and to make experiments for many years. Late in the eighteen fifties, James C. Maxwell formulated the electromagnetic theory of light and Hermann von Helmholtz arrived at similar views about the same time. The discovery of the electromagnet led Samuel F. B. Morse, an American portrait painter, to invent a commercially practicable telegraph about 1835, and a telegraph line had been constructed between Baltimore and Washington by 1844. About 1850 Fizeau and Foucault devised new methods for more accurate determination of the velocity of light. It was the latter who furnished the first experimental proof of the earth's rotation in 1851 by means of a large suspended pendulum, and for his invention of the gyroscope he received the Copley medal of the Royal Society in 1855. By 1850 William Thomson was elaborating the theory of heat engines worked out somewhat earlier by Carnot and the principle of the conservation of energy revealed by James Joule, who had studied the mechanical equivalent of heat in 1847. When Kirchhoff and Bunsen perfected the spectroscope in 1859-1860, they opened a new era in physics, chemistry, and astronomy.

In his *Principles of Geology*, published in 1833, Lyell came forward as the champion of the doctrine of uniformitarianism and sought to account for the major geological changes of the earth's past, not in terms of sudden catastrophic occurrences as had theretofore been commonly held, but rather by postulating that the common agencies of physical force such as the rain, wind, and ice have been at work ceaselessly through countless centuries. Lyell, in his attempts at reading the records of these changes in the rocks, laid the foundations of paleontology, and the stratigraphic deposits of fossils were to furnish invaluable evidence in support of the evolutionary hypothesis enunciated about a quarter of a century later. Other British geologists engaged in similar studies were William Smith, James Hutton, Roderick Murchison, and Adam Sedgwick. William Maclure, Amos Eaton, Thomas Say, James Hall, and James D. Dana did comparable work in America. The explorations of Alexander von Humboldt in Central and South America, extending over a period of years, aroused great interest in the United States, where his travel works were avidly read. In 1840, Louis Agassiz, in Europe, enunciated his theory of ice ages characterized by widespread continental glaciation. Later, he came to reside in America and teach at Harvard, where he won many adherents to his views, which have subsequently become almost universally accepted. Charles Wilkes, an American explorer in polar seas, was perhaps the first to sight the Antarctic Continent, the misty shore of which he appears to have seen in 1840. Matthew F. Maury did pioneer work in the realm of oceanography while in the service of the United States government.

Fortunately for the cause of science, the federal government, actuated by practical motives, lent financial support to certain types of scientific activity. The Sibley expedition explored the Red River in 1803, the Lewis and Clark expedition set forth in 1804, Pike set out in 1805, the Long expedition to the Rockies in 1819, the Schoolcraft expedition to the Indian Country in 1831; and Frémont explored Oregon and California in 1843. These were but the more prominent of numerous exploring parties which opened up new regions to the knowledge of the white man. Moreover, the eastern United States, and even the seaboard region, had yet to be explored in detail. New flora and fauna were found which differed from those of Europe. New strata of rock had to be classified, and topographic maps were needed. Mineral resources were eagerly sought. The first geologic survey in America made at public expense was begun by the state of North Carolina in 1823. Similar surveys were commenced by South Carolina in 1824, Massachusetts in 1830, Tennessee in 1831, New Jersey in 1835, Georgia, Maine, and New York in 1836, Delaware, Indiana, Michigan, and Ohio in 1837, New Hampshire and Rhode Island in 1839, Vermont in 1844, Mississippi in 1850, Illinois, Missouri, and Wisconsin in 1853, Kentucky and Iowa in 1854, Arkansas in 1857, Texas in 1858, Louisiana in 1864, and Nevada in 1865.

Notable advances were made in the biological studies on both sides of the Atlantic. In 1809, Lamarck published his views on evolution and the transmutation of species in his *Philosophie Zoologique*. Although many of his ideas have received subsequent vindication, his views were bitterly assailed by his colleague Cuvier, generally considered as the greater zoological authority at the time and a staunch protagonist of the special creation theory. In America, Thomas Jefferson became interested in the fossil bones of huge extinct prehistoric creatures. Somewhat later, John J. Audubon, a self-taught ornithologist, and Asa Gray, professor of botany at Harvard, were without peers in America in the fields of their chosen labors. In Europe, Schleiden and Schwann were propounding the cell theory by 1838-1839. A classic of biological literature appeared in 1859, when Charles Darwin published his *Origin of Species*. Copies were sent to America, and it was read by a small but select group of scholars. Indeed, Asa Gray, an ardent exponent of the Darwinian hypothesis in later years, had even earlier been corresponding with Darwin on the subject of evolution.[4]

Creditable work in medicine and surgery was performed by such Americans as John C. Warren, Walter Brashear, Valentine Mott, John King, Horatio G. James, William Beaumont, and David I. Rodgers. Caspar Wistar published important studies in anatomy, especially between the years 1811 and 1814. The study of the mind received impetus from the work of Pinel in France, who secured kinder and more rational treatment for the insane, and from that of Benjamin Rush, sponsor of similar ideas in America. The brain physiologists, Franz Joseph Gall and Caspar Spurzheim in Europe, elaborated phrenology, a system alleged capable of enabling its votaries to discover a person's mental aptitudes by noting irregularities of skull structure, supposed to correlate with one's faculties. Phrenology had a wide vogue in America.[5] Anesthesia, or the "death of pain," was given to the world by three Americans working independently. Crawford W. Long used ether successfully in 1842, W. T. G. Morton did likewise in 1846, attracting the notice of the scientific world, and Horace Wells had similarly used nitrous oxide in 1844. Sir James Y. Simpson, a Scotch surgeon, used chloroform successfully in 1847. Oliver Wendell Holmes, generally remembered by posterity for his poetry, was by profession a medical man teaching in the Harvard Medical School, and he antici-

[4] G. H. Genzmer, "Asa Gray," *Dictionary of American Biography*, edited by Allen Johnson and Dumas Malone (20 vols., New York, 1928-1936), VIII, 513.

[5] An interesting account of this pseudo-science is to be found in R. E. Riegel, "Early Phrenology in the United States," *Medical Life*, XXXVII (1930), 361-376.

pated the European gynecologist, Semmelweiss, by several years in explaining the contagiousness of puerperal fever. Joseph Leidy, one of the most versatile scientists America has ever produced, made important discoveries in bacteriology.

In Europe, pioneer work was being done in experimental psychology. Gustav Fechner published his *Psychophysik* in 1860, thereby introducing a new word into the vocabulary of science and new ideas as well, elaborating on the relationships between physical stimuli and psychological perception.

An important factor in the diffusion of scientific knowledge in America was the appearance of an increasing number of scientific periodicals. The American Philosophical Society and the American Academy of Arts and Sciences continued to issue publications to keep their members informed of the scientific topics of the day. These were soon supplemented by the appearance of scientific magazines issued as purely commercial enterprises. *The Medical Repository* (New York, 1797-1822) seems to have been the first purely commercial scientific periodical in the United States. While emphasizing medicine in particular, it devoted space to other scientific subjects as well. The *Philadelphia Medical and Physical Journal* existed from 1804 to 1807. The *American Mineralogical Journal* (New York, 1810-1814) was the first American journal to be devoted to a special branch of natural history, it would seem.[6] The *American Medical and Philosophical Register* was published in New York from 1810 to 1814, and the *Medical and Physical Journal* in the same city from 1822 to 1830. The first strictly scientific journal west of the Alleghanies was the *Western Quarterly Reporter of Medical, Surgical and Natural Science* (Cincinnati, 1822 to 1823). The *Western Journal of the Medical and Physical Sciences* was published in the same city from 1827 to 1836.

The American Journal of Science and Arts,[7] established at New Haven in 1818 under the able editorship of Benjamin Silliman, became the leading scientific journal in America for several decades and still holds a prominent place among such periodicals. Its purpose was set forth by Silliman as follows:

[6] "The object of this work is to collect and record such information as may serve to elucidate the Mineralogy of the United States, than which there is no part of the habitable globe, which presents to the mineralogist a richer or more extensive field for investigation." *American Mineralogical Journal*, I (1814), preface, iii.

[7] It soon became known as *Silliman's Journal*, and after 1880 as the *American Journal of Science*. Its history is traced in E. S. Dana, and others, *A Century of Science in America* (New Haven, 1918), ch. I, "The American Journal of Science."

It is designed as a deposit for original communications; but will contain also occasional selections from foreign journals. Within its plan are embraced

Natural History, in its three great departments of mineralogy, botany, and zoology.

Chemistry and natural philosophy, in their various branches; and mathematics pure and mixed.

It will be a leading part to illustrate American natural history and especially our mineralogy and geology. . . .

Learned societies are invited to make this Journal, occasionally the vehicle of their communications to the public.[8]

The importance of this journal in the development of American science in the pre-Civil War deserves especially to be stressed.[9]

Laboring for the cause of scientific societies for thirty years through the vehicle of his journal, Silliman was able to express his satisfaction at their progress, an advance to which he had himself so largely contributed:

Comparing 1817 with 1847, we mark . . . a very gratifying change. The cultivators of science, were then few—now they are numerous. Societies and associations of various names, for the cultivation of natural history, have been instituted in very many of our cities and towns, and several of them have been active and efficient in making observations and forming collections.[10]

One of the agencies for the popularization of scientific discoveries was the lyceum. The father of the lyceum movement was Josiah Holbrook of Darby, Connecticut, who published in 1826 an article in the *American Journal of Education* entitled "Associations of Adults for the Purpose of Mutual Education." The same year he organized in Massachusetts, "Millbury Lyceum, No. 1, Branch of the American Lyceum," which led the way for the formation of the some three thousand similar lyecums existing in 1834 in every part of the Union, especially in New England. A national convention of lyceums was held in

---

[8] *American Journal of Science*, I, 2d ed. (1819), preface, v-vi.

[9] "Of the services of this Journal to American science, it is not too much to say that more than any other similar publication, it has aided and stimulated our countrymen in their scientific labors, and has made their names and works familiar to men of science abroad while through the variety and weight of its contributions it has not only won a high reputation among contemporary journals but has vindicated for our country an honorable place among the communities in which science is most promoted and esteemed. . . . " American Academy of Arts and Sciences, *Proceedings*, VI (1865), 512.

[10] Benjamin Silliman in *American Journal of Science and Arts*, 1st ser., L (1847), preface, xiv-xv.

1839, and the movement continued to prosper until the Civil War. Many prominent men in letters or science, such as Emerson, Silliman, and Agassiz, made frequent appearances on the platforms of the lyceums.

By the eighteen forties the United States had undergone a transformation on the material side which prepared the way for its scientists to enter upon a period of national integration of scientific endeavor. The nation was being welded together by a network of canals, highways, and railroads. The public was accordingly becoming more accustomed to travel. Scientific gatherings could now be held with greater ease. The telegraph and fast mail made it possible to transmit scientific communications with a rapidity and safety which would have been envied by Newton, Leibnitz, or Franklin, who experienced difficulties with belated communications and transactions. The new facilities for transportation and communication went a long way toward breaking down the local feeling and provincialism of the decades which had preceded.

A rather subtle psychological factor ought also to be mentioned. Americans were becoming a nation of "joiners." Alexis de Tocqueville and other Europeans expressed astonishment at the propensity of Americans for organizing private associations for every conceivable purpose, nor did this significant trend of the times pass unnoticed by native American writers.[11] Political clubs throve, temperance and anti-slavery societies flourished, charitable and educational societies were doing work presently to be assumed by the states as obligations of government, religious societies abounded, labor unions were struggling desperately for recognition, the sponsors of bizarre therapeutic cults and the devotees of novel religious faiths paraded the American scene and drew votaries from every walk of life. College fraternities and fraternal orders were also to be reckoned among the multifarious

[11] "In truth one of the most remarkable circumstances or features of our age is the energy with which the principle of combination, or of action by joint forces, by associated numbers is manifesting itself. It may be said, without much exaggeration, that everything is done now by societies. Men have learned what wonders can be accomplished in certain cases by unions, and seem to think that union is competent to everything. You can scarcely name an object for which some institution has not been formed. Would men spread one set of opinions or crush another? They make societies. Would one class encourage horse-racing and another discourage travelling on Sunday? They form societies. We have immense institutions spreading over the country, combining hosts for particular objects. We have minute ramifications of these societies. . . . This principle of association is worthy of the philosopher, who simply aims to understand society and its most powerful springs." W. E. Channing, *The Works of William E. Channing* (Boston, 1877), 139.

societies flourishing on every hand. Democratic America was *par excellence* the land of the ubiquitous society where organization was unhampered at a time when much of Europe, under the heel of Metternich, the Tsars, and the Bourbons of Spain and Italy, was taking severely repressive measures to stamp out many forms of private organization, going even at times to the length of putting scientific academies under governmental surveillance. Americans, on the other hand, were wholly free to form this or that group as they saw fit. Surrounded then by their countrymen embarking upon every sort of political, social, economic, educational, or religious organization, it was only natural that those interested in the promotion of science should turn more and more to the idea of banding together to further their common pursuit of knowledge. This led, as we shall see in the following pages, to the creation, first of all, of countless local scientific societies, many of them ephemeral and wholly forgotten, and, second, to the establishment of a number of strong, important, and lasting state academies of science and national scientific societies.

Increasing interest in science throughout the nineteenth century led, among other things, to the formation of state academies in most of the states of the Union. Indeed, by the time of the outbreak of the Civil War, such bodies had been established in almost every Northern state and in a few of those of the South. Simultaneously, countless municipal and other local academies of science or general science societies were springing into existence. In our study of these societies it will prove advantageous to classify them according to geographical regions and base our deductions accordingly. Our basis of classification shall be the nine geographic divisions of the United States census: the New England, Middle Atlantic, East North Central, West North Central, South Atlantic, East South Central, West South Central, Mountain, and Pacific.[12]

We shall begin our study with a consideration of the societies of New England. Maine attempted a state academy of science, one being formed in 1836 under the name of Maine Institute of Natural Science.

[12] The detailed classification is as follows: New England Division, comprising Me., N. H., Vt., Mass., R. I., and Conn.; Middle Atlantic Division, comprising N. Y., N. J., and Pa.; East North Central Division, comprising Ohio, Ind., Ill., Mich., Wis.; West North Central Division, comprising Minn., Iowa, Mo., N. D., S. D., Neb., Kan.; South Atlantic Division, comprising Del., Md., D. C., Va., W. Va., N. C., S. C., Ga., Fla.; East South Central Division, comprising Ky., Tenn., Ala., Miss.; West South Central Division, comprising Ark., La., Okla., Tex.; Mountain Division, comprising Mont., Idaho, Colo., N. M., Ariz., Utah, Nev.; Pacific Division, comprising Wash., Ore., and Calif.

It was of brief duration, however, giving place to the Portland Society of Natural History in 1843. Numerous and important local bodies were formed for the general promotion of science. An ambitious enterprise was the formation at Boston of the New England Society for the Promotion of Natural History in 1814. Its name was changed in the following year to Linnaean Society of New England. In 1817 the Linnaean Society issued a report about a supposed sea serpent seen near Cape Ann, with plates illustrating dissection of the supposed young of the monstrosity. The society dissolved about 1822, and its collections ultimately went to the Boston Society of Natural History.

Perhaps the most active of the local scientific organizations founded in New England during this period was the Boston Society of Natural History, which came into existence in the winter of 1830, when "a few gentlemen of scientific attainments conceived the design of forming a society in Boston, for the promotion of natural history."[13] A formal organization was effected on April 28, 1830, and the society was incorporated the following year. Soon it had successfully built up a museum and library, which still remains one of the show places of Boston. Early members of the society took a prominent part in the first geological and botanical surveys of Massachusetts. The *Boston Journal of Natural History,* a publication sponsored by the society, gave stimulus to scientific inquiry and served as a clearing house for the numerous natural history societies of New England. It was but natural that the Academy of Arts and Sciences should suffer a little by the competition of the young and vigorous Boston Society of Natural History. Time has shown, however, that there is ample room for both in Boston, and the Boston Society of Natural History has entered upon its second century of unbroken activity.[14]

Meanwhile similar, but lesser, societies were being formed in small places throughout New England. The Providence Franklin Society (Providence, Rhode Island), founded in 1821, was in a flourishing condition by 1830.[15] In the twenties, a Philophusian Society is reported to have been pursuing similar aims.[16] The Amherst College Natural History Society, begun in 1822, was vitalized in 1826 through the efforts of Professor Edward Hitchcock who arrived on the scene

[13] A. A. Gould, "Notice on the Origin, Progress, and Present Condition of the Boston Society of Natural History," *American Quarterly Register,* XIV (1842), 256.

[14] The most recent of the histories of the society is P. R. Creed, edit., *The Boston Society of Natural History, 1830-1930* (Boston, 1930).

[15] Its structure and work are described in a note in *American Journal of Science and Arts,* XVIII (1830), 195-196.

[16] *American Journal of Science and Arts,* X (1826), 370.

in 1826 and built up collections into a famous museum. The Worcester Lyceum of Natural History, commencing operations in 1825, deposited its collections with the American Antiquarian Society about 1830, when it became inactive.[17] It was revived in 1854 as the Natural History Department of the Young Men's Library Association. Its name was changed in 1866 to Worcester Lyceum and Natural History Association and in 1884 to Worcester Natural History Society. After 1830 the natural history movement in New England gained still more impetus, and was reflected in an abundance of societies. The Essex County Natural History Society (Salem, Massachusetts) was established in 1833. The Yale Institute of Natural Science, formed in 1834, became known as the Yale Natural History Society in the following year. Benjamin Silliman, its first president, published its early papers in the *American Journal of Science*. The Natural History Society of Hartford was founded in 1835, becoming the Connecticut Society of Natural History in 1845. The Williams College Lyceum of Natural History was established in 1835. It conducted expeditions to the Bay of Fundy, Berkshire County, Massachusetts, Florida, Newfoundland, and Greenland. The Cuvierian, or Natural History Society of Wesleyan University (Middletown, Connecticut), was formed in 1836, the Harvard Natural History Society in 1837, the Northern Academy of Arts and Sciences (Hanover, N. H.) in 1841, the Lynn Natural History Society in 1842, the Portland Society (Portland, Maine) of Natural History in 1843, the Concord Natural History Society (Concord, N. H.) in 1846, and the Essex Institute (Salem, Massachusetts) in 1848. The last was formed by a merger of the aforementioned Essex County Natural History Society and the Essex Historical Society. By 1861 the Institute had a library of about 22,000 volumes and, a museum of over 50,000 specimens. The Institute inaugurated field meetings as a part of its educational program.

> At these meetings we bring together those who feel an interest in the works of Nature, and who make them their especial study; and we place them face to face with the various phenomena of creation, as they are exhibited in our fields, our hills and our forests. By these excursions we are relieved from the necessity of studying these things in the dry dead cabinets of home and the student who walks with us has a view of them as Nature has herself arranged them, drawing his conclusions from facts undisguished by the interference of man, and free from the partial and imperfect character which will ever be detected, even in the best ordered and fullest collection.[18]

[17] *Ibid.*, XVIII (1830), 139.
[18] G. D. Phippen, "Field Meeting at Lynn," Essex Institute, *Proceedings*, III (1861), 101.

Meanwhile, as we shall presently see, parallel developments were taking place in other sections of the country, especially in those adjacent to New England, where the study of natural history was especially the vogue. New York City was destined early to become a notable center of scientific organization. The earliest general enterprise was the Society Instituted in the State of New York for the Promotion of Agriculture, Arts, and Manufactures, organized in 1791, and presently becoming known as the Society for the Promotion of Agriculture, Arts and Manufactures. Its president, from 1791 to 1813, was Robert R. Livingston. Headquarters of the Society were transferred to Albany, the new capital, in 1798. The Literary and Philosophical Society of New York was founded in 1814 and incorporated the following year. De Witt Clinton, who served as its president during the entire time of its existence until the late twenties, delivered a memorable address on May 4, 1814, summarizing scientific progress in America and the achievements of scientific societies down to that date.[19] The society's membership gradually passed over to the Lyceum of Natural History of New York, including Samuel L. Mitchill. About 1816 the New York Historical Society (1804) embarked temporarily on the collection of natural history specimens.

The Lyceum of Natural History of New York was launched in 1817. The adoption of the designation "lyceum," instead of the usual words "society or academy," was promptly imitated by other organizations presently springing up in the region and later by numerous societies throughout the country. Commenting on the name, Samuel L. Mitchill, its first president, wrote: "The members called it the Lyceum, in remembrance of the school founded by that sublime genius, Aristotle, at Athens. Disciples of the 'mighty Stagirite,' they determined after his example, to be peripatetics, and explore and expound the arcana of nature as they walked."[20] The society built up an extensive collection of biological and geological specimens and an excellent library. It adopted its present name of New York Academy of Sciences in 1876.[21] The Brooklyn Institute of Arts and Sciences was organized in 1823, incorporated in the following year as the Apprentices' Free Library of Brooklyn, rechartered as the Brooklyn Institute in 1843, and re-

[19] Literary and Philosophical Society of New York, *Transactions,* I (1815), 19-184; also printed separately. Reviewed in *North American Review,* I (1815) , 390-402.

[20] Printed in J. H. Barnhart, "The First Hundred Years of the New York Academy of Sciences," *Scientific Monthly,* V (1917) , 464.

[21] The history of the New York Academy of Sciences is traced in the reference mentioned in the preceding footnote and in H. L. Fairchild, *History of the New York Academy of Sciences* (New York, 1887) .

chartered in 1890 as the Brooklyn Institute of Arts and Sciences. It
has become one of the most active of the local bodies of science, carrying on extensive lecture courses in practically every department of the
arts and sciences and maintaining libraries, museums, and publications. The United States Naval Lyceum was established at Brooklyn
in 1833 by a group of naval officers, many of whom were engaged in
protecting a flourishing Oriental sea trade.[22]

> As mere collectors of specimens in natural history and curiosities, illus-
> trative of the manners and customs of distant nations, which the mem-
> bers in discharge of their official duties to their country, may be called
> upon to visit, they possess some advantages that they are not only will-
> ing but anxious to improve. . . . Indeed the return of every National ves-
> sel, since the institution has been in operation, has contributed more or
> less to its collection.[23]

The valuable museum thus built up was given to the Naval Academy
when the society became inactive in 1891.

A Brooklyn Lyceum of Natural History, formed in 1838, evidently
was unimportant and short-lived.[24] About 1798 the Society Instituted
in the State of New York for the Promotion of Agriculture, Arts, and
Manufactures (founded in New York City in 1791 and incorporated
in 1793) removed to Albany and reincorporated in 1804 as the Society
for the Promotion of Useful Arts. The Society, of which Robert R.
Livingston served as president for many years, maintained a museum
in the Capitol. In 1824 it united with the Albany Lyceum of Natural History (founded the preceding year) and became the Albany
Institute. Upon consolidation with the Albany Historical and Art
Society in 1900, it assumed its present name of Albany Institute of
History and Art.

The United States Military Society, organized at West Point in 1803,
is known to have lasted until at least 1810. It is said to have been the
first society in the United States to hold peripatetic meetings, moving
about from place to place. The Troy Lyceum of Natural History
came into being in 1818 under the inspiration of Amos Eaton. Publications were issued sporadically until 1850. The Catskill Lyceum of
Natural History, formed in 1820, was probably also due to the stimulus

[22] "We the officers of the Navy and the Marine Corps, in order to promote the
diffusion of knowledge, to foster a spirit of harmony, and community of interest in
the service, and to cement the links which unite us as professional brethren, have
formed ourselves into a society, etc." *American Journal of Science and Arts*, 1st ser.,
XXXVIII (1835), 390.

[23] *Ibid.*, 390-391.

[24] Stiles, *History of Brooklyn* (2 vols., Brooklyn, 1884), II, 1302.

of Eaton's lecture tour, and at Hudson, New York, Amos Eaton also stimulated sufficient interest in science in 1819 to cause the Hudson Association for Improvement in Science to be formed. Incorporated in 1821,[25] it lingered on for a few more years. The Newburgh Lyceum of Natural History was formed in 1823, and was in existence for several years; the Utica Lyceum of Natural History, dating from the same year, lasted about a decade. In Western New York, we find the Natural History Society of Geneva College coming into existence in 1840 and lasting for about a decade and a half, despite much "dissention and unpleasantness" within its ranks.[26] The Buffalo Society of Natural Sciences, still active and important, began its career in 1861.[27] Sporadic attempts in this period to organize societies in Rochester and Syracuse, and elsewhere in western New York, are also known to have occurred, though not achieving much success. In this period Pennsylvania, no less than New York, was the scene of considerable activity in scientific organization.

The Academy of Natural Sciences of Philadelphia was started in 1812 by founders who agreed "to contribute to the formation of a Museum of Natural History, a Library of works of Science, a classical experimental philosophical apparatus, and every other desirable appendage or convenience for the illustration and advancement of natural knowledge. . . . "[28] William Maclure presented specimens collected on his travels in Europe and generously made grants of books and money as well. Dr. Adam Seybert's collection of minerals, one of the finest in America, was early acquired by the society. The valuable entomological library of Thomas Say was presented to the society in 1834. Thomas Nuttall gave his herbarium to the society. Ornithological collections

---

[25] *American Journal of Science,* X (1826), 372.

[26] Letter written by J. S. Fowler, of Hobart College, commenting on the minutes for 1843, cited in Max Meisel, *A Bibliography of American Natural History* (3 vols., Brooklyn, 1924-1929), II, 705.

[27] A precursor of the society was described as follows: "The earliest effort in the city of Buffalo to organize a society for the study of natural history dates back to April 10, 1858, when eight or ten boyish enthusiasts agreed to meet every Thursday evening for the interchange of scientific observations . . . where they discussed the first chapter of Genesis, the value to science of Hugh Miller's labors or those of Dr. Edward Hitchcock and other like grave subjects and out of their slender funds subscribed for the Scientific American and the Horticulture Magazine to aid their budding endeavors." R. H. Howland, "The Buffalo Society of Natural Sciences, Historical Sketch." Buffalo Society of Natural Sciences, *Bulletin,* VIII (1907), no. 6, 3.

[28] S. G. Morton, "History of the Academy of Natural Sciences of Philadelphia," *American Quarterly Register,* XIII (1841), 433.

were garnered from the remotest parts of the earth.[29] The *Journal* of
the Academy of Natural Sciences, commencing in 1817,[30] was published for over a century. In the years prior to the Civil War, the
names of many prominent American scientists adorned its pages.
Thomas Say, Thomas Nuttall, Joseph Leidy, S. G. Morton, S. F. Baird,
S. W. Woodhouse, J. L. Conte, Edward Hallowell, Isaac Lea, J. D.
Dana, and Asa Gray were among the most active contributors. Publication of the *Proceedings* was begun in 1841. The Maclurian Lyceum
of Philadelphia was opened in 1826, when "in consequence of an
increasing taste for scientific pursuits, it was thought advisable to
form another institution in this city, which should afford additional
facilities for the acquisition of knowledge."[31] Comprehensive plans for
a library and museum were projected, but the society came to an end
four years later. Thomas Say, its president, and C. L. Bonaparte contributed to its papers.

Meanwhile, societies were springing up at other places in Pennsylvania. The Chester County Cabinet of Natural Sciences, a society
founded at West Chester in 1826, continued in operation until 1869,
and it published a number of papers. Active also in the days down to
the Civil War were the Delaware County Institute of Science, organized at Media, Pennsylvania, in 1833, and the Pottsville Scientific
Association of Schuylkill County, in 1854.[32] Still a third scientific enterprise launched in Philadelphia was the Wagner Free Institute of
Science, founded in 1855 by William Wagner for the purpose of
"Gratuitous instruction in the natural sciences and the arts, the support of a free reading room and library, and of a museum for all objects
designed in the opinion of the trustees to instruct those inhabitants
of the city of Philadelphia whose occupations are such as to deprive

[29] These collections and others are discussed in S. G. Morton, "History of the
Academy of Natural Sciences of Philadelphia," *American Quarterly Register*, XIII
(1841), 437-438.

[30] "In further pursuance of the objects of their institution, the Society has now
determined to communicate to the public such facts and observations as having
appeared interesting to them are likely to be interesting to other friends of natural
science.

"They invite the lovers of science generally and particularly all those who are
anxious for its encouragement in the United States, to aid in promoting the objects
of the institution, so long as the contents of it shall be deserving of public approba-
tion." Philadelphia, *Journal*, I (1817), 1-2.

[31] Maclurian Lyceum of Philadelphia, *Contributions*, I (1827), no. 1, 1.

[32] A. W. Schalch and D. C. Henning, edits., *History of Schuylkill County, Pennsylvania* (1907), 179.

them of proper recreation and instruction."[33] There were no publications until 1887. The study of science in New Jersey developed rather more slowly than in the two larger states on either side of it. A Literary and Philosophical Society of New Jersey was instituted at the town of Princeton in 1825. The aims of this society of brief duration were reported to be "the promotion of useful knowledge, and the friendly and profitable intercourse of the literary and scientific gentlemen of New Jersey.[34] And just prior to the Civil War, in 1860, the Burlington County Lyceum of History and Natural Science commenced operations.

The pride of southerners must have been touched to the quick as they watched their northern brethren forging ahead with their scientific organizations. It was in the proud and cultured city of Charleston that the first stirrings in the South Atlantic states took place. There, in 1813, was formed the Literary and Philosophical Society of South Carolina, having among its objects the "collection, arrangement, and preservation of specimens in natural history; and of things rare, antique, curious and useful; and . . . the promotion and encouragement of the arts, sciences and literature, generally."[35] Joel R. Poinsett was prominent among the donors to the collections of the society. The society lasted on until the early eighteen forties. By 1853 the times were again ripe for a scientific society in Charleston, and in that year was formed the Elliott Society of Natural History, named in honor of the distinguished naturalist Stephen Elliott. The introduction to the By-Laws declared rather grandiloquently:

> The advancement of useful knowledge is undoubtedly the leading spirit of the age, and as the principle of association is one of power, success, utility. Societies have been formed in most of the large cities of the civilized world for this object, and their efforts have been crowned with the most flattering success. . . .
> [And far from over-moderate in its claims, it went on] Among the pioneers for the advancement of science and literature, and especially of natural history in North America, the most prominent were Charlestonians . . . the names of Garden, Elliott, Ramsay, Grimke and others . . . have . . . become the common property of the world.
> [It continued] Influenced by these considerations, and with a view not only to the improvement of ourselves, but also to unite our feeble in-

[33] S. B. Weeks, comp., "A Preliminary List of American Learned and Educational Societies," *Report of the Commissioner of Education, for the year 1893-1894* (2 vols., Washington, 1896) , II, 1551.

[34] Quoted in *American Journal of Science*, V (1826) , 374.

[35] "The Literary and Philosophical Society of South Carolina," J. Shecut, in *Shecut's Medical and Philosophical Essays* (Charleston, 1819) , 49.

dividual efforts, more effectually to extend the knowledge we may thereby acquire, . . . do pledge ourselves . . . to follow the rules of an association hereby formed, for developing the natural history of our country, and especially of our own state, and for cultivating and promoting the study of science.[36]

The society survived the Civil War, despite the fact that Sherman's Army destroyed its collections and papers.

In 1816 a group of residents of Washington, who were "impressed with the importance of collecting and distributing the various productions of this and other countries, . . . determined to form themselves into . . . the 'Metropolitan' Society and to connect with it a cabinet of the minerals of the United States and other parts of the world."[37] A few months later the name was changed to the Columbian Institute for the Promotion of Arts and Sciences. In 1818 Congress granted it a twenty-year charter. John Quincy Adams, Andrew Jackson, John C. Calhoun, Henry Clay, Joel R. Poinsett, Richard Rush, and many other prominent men became members, finding in its sessions a common meeting ground where political animosities could be laid aside. John Quincy Adams and John C. Calhoun both served as presidents. A tract of five acres of land, granted to the society for a garden, later, in 1852, became the National Botanic Garden. Upon the disintegration of the society in the eighteen thirties and the expiration of its charter in 1838, the activities of the Columbian Institute passed over to the National Institution for the Promotion of Science, organized in 1840, which invited the members of the Columbian Institute to affiliate with it. The actual record of achievement of the Columbian Institute is not particularly impressive. It deserves, however, more than mere passing mention for it represents the first American attempt to "unite the forces of science and government," and is the spiritual antecedent of the National Institute and the National Academy of Sciences. Although it was a premature and abortive attempt, it shows that there were men of science in the young republic who sought to make Washington the scientific capital of the nation.[38]

[36] *Charleston Medical Journal,* VIII (1853) , 60-61.

[37] Richard Rathbun, "The Columbian Institute for the Promotion of Arts and Sciences," Smithsonian Institution, U. S. National Museum, *Bulletin* 101 (Washington, 1917) , 10. The constitution, charter, and other documents pertaining to the history of the Columbian Institute are reproduced in this admirable monograph which traces the history of the society and shows its relation to the growth of later scientific organizations in Washington, particularly the National Institution and the Smithsonian Institution.

[38] J. W. Oliver, "America's First Attempt to Unite the Forces of Science and Government," *Scientific Monthly,* LIII (1941) , 253-257.

The Potomac-Side Naturalists' Club was formed early in 1858. It lasted through the Civil War, only to go to pieces in 1866 "on the rock . . . of expensive entertainments."[39] Resuscitated in 1873, it passed into "innocuous desuetude"[40] at the close of the eighteen seventies, and its place was taken in the eighties by other Washington organizations. A Delaware Academy of Natural Sciences was established in 1827 at Wilmington, Delaware.[41] By 1833 it had a prized museum,[42] and the society continued an active existence until 1842, when its collections and members were absorbed by the Botanical Society of Wilmington. The Wilmington Institute, organized in 1859, functioned successfully throughout the troublous days of the Civil War. It inherited the books and funds of the Botanical Society, which had become dormant some eight years previously. In 1831 a Virginia Historical and Philosophical Society was formed which attempted to carry on natural history work for several years. In the neighboring state of Maryland persistent efforts were needed to bring about a successful state Academy of Science. In 1819 a few Baltimore gentlemen "formed a society which met in a humble room over a stable."[43] The project collapsed, leaving its funds, which amounted to over $1200, to an Academy formed in 1822. This latter body was incorporated in 1826 under the name of Maryland Academy of Science and Literature. In 1834, when a fire destroyed its buildings and collections,[44] it languished, but was revived in 1836, only to be dissolved in 1844. In 1855 the Maryland Historical Society established a committee on natural history headed by P. T. Tyson. In 1863 another Academy was launched, with Tyson as president. The new constitution declared: "The object of the academy shall be to promote scientific research, and to collect, preserve, and diffuse information relative to the sciences, especially those which are connected with the natural history of Maryland."[45] In 1867 it was incorporated under its present title of Maryland Academy of Sciences.

The first concerted effort to establish a scientific society in the region comprising the East North Central states was made at Cincinnati, in 1818, with the formation of the Western Museum Society. "The price

[39] J. W. Chickering, "The Potomac-Side Naturalists' Club," *Science*, n. s., XXIII (1906), 265.

[40] *Ibid.*

[41] *American Journal of Science*, XIV (1828), 3, footnote.

[42] *Ibid.*, XXIV (1833), 177.

[43] S. B. Weeks, *Report of the United States Commissioner of Education for 1893-1894*, vol. 2, 1521.

[44] *American Journal of Science*, XXX (1834), 192-194.

[45] *Ibid.*, 1522.

of membership," we are told, "was fifty dollars, which was transferable and which secured admission for the subscriber's whole family. The Museum collection was placed in the building of the Cincinnati College. Decent strangers were cheerfully admitted."[46] The Society functioned actively throughout the twenties, but, becoming extinct in the early thirties, turned its collections over to the newly formed Western Academy of Natural Sciences. The latter organization, started in 1835,[47] built up extensive collections of books and specimens which were turned over to the present Cincinnati Society of Natural History in 1871. The devotion to science manifested in Cincinnati did not pass unnoticed in Cleveland, and in 1834 announcement was made of a Society of Natural History recently formed there.[48] It came to a speedy demise, however. Better success was achieved with the formation of a Cleveland Academy of Natural Science ten years later. The women of Cleveland raised some $400 for the support of the project.[49] The society continued to function actively and to publish until 1860. Chicagoans, too, were swept into the movement, establishing the Chicago Academy of Natural Sciences in 1856. Despite the fact that 1500 dollars was promptly raised by subscription, the panic of 1857 made hard sledding for the organization. It was incorporated in 1859 under the name of Chicago Academy of Sciences.

About the same time Robert Kennicott began to display great energy in building up the society's museum, especially doing so in 1862 after he returned laden with scientific treasures garnered on an expedition to northwestern arctic America financed by the Smithsonian, the Hudson's Bay Company, and wealthy residents of Chicago. In 1864, Louis Agassiz was summoned to Chicago for consultation and lecturing on museum problems. In 1865, Kennicott was off again in quest of specimens, this time with the Western Union Telegraph Company's survey of the Western coast of North America, for the purpose of establishing communications with Asia. The first Illinois Natural History Society was founded at Bloomington in 1858 and received its charter three years later. It continued meetings and publications until 1864, the time of its tenth and last meeting.[50] The Grand Rapids

[46] C. T. Greve, *Centennial History of Cincinnati* (Cincinnati, 1904), 524.

[47] Notice of Organization in *Western Journal of the Medical and Physical Sciences*, IX (1835), 155-156.

[48] *American Journal of Science*, XXVIII (1835), 119.

[49] *Annals of Science*, II (1852), 1.

[50] The work of this society is described in detail in S. A. Forbes, "History of the Former State Natural History Societies of Illinois." *Science*, n. s., XXVI (1901), 892-898.

Lyceum of Natural History (1854) was the forerunner of several scientific bodies which sprang up in western Michigan in the post Civil War period. The Wisconsin Natural History Society began in 1855 as the Naturhistorische Verein von Wisconsin. During the winter of 1851 an organization known as the Society for the Advancement of Natural Sciences came into existence in Louisville, Kentucky, in the heart of the East South Central states. A rather amusing account of the first election of officers held by the society shows that, as in these days, strife was not wholly absent among the apostles of science:

> Immediately upon this election, a most decided opposition to the society sprang up, and for some weeks a fierce newspaper war was waged against it. It was denominated "the mutual admiration society" and made the subject of daily ridicule. The cause of this determined opposition was the jealousy existing between the rival medical schools of that period. These rancorous bickerings exhibited a spirit wholly unworthy of gentlemen occupying high positions in the scientific world, and resulted only in directing attention to the society and its operations.[51]

Like all previous efforts to maintain a scientific society in Louisville, the enterprise soon collapsed. The society was reported dormant by 1854, although a brief revival was staged in 1858. The year 1836 witnessed the first important attempt to establish a scientific society in the region now occupied in West North Central states, when the Western Academy of Natural Sciences was established at St. Louis, at the time with a population of "14,253 . . . little more than a frontier town."[52] We are told, however, that "the success of these pioneers in western science did not equal their anticipations and the society dropped into a condition of somnambulism. . . . The books . . . became mouldy . . . the museum specimens dusty. . . ."[53] The academy slumbered for almost a score of years with only occasional and informal gatherings of scientists taking place. Meanwhile the population of St. Louis increased to over 122,000 by 1856, when the present Academy of Science was established.[54] Fifteen persons were present at the initial meeting, but four years later, in 1860, the academy had a membership of 150.[55] The charter members evidently had big things to

51 Society for the Advancement of Natural Sciences of Louisville, *Constitution and By-Laws of the Society for the Advancement of Natural Sciences of Louisville, Including an Account of a Former Natural History Society* (Louisville, 1859) , 3.

52 H. M. Whelpley, "Sketch of the History of the Academy of Science of St. Louis," Academy of Science of St. Louis, *Transactions*, XVI (1906) , no. 1, Celebration of the Fiftieth Anniversary of the First Meeting, preface, xxii-xxiii.

53 *Ibid.*, xxii.

54 *Ibid.*

55 *Ibid.* and xxiv.

talk about, for their first discussions were concerned "with objects of magnitude. . . . The zeuglodon, a gigantic fossil whale, and the . . . Mastodon giganteus were the subjects of the first debates."[56] The academy devoted its early activities largely to the study of natural history, devoting almost all its fourteen committees to branches of biology or geology. The following year the society subdivided its activities and created committees to deal with botany, chemistry, paleontology, and physics.[57] The Missouri Historical and Philosophical Society, founded in 1844, issued an appeal four years later for public participation in its scientific work.

> Citizens of every county are requested to furnish specimens of the coal, ores, rocks and petrifactions of their respective neighborhoods. . . . Such a collection would enable the people of the state to acquire more accurate knowledge of the geology and mineralogy of Missouri, and would open the eyes of . . . public men to the extent and value of our mineral resources. . . .[58]

Of the West South Central States, Texas and what is today Oklahoma seem to have been without scientific societies of any prominence whatsoever in the years prior to the Civil War. At Little Rock, however, in 1832, on what was then the frontier, the Antiquarian and Natural History Society of Arkansas Territory was founded. An ephemeral Lyceum of Natural History is reported to have been formed at New Orleans in 1825.[59] The principal scientific society of the region was the New Orleans Academy of Sciences, which, formed in 1853, was actively engaged in publication in the period prior to the Civil War.

The cultural frontier, as far as scientific organization was concerned, had hardly crossed the Mississippi, as has been indicated, and it lay far to the east of the Mountain states. Indeed, in all the vast, but then sparsely settled, region which is now occupied by Montana, Idaho, Wyoming, Colorado, New Mexico, Arizona, Utah, and Nevada, there seems not to have been a single general scientific society. It may be, of course, that a detailed study of local histories would reveal some groups of minor significance. On the Pacific slope, the California Academy of Natural Sciences was organized in 1853 at San Francisco. Its present name, the California Academy of Sciences, was adopted in 1868. The territory, and later the state of Oregon, and also Washington

---

[56] *Ibid.*, xxiv.

[57] *Ibid.*

[58] W. J. Seever, "Missouri Historical and Philosophical Society, 1844-1851," *Missouri Historical Society, Collections,* II (1900), 8.

[59] *American Journal of Science and Arts,* X (1826), 376.

territory had no scientific societies till after the Civil War, or certainly possessed none of considerable or lasting importance. The accompanying table shows the number and geographical distribution of academies of science and other societies for the general promotion of science at ten-year intervals from 1785 to 1865.

In the early decades scientific activity was centered in the area between Philadelphia and Boston. By 1825 the federal capital was recognized as an important scientific center, and each decade saw its position, in that connection, enhanced. By 1825 trans-Appalachian societies had begun to appear, and twenty years later a few trans-Mississippi societies had been formed. By the close of the period, a number of societies had sprung up in East North Central division in such cities as Cleveland, Cincinnati, Chicago, and Milwaukee. The rapid settlement of the region around the Golden Gate, in the fifties during the gold rush, had called forth a society at San Francisco. Aside from societies in Charleston and New Orleans, the table reveals very clearly the paucity of scientific societies in the south.

THE NUMBER AND GEOGRAPHICAL DISTRIBUTION OF ACADEMIES OF SCIENCE AND OTHER SOCIETIES FOR THE GENERAL PROMOTION OF SCIENCE AT TEN-YEAR INTERVALS FROM 1785-1865

	1785	1795	1805	1815	1825	1835	1845	1855	1865
New England Division	1	1	2	3	6	9	14	13	9
Middle Atlantic Division	1	1	2	3	12	9	11	12	12
South Atlantic Division	1	0	0	1	3	5	2	3	7
East North Central Division	0	0	0	0	1	2	2	3	5
East South Central Division	0	0	0	0	0	0	0	1	0
West North Central Division	0	0	0	0	0	0	2	1	1
West South Central Division	0	0	0	0	1	1	1	1	1
Mountain Division	0	0	0	0	0	0	0	0	0
Pacific Division	0	0	0	0	0	0	0	1	1
Total	3	2	4	7	23	26	32	35	36

A parallel and enriching development took place in the formation of specialized scientific societies. The only national society for the study of mathematics that was established during the years covered by this chapter was the American Statistical Association, formed in 1839, an organization primarily for the study of applied rather than pure mathematics.[60] It still functions actively.

In the field of astronomy, the Cincinnati Astronomical Society, launched in 1842, was especially active in securing a large observatory a little later for Cincinnati. (E. I. Yowell, "One Hundred Years at the Cincinnati Observatory," *Sky and Telescope,* III, No. 2 [1943], 3-5.) It was dedicated by impressive ceremonies, with John Quincy Adams as main speaker. The Allegheny Astronomical Society, formed in 1859 (?) at Pittsburgh, was ultimately instrumental in bringing large telescopes to that city. The Cambridge Branch of the American Astronomical Society was formed in 1854 under the guidance of Harvard astronomers.[61] This group, which it was hoped would be the beginning of a genuinely national American astronomical society, lasted for about three years. More than forty years were to elapse before the present American Astronomical Society began in another Harvard conference. The Chicago Astronomical Society was organized in 1862 by amateurs desirous of purchasing a large telescope.

No society for the especial purpose of studying physics was established during the period covered by this chapter, nor indeed for many years after 1865, although the American Association for the Advancement of Science and many of the academies of science formed sections to deal with it.

Similarly, the American Association for the Advancement of Science and many academies of science formed sections for the study of chemistry. In addition, several chemical societies came into existence in this period. The Chemical Society of Philadelphia, founded in 1792 and lasting for several years, is credited with being the "earliest organized body of chemists in either hemisphere."[62] Dr. James Woodhouse served as its president for a time. Joseph Priestley, Robert Hare, and Adam Seybert were among the most active members of the society. On October 24, 1801, a committee was appointed for the "discovery of

[60] Its object, according to its constitution, was "To collect, preserve, and diffuse statistical information in the different departments of human knowledge." S. B. Weeks, "A Preliminary List of American Learned and Educational Societies in the United States," *United States Commissioner of Education, Report for 1893-1894,* II, 1554.

[61] Solon Bailey, *History and Work of Harvard Observatory, 1829-1927* (New York and London, 1927), 31.

[62] H. C. Bolton, "Early American Chemical Societies," *Popular Science Monthly,* LI (1897), 819.

means by which a greater concentration of heat might be obtained for chemical purposes."[63] On December 20th, one of the committee, Robert Hare, at the time only twenty years of age, reported to the society his remarkable invention of the "hydrostatic blowpipe."[64]

The Columbian Chemical Society of Philadelphia was founded in 1811 by a "number of persons desirous of cultivating chemical science and promoting the state of philosophical inquiry." Thomas Jefferson was elected patron, James Cutbrush president, and Franklin Bache vice-president. Some provisions of the constitution of the society sound a bit amusing, for of it one reads that

> . . . it undertook to control the members' actions by a series of fines: twelve and a half cents for absence each roll, and one dollar for refusing to accept an office or declining to read an original chemical essay when appointed to do so. To insure against members withdrawing early from a dull meeting, the secretary was directed to call the roll at the opening and close of each meeting, and to fine absentees twelve and a half cents. Candidates for membership were required to read an original essay on some chemical subject to be discussed by the members, and a two-thirds vote was required to insure election. It seems to have been easier to be put out of the society than to get into it, for 'any member behaving in a disorderly manner shall be expelled by consent of two thirds of the members present.'[65]

Samuel L. Mitchill, Benjamin Rush, who was the first to hold a professorship of chemistry in America, and Benjamin Silliman were prominent members. In 1813 one volume of *Memoirs* was issued, containing essays on a wide variety of topics. The *Medical Repository* reviewed this volume in the following quaint phraseology:

> It is highly gratifying to behold a band of worthies like those before us laboring to analyze the compounds which they find ready made, to form by synthesis new combinations in the laboratory, and thereby to deduce correct doctrines from the facts which are disclosed. We cordially congratulate them on their noble occupation and on the progress they have made. We hope they will be persevering and undaunted; and if from this beginning there shall arise great improvements in theoretical disquisition, as well as in economical exercise, we shall rejoice with a mingled glow of amicable and patriotic sentiment.[66]

In 1811 Dr. F. Aigster, one of the honorary members of the Columbian Chemical Society, gave a course of lectures in the frontier town of Pittsburgh. This may well have afforded the stimulus for the found-

[63] Quoted in *ibid.,* 820.
[64] *Ibid.*
[65] *Ibid.,* 821.
[66] Quoted in *ibid.,* 825.

ing of the Pittsburgh Chemical and Physiological Society in 1813,[67] which appears to have been modeled closely after the Philadelphia Society.[68] A library, chemical and philosophical apparatus, and a geological collection were acquired by the following year, but "it is almost certain that the Society was disbanded at the special meeting of December 14, 1814. . . . "[69] The Delaware Chemical and Geological Society, formed in 1821, at Delhi, New York, was composed of "between forty and fifty well-informed and respectable inhabitants of the county."[70] Despite well-laid plans for the study of chemistry and mineralogy of the region, the society had soon to disband.

Toward the close of the period the engineers began to organize. The oldest engineering society in America, the Boston Society of Civil Engineers, was formed in 1848. Four years later the earliest engineering society of national scope, the American Society of Civil Engineers, was organized under the name of American Society of Engineers and Architects. The American Institute of Architects dates from 1857. Technological associations, such as the National Association of Cotton Manufacturers (1854) and the American Iron and Steel Association (1855), were springing up, but they lie outside the scope of this study, as does also the American Institute of the City of New York, an important industrial and commercial body which has incidentally lent encouragement to technological research.

Passing mention might be made of the formation of a few typical mechanics' associations of the period. The first of these was the Boston Mechanick Association formed in 1795. The most important was the Franklin Institute of the State of Pennsylvania for the Promotion of the Mechanic Arts, commonly known simply as the Franklin Institute (1824). The circumstances of its founding were somewhat unusual. A fire-engine company failed and fell into the hands of John Vaughan, a Philadelphia wine merchant, who gave it to his nephew, Samuel Vaughan. This young man, aged twenty-one, knew nothing of the "mechanick arts" and speedily discovered to his sorrow that he ought to have a knowledge of several trades in order to operate the business with which he had been presented. Since apprenticeship in these trades was normally a time-consuming affair, he conceived the idea that there

[67] S. H. Killikelly, *History of Pittsburgh* (Pittsburgh, 1906).

[68] John O'Connor, Jr., "Pittsburgh's First Chemical Society," *Science*, n. s., XLIV (1916), 12.

[69] *Ibid.*, 14.

[70] Quoted in *ibid.*, 825.

might be an institution where tradesmen, businessmen, and scientists might meet for the practical application of science. Aided by William Kneass, a copper engraver, and William H. Keating, a young professor at the University of Pennsylvania, he compiled a list of 1200 persons whom he invited to a meeting that filled the Philadelphia courthouse and at which the Franklin Institute was born. Nicholas Biddle, tycoon of finance in his own generation and president of the Bank of the United States, was among the organizers, whose ranks also included such diverse persons as a brewer, a merchant, a teacher, a clothier, a lawyer, a druggist, a farmer, a shuttlemaker. The constitution of the Franklin Institute stated that the new organization would "consist of manufacturers, mechanics artisans, and persons friendly to the mechanic arts," and its object was described to be the "encouragement of manufactures and the Mechanic and Useful Arts, by the establishment of Popular Lectures . . . by offering Premiums on all objects deemed worthy of encouragement, by examining all new Inventions submitted to them, and by such other measures as they may judge expedient." Since its inception more than a century ago, the Institute has published considerable scientific material in its *Journal*. Its museum, planetarium, etc., have exerted a tremendous local educational influence.

The Maryland Institute for the Promotion of the Mechanic Arts, formed in 1825, included among its objects the promotion of manufactures and useful arts, the establishment of popular lectures on the sciences, a library, and a museum of minerals. The Middlesex Mechanics' Association at Lowell, Massachusetts, was formed the same year. The most prominent of the Western mechanics' associations was the Mechanics Institute of Chicago, formed in 1837.

In all, three attempts were made to establish national geological societies in the period covered by this chapter. The first in the United States was an American Mineralogical Society formed at New York in 1798.

> An Association has been formed in the city of New York for the investigation of mineral and fossil bodies which compose the fabric of the globe; and more especially for the natural and chemical history of the minerals and fossils of the United States by the name and style of American Mineralogical Society. . . .[71]

The society collapsed after a few years. The second enterprise was the American Geological Society organized at New Haven, Connecticut, in 1819. Receiving the support of Benjamin Silliman and his satellites

[71] *Medical Repository*, I (1798), 114.

at Yale, the society flourished for a few years, but ultimately failed to survive. The third enterprise was the Association of American Geologists and Naturalists (1840) which broadened into the American Association for the Advancement of Science.[72] A few scattered state and local geological societies were also formed. The first state geological society was the Geological Society of Pennsylvania in 1832. A geological society existed in Baltimore in the 1840's. The geological societies played an important role in fostering geological surveys in many parts of the union. The Iowa Historical and Geological Institute was organized in 1843, and the Mississippi State Geological Society in 1850. The Wyoming Historical and Geological Society was formed at Wilkes-Barre, Pennsylvania, in 1858.

Impetus to the study of geography, broadly conceived, was given by the formation of the American Geographical and Statistical Society, in New York in 1851. Prominent among the founders were Henry Grinnell, its president, Joshua Leavitt, H. E. Pierrepont, and C. A. Dana.[73] The *Journal*, begun by the Society the following year, was for almost forty years "the only American periodical devoted to scientific geography."[74] In 1871 the society adopted its present name of American Geographical Society of New York. The American Oriental Society, formed ten years earlier, augmented geographical knowledge of the Far East among other things it accomplished.

With the accumulation of knowledge in other fields, it was natural that similar developments should take place. The biological and related sciences, including botany, entomology, and agriculture, were, by the close of the period, receiving the attention of specialists and specialized societies.

On July 20, 1794, Washington, then president of the United States, addressed a letter to Sir John Sinclair, in which he said:

> It will be some time, I fear, before an Agricultural society, with Congressional aid, will be established in this country. We must walk, as other countries have, before we can run; smaller societies must prepare the way for greater; but, with the lights before us, I hope we shall not be so slow in maturation as older nations have been. An attempt, as you will perceive by the enclosed outlines of a plan, is making to establish a state society in Pennsylvania for agricultural improvements. If it suc-

---

[72] *Quod vide.*

[73] American Geographical and Statistical Society, *Bulletin,* I (1852), "Origin of the Society," 3-4.

[74] Index to the *Bulletin of the American Geographical Society, 1852-1915* (New York, 1918), "Historical and Bibliographical Note," v.

ceeds, it will be a step in the ladder; at present, it is too much in embryo to decide upon the result.[75]

In his annual address on December 7, 1796, when he met the two houses of Congress for the last time, he said:

It will not be doubted that, with reference to either individual or national welfare, agriculture is of primary importance. In proportion as nations advance in population, and other circumstances of maturity, this truth becomes more apparent, and renders the cultivation of the soil more and more an object of public patronage. Institutions for promoting it grow up, supported by the public purse; and to what object can it be dedicated with greater propriety? Among the means which have been employed to this end, none have been attended with greater success than the establishment of boards, composed of proper characters, charged with collecting and diffusing information, and enabled, by premiums and small pecuniary aids, to encourage and assist a spirit of discovery and improvement.[76]

A few individuals had already felt the desirability of forming such associations. The Philadelphia Society for Promoting Agriculture, formed in 1785, was the first purely agricultural society in the United States. It was followed a few months later, in the same year, by the South Carolina Agricultural Society. The Society Instituted in the State of New York for the Promotion of Agriculture, Arts, and Manufactures was formed in 1791 and incorporated in 1793. The Massachusetts Society for Promoting Agriculture was incorporated in 1792 and it began to publish the *Agricultural Repository* a few years later. The Western Society of Middlesex Husbandmen was formed in Massachusetts in 1794, and the Kennebec Agricultural Society in 1800. Thereafter, agricultural societies were founded in such abundance that it is impossible to take notice of them individually except to mention a very few of the more active.

The earliest agricultural societies in America were what might be termed learned agricultural societies.[77] As with early American philosophical organizations, their membership was limited to the intellectually elite. Doctors, lawyers, and other professional and public men

[75] *The Writings of George Washington*, W. C. Ford, edit. (14 vols., New York and London, 1889-1893) , XII, 442-443.

[76] *Ibid.*, XIII, 348.

[77] See N. C. Neely, *The Agricultural Fair* (Columbia University Studies in the History of American Agriculture, no. II, New York, 1935) , ch. II, "The Learned Agricultural Societies in England and America"; and R. H. True, "The Development of Agricultural Societies in the United States," American Historical Association, *Annual Report for the Year 1920*, 305.

composed the membership rather than the rank and file of ordinary farmers.[78] The premiums offered by the societies were for the solution of general problems.[79]

A new and practical turn was given to the movement through the activities of Elkanah Watson. Often called the father of the agricultural fair in America, Watson exhibited two Merino sheep in the public square of Pittsfield, Massachusetts, in 1807. The interest of the neighboring farmers being aroused, Watson began to dream of cattle shows and agricultural societies.[80] By 1810 he had organized the first Berkshire Cattle Show, and the following year the Berkshire Agricultural Society was incorporated. Under its aegis a larger fair was held, with elaborate and festive ceremonies.[81] Watson traveled widely, giving lectures and writing pamphlets extolling the merits of his institution and expounding the need for practical societies. The principal features of the Berkshire system were agricultural fairs operated by societies of working farmers. Prizes were given, plowing matches and balls held. Watson contrasted the European and earlier American societies with his own enterprises, condemning the efforts of "scientific and literary gentlemen, moving within the walks of *classic ground*" for failing to appreciate the real state and needs of contemporary America. Watson, of course, encountered the expressed opposition of older learned agricultural societies, receiving, for example, in 1812, a squelching rejoinder from John Adams,[82] then president of the Massachusetts Society for the

---

[78] "It should be stated, however, that the prime movers in the formation of these societies were not men actually engaged in farming, though many of them were owners of fine estates. The mass of farmers were not, as yet, fully prepared for this progressive effort, and all the agricultural teachings of educated and scientific men prove unavailing, unless the people themselves, the actual tillers of the soil, are prepared to receive and profit by their teachings. Many years elapsed after these efforts were made, before the habit of reading became sufficiently common among the masses of practical farmers to justify the expectation that any general benefit would arise from the annual publication of the transactions of these societies.

"There was little or no disposition in the community to examine the subject, and they failed to excite any spirit of emulation in the public mind. The improvements proposed fell almost dead upon the people who rejected 'book farming' as impertinent and useless, and knew as little of the chemistry of agriculture as of the problems of astronomy. A quarter of a century, however, effected some change, and in 1816 the Massachusetts society held its first exhibition." C. L. Flint, *Eighty Years of Progress in the United States* (New York, 1861) , 24.

[79] Neely, *op. cit.*, 43-45.

[80] Elkanah Watson, *History of Agricultural Societies on the Modern Berkshire System* (Albany, 1820) , 116.

[81] *Ibid.*, 119.

[82] *Ibid.*, 133.

Promoting of Agriculture. Four years later, however, when the Massachusetts society adopted the plan of holding an annual cattle show and exhibition, John Lowell, in his address to the society, expressed indebtedness to the Berkshire plan,[83] and even Adams came to the same point of view.[84] Progressively, the older societies became democratic under the impact of the newer ones.

By 1819, according to Watson, all the counties in New England, except those of Rhode Island, had organized societies by his plan, and the movement was spreading into Pennsylvania, North Carolina, South Carolina, Virginia, Maryland, Kentucky, Ohio, and Illinois. The crest was reached in 1820-1825. But the hard times of the later twenties saw the withdrawal of state aid and the temporary collapse of the movement in New Hampshire, New York, Pennsylvania, Connecticut, and elsewhere. A revival took place in the eighteen thirties. After 1840 American agriculture entered upon a new period of prosperity. New lands were opened, railroads were built, machinery was improved, and European markets were entered. In 1841 New York granted state aid, which helped revive many local societies. Presently, hundreds of local associations sprang up throughout the union. In 1858 the Commissioner of Patents compiled a list of 912 state and county agricultural and horticultural societies, five-sixths of which Butterfield estimated were established after 1849.[85] In 1868 the Department of Agriculture listed 1367 state, district, and other agricultural and horticultural societies, 86 per cent of which, according to Neely, were formed after 1850.[86] Neely makes an interesting general observation on the westward movement of the societies: "The agricultural society spread from east to west, but pretty largely through the northern states. Initiated along the Atlantic seaboard, it turned westward principally through western New York and Pennsylvania into the great valleys of the Ohio, the Upper Mississippi and the Missouri."[87] His studies also led to another interesting observation: "The Civil War put a sudden end to a great many of the agricultural societies, but it was more than offset by the increase in New England and the western states."[88]

Several tables are here reproduced, showing Neely's studies of the

[83] *Ibid.*, 179.

[84] Elkanah Watson, *Men and Times of the Revolution* (New York, 1857), 499.

[85] K. L. Butterfield, "Farmers Social Organizations," L. H. Bailey, edit., *Cyclopedia of American Agriculture* (New York, 1907-1909), IV, 292.

[86] Neely, *op. cit.*, 83.

[87] *Ibid.*, 85. Reprinted from *The Agricultural Fair*, by permission of The Columbia University Press.

[88] *Ibid.*, 87.

geographical distribution of agricultural societies in the United States in 1858 and 1868, respectively.

DISTRIBUTION OF AGRICULTURAL AND HORTICULTURAL SOCIETIES, 1858*

Numbers and Percentages of the State and Local Agricultural and Horticultural Societies Reported to the United States Commissioner of Patents in 1858 and the Numbers and Percentages of the Total Population in 1860, by Geographic Division †

Geographic Division	Numbers		Percentages	
	Population, 1860 ‡	Societies, 1858	Population, 1860	Societies, 1858
New England	3,135,283	95	9.9	10.4
Middle Atlantic States	7,458,985	184	23.7	20.1
East North Central States	6,926,884	308	22.0	33.7
Iowa, Minnesota, Missouri	2,028,948	135	6.4	14.8
South	11,133,361	165	35.4	18.0
Pacific States	444,053	12	1.4	1.3
All other territory	315,807	13	1.0	1.4
Total	31,443,321	912	100.0	100.0

* Reprinted from Neely, *The Agricultural Fair*, 84, by permission of the Columbia University Press.
† Compiled from Commissioner of Patents, *Report*, 1858, *Agriculture* (Washington), 91, and United States *Census*.
‡ The total population rather than the rural population is used here, since, in matters of initiative, financial support, and attendance, the fair was, to a very appreciable extent, an urban as well as a rural enterprise.

A few of the leading state agricultural societies may be mentioned. The Pennsylvania Agricultural Society was established in 1823, the New York State Agricultural Society in 1832, the Louisiana State Fair Association in 1841, the Kentucky Agricultural and Mechanical Association in 1850, the Virginia State Agricultural Society in 1850, the Wisconsin State Agricultural Society in 1851, the Connecticut State Agricultural Society in 1852, the Illinois State Agricultural Society in 1853, the California State Agricultural Society in 1854, the Indiana State Agricultural Society in 1857, and the New Jersey State Agricultural Society in 1865. The New England Agricultural Society dates from 1864.

The Columbian Agricultural Society, organized at Washington in 1809 by men interested in agriculture in the District of Columbia, Virginia, and Maryland, was the first attempt to establish a national agricultural society in the United States. It held a fair or exhibition on May 10 at Georgetown, a gala event for Washington society. A similar fair was held in the autumn of the same year. The society declined with the War of 1812, and expired.

DISTRIBUTION OF AGRICULTURAL AND HORTICULTURAL SOCIETIES, 1868*

Numbers and Percentages of the State and Local Agricultural and Horticultural Societies Reported to the United States Commissioner of Agriculture in 1868 and the Numbers and Percentages of the Total Population in 1870 with the Percentage Increase in Numbers of Societies and Population Over 1858 and 1860, Respectively, Geographic Division †

Geographic Division	Numbers		Percentages		Percentage Increase	
	Population, 1868	Societies, 1868	Population, 1870	Societies, 1870	Population, over 1860	Societies, over 1858
New England	3,487,924	139	9.0	13.4	11.2	46.3
Middle Atlantic States	8,810,806	186	22.8	18.0	18.1	1.0
East North Central States	9,124,517	390	23.6	37.7	31.7	26.6
Iowa, Minnesota, Missouri	3,355,021	181	8.7	17.5	65.3	34.0
South	12,288,020	62	31.8	6.0	10.3	−62.4
Pacific States	675,125	6	1.7	0.5	52.0	−50.0
All other territory	816,958	68	2.1	6.5	158.6	423.0
Total	38,558,371	1,032	100.0	100.0	22.6	13.1 ‡

* Reprinted from Neely, *The Agricultural Fair*, 86, by permission of Columbia University Press.
† Compiled from Commissioner of Agriculture, *Report*, 1867, 364 *et seq.* and United States *Census*.
‡ The Department of Agriculture had on its records 1367 organizations, of which 1052 had made reports in response to a questionnaire sent out by the department. The published list, however, contains only 1032 with the information necessary for inclusion in this tabulation.

The following table shows Neely's tabulation of the founding of agricultural societies for five-year intervals.

DATE OF ORGANIZATION OF AGRICULTURAL SOCIETIES*

Number and Percentages of Agricultural Societies Reporting Date of Organization to the United States Department of Agriculture in 1868, by Five-Year Periods †

	Numbers	Percentages
Before 1841	40	4.0
1841-1845	30	3.0
1846-1850	68	6.8
1851-1855	185	18.6
1856-1860	250	25.1
1861-1865	178	17.9
1866-1868‡	242	24.3
Total	993	100.0

* Reprinted from Neely, *The Agricultural Fair*, 87, by permission of Columbia University Press.
† Compiled from Commissioner of Agriculture, *Report*, 1867, 364 *et seq.*
‡ A period of about two and a half years.

In 1841 a group of gentlemen, desiring "to elevate the character and standing of the cultivation of the American soil,"[89] met in Washington and organized a national agricultural society. Failure to secure part of the Smithsonian fund, as had been hoped, caused the society to remain dormant until 1862, when another convention of over 150 delegates met at Washington upon the call of twelve state agricultural associations and organized the United States Agricultural Society. The society was incorporated by an act of Congress in 1860.[90] Annual meetings were held until the beginning of the Civil War, with delegates attending from agricultural associations from all over the union. The United States Agricultural Society held national exhibitions and field trials at Springfield, Massachusetts, Springfield, Ohio, Boston, Philadelphia, Syracuse, Louisville, Richmond, Chicago, and Cincinnati. The society was instrumental in securing the establishment of the national Department of Agriculture.

Societies of fruit growers were also coming into existence throughout the union. The oldest of these, the Pennsylvania Horticultural Society, was organized in 1827.[91] The Massachusetts Horticultural Society was formed in 1829, the Kentucky Horticultural Society in 1840, the Rhode Island Horticultural Society in 1845, the Delaware Horticultural Society two years later, the Illinois State Horticultural Society in 1856, the Missouri State Horticultural Society in 1859, the Indiana Horticultural Society in 1860, and the Wisconsin State Horticultural Society in 1865. A convention of fruit growers organized the National Pomological Society in 1848.

Commenting on the importance of agricultural societies for American agriculture, one authority writes:

> I do not think it is claiming too much for the agricultural societies throughout the country, to say that the general spirit of inquiry in relation to farm improvements, and much of the enterprise manifested by farmers of the present day, is due to their efforts. The most impartial judgment would, in fact, go much further than this, and say that a large proportion of the actual improvement that has been made in farm stock,

[89] Quoted in B. P. Poore, "History of Agriculture in the United States," *Report of the U. S. Commissioner of Agriculture for the Year 1866* (Washington, 1867), 525.

[90] The preamble to the constitution stated the objects of the society to be to "improve the agriculture of the country, by attracting attention, eliciting the views, and confirming the efforts of that great class composing the agricultural community, and to secure the advantages of a better organization and more extended usefulness among all state, county, and other agricultural societies." Quoted in *ibid.*, 525.

[91] A now extinct New York Horticultural Society had been formed as early as 1818. *Science*, n. s., LXXXIV (1936), 360.

farm implements, and farm products, may be traced, directly or indirectly, to the influence of the agricultural associations of the country.

To appreciate this influence it is only necessary to consider the immense facilities which a well conducted exhibition gives, not only to the agricultural mechanic for making known the nature and value of his improvements, but to the farmer for becoming acquainted with them.[92]

A number of botanical societies came into existence in this period. The first of them was the Charleston (South Carolina) Botanic Garden and Society, instituted in 1805, which declared

> Our first concern will be the examination and arrangement of our indigenous plants, among which are doubtless many non-descripts. To extend the knowledge of our favorite pursuit, we beg leave to propose . . . an exchange of our indigenous plants. . . .[93]

The Linnaean Society of Philadelphia was formed in 1806, the Linnaean Society of New England in 1814, the Washington Botanical Society in 1817 (this became defunct and was replaced by the Botanic Club of Washington in 1825), the Botanical Society of Wilmington, Delaware, in 1842, the Linnaean Association of Pennsylvania College (Gettysburg, Pennsylvania) in 1844, the important Torrey Botanical Club (New York) in 1858, and the Linnaean Society of Lancaster, Pennsylvania, in 1862.

The East India Marine Society was established in 1799 at Salem, Massachusetts, "for the purpose of investigating and recording facts relative to the natural and physical history of the ocean."[94] Its original

---

[92] C. L. Flint, *op. cit.*, 26. Daniel Webster, often referred to by his contemporaries as the "farmer of Marshfield," is said to have held similar views: "The great practical truth and characteristic of the present generation is that public improvements are brought about by a voluntary association and combination. The principle of association—the practice of bringing men together for the same general object, pursuing the same general end, and uniting their intellectual and physical efforts to that purpose—is a great improvement in our age and the reason is obvious. Here men meet together that they may converse with one another—that they may compare with each other their experience, and thus keep up a constant communication. In this practical point of view, these agricultural associations are of great importance. . . . it is the meeting of men face to face, and talking over what they have in common interest—it is this intercourse that makes men sharp, intelligent, ready to communicate to others, and ready to receive instruction from them." Quoted in B. P. Poore, "History of the Agriculture of the United States," *Report of the U. S. Commissioner of Agriculture for the year 1866* (Washington, 1867), 525. I have been unable to locate the quotation in any of the published writings of Daniel Webster, although as is well known he made a number of addresses before agricultural societies and was interested in their spread.

[93] *Medical Repository*, IX (1806), 434-436.

[94] A notice in the *American Journal of Science*, X (1826), 369.

membership included only those who navigated the waters near either the Cape of Good Hope or Cape Horn. Under its able president, Nathaniel Bowditch, it built up a splendid collection of marine specimens that was ultimately transformed into the Peabody Museum.

The Entomological Society of Pennsylvania, formed at York in 1842, led two or three years of active existence. At a meeting held in Philadelphia in 1859 for forming a society for "the advancement of entomological science,"[95] the Entomological Society of Philadelphia was organized. Eight years later it became the American Entomological Society. It is the oldest of the national specialized biological societies, and no general society to deal with biology as a whole was founded in the period covered by this chapter.

The American Ethnological Society was organized in 1842 for making "inquiries into the origin, progress, and characteristics of the various races of men."[96] This society published discoveries relating to the lands and peoples of the Spanish American countries and Africa, which were at the time being explored by adventurous scientists. Albert Gallatin, secretary of the treasury under Jefferson and Madison, one of the founders of the society, contributed as the first paper in its *Transactions* his celebrated "Notes on the Semi-Civilized Nations of Mexico, Yucatan, and Central America." In the 1860's the society be-

[95] E. T. Cresson, *A History of the American Entomological Society* (Philadelphia, 1909) , 13.

[96] American Ethnological Society, *Transactions,* I (1845) , iii. Its aims were further discussed as follows: "The American Ethnological Society was established for the promotion of a most important and interesting branch of knowledge, that of Man and the Globe he inhabits, as comprised in the term Ethnology in its widest acceptation . . .

"To its native and resident members, the American Ethnological Society feels it has but to indicate the field presented for their exertions, and the immense extent and variety of subjects that call for their investigation. The mystery that still envelops the history and origin of the American races of man—the phenomena connected therewith—the diversity of languages—the remains of ancient art and traces of ancient civilization among the aborigines of Peru, Mexico, and Central America —the spontaneous growth or imported origin of arts, science, and mythology—the earth-works of the Ohio and Mississippi valleys and their founders:—these are amongst the topics for inquiry which the most cursory view suggests; and there are few individuals in our western country who may not obtain interesting materials for their elucidation." *Ibid.,* ix, x.

See also Franz Boas, "The American Ethnological Society," *Science,* n. s., XC (1943) , 7-8, for a centenary account. Schoolcraft, one of the principal founders, had advocated establishment of such a society as early as 1819; see C. S. Osborn and Stellanova Osborn, "Schoolcraft and the American Ethnological Society," *Science,* n. s., LXXXXVII (1943) , 161-162.

came inactive, but it was revived and reorganized in 1871 as the Anthropological Institute, returning, however, soon afterwards to its original name of American Ethnological Society, which it still retains. The American Antiquarian Society, established in 1812, had done some ethnological work in the thirty years prior to the establishment of the American Ethnological Society.

As we have already seen, the medical profession had begun to organize in the latter part of the colonial period, and by 1789 a number of state and local medical societies were in existence.[97] Others were presently springing up in abundance. It is convenient to study these by geographical regions. In New England, the New Hampshire Medical Society was formed in 1791, the Connecticut State Medical Society in 1792, the Rhode Island Medical Society in 1811, the Vermont State Medical Society in 1813, and the Vermont Medical Association in 1852. In the Middle Atlantic States, the Medical Society of the State of New York appeared in 1807 and the Medical Society of the State of Pennsylvania in 1848.

Among the local groups were the Medical Society of the Eastern District of New Jersey in 1790, the Academy of Medicine of Philadelphia in 1797, the Medical Society of the County of Albany, New York, in 1806, the Physico-Medical Society of New York in 1815, the Medical Society of the County of Kings, New York, in 1822, the New York Pathological Society in 1844, the New York Academy of Medicine in 1847, and the Pathological Society of Philadelphia in 1857. In the East North Central division, the Michigan State Medical Society was organized in 1820, the State Medical Society of Wisconsin in 1841, the Ohio State Medical Society in 1846, and the Indiana State Medical Association in 1849. The Chicago Medical Society dates from 1850. In the West North Central division, the Iowa State Medical Society and the Missouri State Medical Association were established in 1850, and the Kansas Medical Society in 1859. In the South Atlantic division, the Medical and Chirurgical Faculty of the State of Maryland was formed in 1799, the Medical Society of Virginia in 1821, the South Carolina Medical Association in 1848, and the Georgia Medical Association in 1849.[98] The Medical Society of the District of Columbia was formed in 1819. In the East South Central division, the Medical Association of the State of Alabama was organized in 1847, the Kentucky State Medical Society in 1851, and the Mississippi State Medical Association in

[97] A list of early American medical societies is given in F. R. Packard, *History of Medicine in the United States* (Philadelphia and London, 1901), 525-526.
[98] A transitory Georgia Medical Society was formed in 1804.

1858. In the West South Central division, the State Medical Association of Texas dates from 1853. The Société Médicale de la Nouvelle Orléans was formed in 1812. No medical society in the mountain division dates from this period. On the Pacific Coast the California Medical Association was formed in 1856.

The crowning development in the matter of medical organization came in 1847, with the formation of the American Medical Association,[99] which organized as a body of representatives composed of delegates from medical colleges, hospitals, and state, district, county, or other medical societies which adopted its code of ethics. The American Medical Association has labored increasingly to raise the standards of the medical profession and medical education in the United States. The American Institute of Homeopathy was formed two years prior to the American Medical Association. The American Veterinary Medical Association dates from 1863. The American Ophthalmological Society came into existence in 1864.

The American Society of Dental Surgeons, formed in 1840, took the lead in establishing the first dental school in the United States, the Baltimore College of Dentistry. The American Dental Association,[100] launched in 1859, was expressive of the consciousness of the emergence of dentistry as an independent profession.

Akin to the study and practice of medicine is the apothecary's pro-

[99] It was the Medical Society of the State of New York that took the lead in summoning the convention which formed the American Medical Association: "Whereas, It is believed that a national convention would be conducive to the elevation of the standard of medical education in the United States, and

"Whereas, there is no mode of accomplishing so desirable an object without concert of action on the part of the medical societies, colleges and institutions of all the states, therefore be it

"*Resolved,* That the Medical Society of the State of New York earnestly recommends a national convention of delegates from medical societies and colleges of the whole union, to convene in the City of New York, on the first Tuesday in May, in the year 1846, for the purpose of adopting some concerted action on the subject, set forth in the foregoing preamble." Quoted in L. S. McMurtry, "The American Medical Association: Its Origin, Progress and Purpose," *Science,* n. s., XXII (1915), 103.

[100] Its objects are "to cultivate and promote the art and science of dentistry and of its collateral branches; to conduct, direct, encourage, support or provide for exhaustive dental and oral research; to elevate and sustain the professional character and education of dentists; to promote among them mutual improvement, to disseminate knowledge of dentistry and dental discoveries, to enlighten and direct public opinion in relation to this science." C. J. West and Callie Hull, comps., *Handbook of Scientific and Technical Societies of the United States and Canada* (National Research Council, *Bulletin* 76, Washington, 1930), 39.

fession. In 1820 the United States Pharmacopoeial Convention held its first meeting, presaging better organization among the pharmacists of the nation, and in 1852 the American Pharmaceutical Association was formed.[101]

No societies were formed for the express purpose of studying psychology during the period covered by this chapter, although the American Psychiatric Association was organized in 1844 by the superintendents of thirteen insane asylums. About the same time, societies and less formal groups were organized in Philadelphia, Boston, and other cities for the study of phrenology.

The crowning development of the period 1790-1865 was, of course, the formation of important new agencies of general scope, the Smithsonian Institution, the American Association for the Advancement of Science, and the National Academy of Sciences. Before turning to a consideration of these, however, one ought first to discuss the work of the two older national bodies of general character, the American Philosophical Society and the American Academy of Arts and Sciences.

The American Philosophical Society continued to promote actively the cause of science. David Rittenhouse became president in 1791, succeeding the venerable Franklin, who had died in the preceding year. The next president, Thomas Jefferson, served from 1797 to 1815, and he was followed by Caspar Wistar, who was in office from 1815 to 1819.[102] Peale's Museum, organized in 1784 by Charles Wilson Peale, was moved to Philosophical Hall in 1794, where it remained for several years. Some decades later the financial failure of the museum was to involve the American Philosophical Society in difficulties, but its sponsorship of the museum served to stimulate public interest in natural history.

On March 6, 1795, a committee of the society offered a prize for "an essay on a system of liberal education & literary instruction, adapted to the genius of the government, & best calculated to promote the general welfare of the United States:—comprehending also, a plan for instituting & conducting public schools in this country on principles of the

[101] Its objects are "To improve the science and art of pharmacy by diffusing scientific knowledge among apothecaries and druggists, fostering pharmaceutical literature, developing talent, stimulating discovery and invention, and encouraging home production and manufacture in the several departments of the drug business." S. B. Weeks, comp., "A Preliminary List of American Learned and Educational Societies," United States Commissioner of Education, *Report for 1893-1894*, II, 1534.

[102] Succeeding presidents in the period covered by this chapter were: Robert M. Patterson (1819-1825), William Tilghman (1825-1828), Peter S. DuPonceau (1828-

most extensive utility."[103] Many essays were received and the prize was ultimately divided evenly between Samuel Knox for "An Essay on Education" and Samuel H. Smith for "Remarks on Education: Illustrating the close connection between Virtue and Wisdom: to which is annexed a System of liberal Education." Both plans called for state systems of education from the primary school to the university under democratic control, and both emphasized the need for an education based upon science instead of upon superstitution and prejudice.[104] Thomas Jefferson succeeded in having the manuscript records of the Lewis and Clark expedition, the Pike expedition of 1805-1806, and the Dunbar expedition to Louisiana deposited with the society. The establishment of the Academy of Natural Sciences of Philadelphia in 1812 drew away some of the support which otherwise would have gone to the American Philosophical Society. In 1849 the curators of the American Philosophical Society deposited its collection of fossils with the Academy of Natural Sciences of Philadelphia. Mühlenberg's herbarium was given to the American Philosophical Society in 1818. A collection of Mexican minerals was secured for the society in 1830 by Joel R. Poinsett. The society gradually gained possession of those Franklin papers which had not disappeared in the course of the years. The hundredth anniversary of the original Junto was celebrated in 1843 with appropriate addresses, including an able historical sketch by President Peter S. Du Ponceau. Among those who contributed scientific papers to the American Philosophical Society in the years prior to the Civil War were Thomas Nuttall, Thomas Say, and Joseph Leidy.

The American Academy of Arts and Sciences also continued its activities and prospered even in spite of severe competition and trying vicissitudes. In 1796 it received the Rumford fund, a bequest from Benjamin Thompson for the advancement of knowledge of light and heat, which it still administers. Nathaniel Bowditch was president of the academy from 1829 to 1838, Joseph Bigelow from 1846 to 1863, and Asa Gray from 1863 to 1873. Augustus A. Gould was corresponding secretary from 1850 to 1852. With the establishment of the Boston Society

---

1844), Robert M. Patterson (1845), Nathaniel Chapman (1846-1849), Robert M. Patterson (1849-1853), Franklin Bache (1853-1855), Alexander D. Bache (1855-1857), John K. Kane (1857-1859), and George B. Wood (1859-1880).

[103] American Philosophical Society, *Proceedings*, XXII (1884), pt. 3, *Minutes, 1744-1838*, 229.

[104] See also A. O. Hansen, *Liberalism and American Education in the Eighteenth Century* (New York, 1926), ch. IV, for a discussion of these plans of education.

of Natural History in 1830, there was some diminution of activity on the part of the American Academy, but a revival of interest in it took place in the fifties and sixties. In 1854 new statutes of the academy were issued which recognized a somewhat earlier division of it into three classes, to deal respectively with (1) the mathematical and physical sciences, (2) the natural and physiological sciences, and (3) the moral and political studies.[105] Thomas Nuttall, William S. Sullivant, Louis Agassiz, James Hall, Edward Hitchcock, Jeffries Wyman, James D. Dana, and others contributed important papers to the publications of the society in the years prior to the Civil War.

The way was paved for the Smithsonian Institution and the National Academy of Sciences by the now all but forgotten National Institute, which was, for a very brief period, the nation's foremost scientific society. The death of the Columbian Institute, with the expiration of its charter in 1838, left the capital devoid of a central scientific agency. As a result the National Institution for the Promotion of Science was organized in Washington on May 15, 1840. An act incorporating it under the name of the National Institute passed Congress in 1842. Joel R. Poinsett became its president, John Quincy Adams was an especially active member, and many high governmental officials became affiliated with it. We are told that bitter congressional enemies cast aside political animosities to enjoy its meetings and to work in common for its welfare. When President Tyler, in April, 1844, presided over the opening session of the first national scientific congress ever held in the United States, one which had been called under the auspices of the society, it could be truly said that the National Institute had become the leading national scientific organization of the Republic. Its high aspirations were expressed by Senator Robert J. Walker, who spoke at this congress as follows:

> This institute is located at the home of the Federal Government, and its operations are designed to embrace the whole Union. Rising above local and sectional influences, it appeals to the friends of science throughout the nation, and asks the support of all, with a view to the general diffusion of knowledge, and the advancement of American science. It is not designed to impede the progress or impair the usefulness of any present or future scientific institutions or societies in any of the States, but would desire to establish between them and this Institute the most cordial relations, together with reciprocal aid and encouragement.[106]

Notwithstanding the fact that the National Institute performed val-

---

[105] American Academy of Arts and Sciences, *Memoirs*, XI (1882), pt. 1, 91.
[106] National Institute, *Bulletin* 3 (1845), 439.

uable scientific work, helped exploring expeditions, and aided in the reorganization of the Coast Survey, it, despite the roseate hopes of its promoters, was doomed to sudden decay. Poinsett declined re-election, meetings and publications ceased, and its list of three hundred and fifty resident and twelve hundred and fifty corresponding members soon dwindled from these impressive figures. An attempt to resuscitate the National Institute in 1847 ended in failure, but it was revived in 1855 as a local scientific society. Upon the expiration of its charter in 1862, its museum and library were turned over to the Smithsonian Institution. Although the National Institute thus perished, it is to be remembered for having exerted considerable influence on the development of the bodies which followed it: the Smithsonian Institution, the American Association for the Advancement of Science, the United States National Museum, the Coast Survey, the Naval Observatory, and the National Academy of Sciences. The National Institute seems to have been the first American society to adopt the adjective "national" in its name, a practice imitated by numerous subsequent societies in the United States. The rising tide of nationalism led scientific societies, no less than political parties, to style themselves with names which indicate to the historian the passing of provincialism.

Although the Smithsonian Institution is not a scientific society in the same sense as the other organizations considered in this work, its influence on the development of American scientific organizations has been so important as to justify the inclusion of a considerable discussion of its formation and work. James Smithson, an English scientist of high achievements, died at Genoa, Italy, on June 27, 1829, leaving his property as a bequest "to the United States of America, to found at Washington, under the name of the Smithsonian Institution, an Establishment for the increase & diffusion of knowledge among men."[107] Owing to legal difficulties in connection with the will, and to a prolonged dispute as to what should be the character of the new agency, it was not until 1846 that the Smithsonian Institution was chartered by Congress and came into actual existence. The act created it as a corporation or establishment and included among its statutory members the president of the United States, the vice-president, the chief justice, the secretary of state, the secretary of war, the attorney-general, the postmaster-general, the secretary of the navy, the secretary of the interior, and the secretary of agriculture. The task of conducting the Smithsonian Institution was intrusted to a board of regents, including the president

[107] Reprinted in G. B. Goode, *The Smithsonian Institution, 1846-1896* (Washington, 1897), 20.

of the United States, the vice-president, the chief justice, three senators, and three representatives. This board selects the secretary of the
Smithsonian Institution, who is its principal executive officer, responsible for all staff appointments and all expenditures, and who acts as
the custodian of the property of the Institution. He transmits an annual
report to Congress.

It was Joseph Henry, renowned as an investigator in the field of
physics, who was selected to be the first secretary of the Smithsonian
Institution. For more than three decades his tireless labors were devoted to the task of organizing and directing the work of the Institution, amply demonstrating his capacity as an administrator whose
catholicity of interest reached to the work carried on in every branch
of science. From the first he held that the Smithsonian should interpret
the terms of Smithson's bequest literally and concentrate its funds on
the actual increase and diffusion of knowledge—the increase of knowledge to be promoted by original research of the highest character to
discover general principles or laws of nature, and the diffusion of
knowledge to be accomplished through the publication of a series of
memoirs. He maintained that the work of the Institution should be
of lasting and world-wide import, holding that museums, libraries, and
lectures were local in character and should be supported by government revenues rather than from the funds of the Smithsonian. After
a hard fight stretching over several decades, Henry had the satisfaction
of seeing his views adopted. The Library of Congress and the National
Museum came to be supported by direct appropriation, and still other
changes were made which freed the funds of the Smithsonian for pure
research. It was his firm conviction that the Institution should not do
any piece of work which could be performed equally well by any
other agency. In 1847, a year after he was chosen as secretary, Henry
submitted a "Programme of Organization" providing for the details of
the plan that Congress had prescribed, including the collection and
preservation of objects of nature and art, the plan of the library,
public lectures, and a meteorological service.

The work of the Smithsonian Institution, from the time of its organization down to 1865, was one of decidedly significant accomplishment. Its library contained one of the principal collections of scientific
publications in the country. Its buildings housed the National Museum,
and, later, the National Academy of Sciences. In 1849 the Smithsonian
inaugurated a system of international exchange of the periodicals of
learned and scientific societies which soon superseded all other plans
for foreign exchange, and, continued until the present, has rendered

inestimable service to American scientific societies. The records of several exploring expeditions were published by the Smithsonian in the period prior to the Civil War. *Annual Reports* were issued from the outset, and, beginning in 1848, *Smithsonian Contributions to Knowledge* appeared, from time to time, containing additions to human knowledge based upon exhaustive research. The *Smithsonian Miscellaneous Collections,* begun in 1862, were designed for the publication of papers summarizing existing states of knowledge in particular branches of science at given times. Lists, catalogues, and bibliographical guides have been forthcoming through this medium.

A few of the more significant pieces of investigation published by the Smithsonian Institution down to 1865 may be mentioned. The *Smithsonian Contributions to Knowledge* contained papers on electromagnetism and the undulatory theory of light. Copies of Benjamin Peirce's *Analytical Mechanics,* published in 1855, were distributed in part by the Institution, and C. H. Davis's translation of Gauss's *Theoria Motus Corporum Coelestium,* appearing in 1857, was financed in part by Smithsonian aid. Reports of solar eclipse expeditions were printed in the *Contributions to Knowledge,* as well as lists of lunar occultations, and monumental studies of the orbits of the planets Uranus and Neptune by Peirce and Gould. In 1864 Henry Draper's paper "On the Construction of a Silvered-Glass Telescope . . . " appeared in the *Contributions.* It was destined long to be the bible for those interested in constructing reflecting telescopes. Robert Hare and Wolcott Gibbs contributed weighty discussions to the publications of the Institution. Lester F. Ward, Fielding V. Meek, F. V. Hayden, and others were indefatigable classifiers of paleontological specimens whose catalogues filled many pages of the *Contributions.* Jeffries Wyman, John Torrey, Asa Gray, and Joseph Leidy were leaders in the realm of botanical studies; Spencer F. Baird and Charles Girard were eminent among a host of contributors on zoological topics. E. G. Squire and E. H. Davis furnished an epochal essay to the literature of American anthropology by their "Ancient Monuments of the Mississippi Valley," appearing in 1848. Edward Hitchcock contributed an important paper, "Illustrations of Surface Geology." Papers on American geography and explorations conducted in Asia and Africa found a place in the *Contributions.* The Smithsonian also concerned itself with the task of publishing bibliographies on scientific subjects and preparing tables of physical and meteorological constants.

Certain miscellaneous activities of the Smithsonian in this period are also worthy of mention. Benjamin Silliman, Jr., was sponsored for a course of lectures on the four elements of the ancients, "Earth, Air, Fire, and Water." James D. Dana, Edward Hitchcock, and Louis Agassiz were likewise engaged to lecture. In 1849 the Smithsonian inaugurated a daily meteorological service, the predecessor of the Weather Bureau. Its success rested largely on telegraphic reports of simultaneous weather conditions in many localities, a fact which must have given peculiar satisfaction to Joseph Henry, whose experiments with the electromagnet had paved the way for Morse's invention of the telegraph. The relations of the Smithsonian with the American Association for the Advancement of Science were marked by cordiality from the outset.

For almost a century the American Association for the Advancement of Science has been a potent factor in stimulating scientific research in America. The successive steps leading to its formation deserve to be noted rather carefully. During the latter part of the eighteen thirties the members of the New York State Geological Survey and those engaged in similar surveys in adjacent states found themselves experiencing difficulty in keeping up their mutual professional contacts. Suggestions for calling a conference of American geologists were presently forthcoming in the correspondence of Lardner Vanuxem, Edward Hitchcock, William Mather, Ebenezer Emmons, and others.[108] William Mather, broaching the subject in a letter to the Board of Geologists of New York, well expressed what was in the minds of the advocates of such a conference when he wrote:

> Would it not be well to suggest the propriety of a meeting of the geologists and other scientific men of our country at some central point next fall, say in New York or Philadelphia. . . ? It would undoubtedly be an advantage not only to science, but to the several surveys that are now in progress. . . . It will tend to make known our scientific men to each other. . . . More questions may be satisfactorily settled in a day by oral discussion, than in a year by writing and publication.[109]

The first meeting of the Association of American Geologists was

[108] There has been some dispute among historians as to who first suggested the idea of calling a conference of American geologists. The correspondence of Vanuxem, Hitchcock, Mather, and Emmons shows that they were all interested in such a project. It was probably at the home of the last in Albany, New York, that the first definite steps were taken, a tablet of commemoration being placed there in 1901 by the American Association for the Advancement of Science. New York State, *Museum Bulletin* 52, 452-456.

[109] Quoted by Edward Hitchcock in an "Address Delivered at the Inauguration of the State Geological Hall at Albany, N. Y., August 27, 1856." New York State, Regents of the University of New York, *Tenth Annual Report on the Condition of the State Cabinet of Natural History*, printed as New York State Senate Document 109 (Albany, 1857), 24.

held on April 2, 1840, in the Hall of Franklin Institute in Philadelphia, with Edward Hitchcock acting as chairman.[110] Geologists attended from Massachusetts, New York, Pennsylvania, Delaware, Virginia, and Michigan. From the last state came Bela Hubbard and Douglas Houghton. It "took them an entire week, traveling day and night by the most direct route, and the roads in Ohio were so muddy that the passengers often had to alight and assist in pulling the stage out of the mud."[111] At the meeting, Professor Hitchcock gave an account of the gigantic fossil footprints which he had recently called to the attention of the scientific world. In the following year another meeting was held in Philadelphia. At the Boston meeting in 1842, Henry Darwin and William Barton Rogers collaborated on a monumental paper on the structure of the Appalachian system. Several naturalists attended this latter meeting, with the result that, largely through the influence of A. A. Gould and Amos Binney, the name was changed to "The Association of American Geologists and Naturalists." A report of the first three meetings lists the names of seventy-seven persons who had either attended sessions or presented communications. Four of the names, including Hitchcock, carry the prefix "Rev.," so that it "appears that in those days geology included about 5 per cent of theology."[112] Meetings in subsequent years were held in Albany, Washington, New Haven, New York, and Boston. Progressively, chemists, physicists, and other men of science came into the membership.

As early as 1845, the Association of American Geologists and Naturalists had announced that "a constant effort has been made to counteract the impression that the objects of the Association are exclusively geological or directed to those cognate subjects only which have a direct bearing upon that subject."[113] By 1847, when thirty-seven papers were read, nine related to geology, seven to paleontology, seven to zoology, five to chemistry and physics, three to anthropology, three to meteorology, and one to institutions.[114] Meanwhile, the British Asso-

[110] An account of this meeting is given in *American Journal of Science*, XXXIX (1840), 189-191.

[111] W. H. Hale, "Early Years of the American Association," *Appleton's Popular Science Monthly*, XLIX (1896), 562.

[112] H. L. Fairchild, "The History of the American Association for the Advancement of Science," *Science*, n. s., LIX (1924), 366.

[113] Quoted in G. B. Goode, "The First National Scientific Congress (Washington, April, 1844) and Its Connection with the Organization of the American Association," American Association for the Advancement of Science, *Proceedings*, XL (1891), 46.

[114] *American Journal of Science*, 2d ser., IV (1847), 428-429.

ciation for the Advancement of Science, formed at York, England, in 1832, had been conspicuously successful, and there was ever present before the eyes of American scientists the example of that active and rapidly growing enterprise. It became increasingly apparent that a similar comprehensive and peripatetic society was needed in the United States. At the Boston meeting of the Association of American Geologists and Naturalists in 1847, it was decided that the usefulness of the society would be increased by a broadening of its scope, and the body accordingly resolved itself into the American Association for the Advancement of Science. In order to prepare the way for the change and to explain the purpose of the American Association for the Advancement of Science, a committee composed of H. D. Rogers, Benjamin Peirce, and Louis Agassiz had meanwhile issued a circular to scientists throughout the country explaining the reasons for the change, containing a draft of the proposed constitution, and giving an account of the new plan of organization.

The first meeting of the American Association for the Advancement of Science was held on September 20, 1848, at Philadelphia. W. C. Redfield became the first president. It adopted a brief constitution called "Objects and Rules of the Association," which set forth the aims of the society as follows:

> The objects of the Association are, by periodical and migratory meetings, to promote intercourse between those who are cultivating science in different parts of the United States; to give a stronger and more general impulse, and a more systematic direction to scientific research in our country; and to procure for the labours of scientific men, increased facilities and wider usefulness.[115]

The "Rules" were the work of a committee composed of H. D. Rogers, Benjamin Peirce, and Louis Agassiz, and were modeled after those of the British Association. It is inferred that Agassiz, having worked in England and having attended the British Association before arriving in America in 1846, was influential, if not dominant, in the organization of the broader association, and in the formulation of its "Objects and Rules."[116] It is highly probable that the name American Association for the Advancement of Science was simply an adaptation from British Association for the Advancement of Science.[117] In

[115] American Association for the Advancement of Science, *Proceedings*, I (1849), 8.
[116] H. L. Fairchild, *loc. cit.*, 366.
[117] "It is quite certain that the name and rules were borrowed from the British Association." H. L. Fairchild, *loc. cit.*, 366. G. B. Goode, *loc. cit.*, 46-47, on the other hand, maintained that the name American Association for the Advancement of Science was compounded from the names of the Association of American Ge-

1851, the "Objects and Rules" were slightly changed and called "Constitution."

The American Association for the Advancement of Science early began to recognize the need for providing for specialization and divisions of effort within its own ranks. Accordingly, on September 20, 1848, it was "Resolved, That the Association be divided into two Sections, one to embrace General Physics, Mathematics, Chemistry, Civil Engineering, and the Applied Sciences generally, the other to include Natural History, Geology, Physiology and Medicine."[118]

The American Association for the Advancement of Science had arisen in response to new and definite needs which it could meet better than any existing organization could hope to do. It could transcend the interests of state and local societies. Equally fundamental, it opened its membership to all who were genuinely interested in science, a thing which the American Philosophical Society and the American Academy of Science did not condescend to do. It possessed an additional desideratum in the fact that its meetings were peripatetic. Under the presidency of such outstanding men of science and learning as Joseph Henry, Alexander D. Bache, Louis Agassiz, Benjamin Peirce, and James Hall, the American Association for the Advancement of Science was very

---

ologists and Naturalists and the National Institution for the Promotion of Science. The latter, he argues, had a "direct influence upon the origin of the American Association for the Advancement of Science" inasmuch as several of the leaders in the American Association for the Advancement of Science took part in the scientific congress held in Washington in 1844 under the influence of the National Institution. He therefore maintains that the influence of the National Institution should receive "due recognition" and that it was one of the "parents" of the American Association for the Advancement of Science, the Association of Geologists and Naturalists having been the other. If Fairchild's conclusions about the origin of the name of the American Association for the Advancement of Science are sound, then Goode's statements contain two errors (H. L. Fairchild, *loc. cit.*, 366) for the name came from neither of the "parents" assumed by him. Goode mentions in support of his contention the fact that the name was given as the American Association for the Promotion of Science in the *American Journal* giving a preliminary notice of the September meeting (*American Journal of Science*, VI [1848], 294) and in an account of the first meeting (*ibid.*, 393). But, as we have seen, Agassiz had a hand in the drafting of the "Objects and Rules of the Association," which definitely state, "The Society shall be called the American Association for the Advancement of Science." It seems to me that, on the whole, the theory of adaptation of the British name is the more tenable explanation of the name actually adopted at the first meeting.

[118] American Association for the Advancement of Science, *Proceedings*, I (1848), 26.

active from the time of its inception down to the close of the period with which we are dealing.[119]

The American Association for the Advancement of Science immediately began issuing *Proceedings,* which contained notices and important scientific papers from the outset down to 1861, when it was interrupted for the duration of the Civil War, only to be resumed on an even larger scale in 1866 and the succeeding years. Venerable names in the history of American science are to be found in abundance in the early volumes. Joseph Henry and Wolcott Gibbs occasionally found time for discussion of physical topics. Benjamin Peirce furnished mathematical disquisitions. Matthew F. Maury discussed problems in his chosen field of oceanography, C. H. Davis and Alexander D. Bache explained geodetic methods of survey, and William Chauvenet those of astronomical measurement. Other astronomers who were especially active were Benjamin Gould, Daniel Kirkwood, and Simon Newcomb. Asa Gray as a botanist and Louis Agassiz, Joseph Leidy, and Jeffries Wyman as zoologists were leaders among the biological contributors. Edward Hitchcock, James D. Dana, James Hall, and Louis Agassiz were among the more prolific of the geological writers. Lewis H. Morgan was already embarked upon anthropological researches destined to bring him world renown.

Although Washington, Adams, Jefferson, Barlow, and others of the early Republic had considered the feasibility of a national academy of sciences,[120] nothing had come of projects in that direction. Later, the Jacksonian Democrats looked with suspicion upon all plans for strengthening the federal government in the intellectual sphere, and attempts to establish a national university, a national observatory, and a national academy of sciences were alike unfavorably received, especially since the proponents of such schemes were usually from the northern and eastern portions of the country.[121] It was not to be expected, however, that the great American republic would be permanently without a national academy of sciences under its sponsorship.

---

[119] The history of the American Association for the Advancement of Science has been ably told in H. L. Fairchild, "The History of the American Association for the Advancement of Science," *Science,* n. s., LIX (1924), 365-369, 385-390, 410-415. Many other references are given in my Bibliography.

[120] G. B. Goode, "The Origin of the National Scientific and Educational Institutions of the United States," *Annual Report of the Smithsonian Institution, 1897, U. S. National Museum,* II (Washington, 1901), 312-313, in footnote.

[121] "During Jackson's Presidency all ideas of centralization, even in scientific matters, appear to have fallen into disfavor, and the Columbian Institute and the Columbian College were forced to abandon their hopes for governmental aid." *Ibid.,* 286.

The example of European countries—Great Britain with its Royal Society, France with l'Académie des Sciences, the royal academies of Russia, Prussia, and Austria, and the quasi-governmental scientific bodies of the lesser states—was constantly before the eyes of American men of science. The growing feeling that there was a distinct place for a similar scientific body here in America was well expressed by Alexander D. Bache in an address in 1851 before the American Association for the Advancement of Science.

> Our country is making such rapid progress in material improvement, that it is impossible for either the legislative or executive departments of our Government to avoid incidentally, if not directly, being involved in the decision of such questions. Without specification, it is easy to see that there are few applications of science which do not bear on the interests of commerce and navigation, naval or military concerns, the customs, the light-houses, the public lands, post-offices and post-roads, either directly or remotely. If all examination is refused, the good is confounded with the bad, and the Government may lose a most important advantage. If a decision is left to influence, or to imperfect knowledge, the worst consequences follow.
>
> Such a body would supply a place not occupied by existing institutions, and which our own is, from its temporary and voluntary character not able to supply.[122]

It was over a decade later, however, before the opportunity presented itself to American scientists to realize the long-cherished dream. The National Academy of Sciences owed its immediate origin to the need of the Federal Government for technical advice of a scientific character in the prosecution of the Civil War. During the early years of the war various temporary boards composed of scientific experts were established. This fact gave Charles H. Davis, who had served on several of them, the idea of suggesting to the Navy Department the creation of a permanent scientific commission. The suggestion of such a commission was accordingly made through the instrumentality of Joseph Henry with the gratifying result that the "Permanent Commission," consisting of Joseph Henry, secretary of the Smithsonian Institution, Alexander D. Bache, superintendent of the Smithsonian Institution, and Charles H. Davis, chief of the Bureau of Navigation, Navy Department, was inaugurated on February 11, 1863, to report on such "questions of science and art"[123] as the Navy Department should desire.

[122] A. D. Bache, "Address . . . on Retiring from the Duties of President," American Association for the Advancement of Science, *Proceedings*, VI (1852), xlvii-li.

[123] Letter of appointment, signed by Gideon Welles. Navy Archives, reprinted in F. W. True, edit., *A History of the First Half-Century of the National Academy of Sciences, 1863-1913* (Washington, 1913), 1.

These men, as well as many others, had come to feel the need for a national scientific body to which the government could turn for advice and they had already begun to look forward to the establishment of a national academy of sciences. Even while action of the Navy Department on the proposal for the permanent commission was awaited, it occurred to Davis that the academy of sciences might be established by the simple expedient of asking Congress to incorporate such a body "in the name of some of the leading men of science from different parts of the country."[124] About the same time, Louis Agassiz had been nominated a regent of the Smithsonian, went to Washington, and, after discussion of the proposal for an academy by Davis, Bache, Agassiz, and others, Senator Henry Wilson of Massachusetts introduced a bill (S. No. 555) on February 25, 1863, authorizing a national academy of sciences.[125] This bill passed both houses of Congress on March 3, 1863, and was signed by President Lincoln on the same day. Thus, there quickly came into existence such a body as Jefferson, Barlow, and Poinsett had dreamed of and labored for in vain. Davis was not backward in taking the credit for the action which he, probably more than any other, had brought about. And yet, as has been shown, the seed for such an idea had long been planted in the minds of men.

The historian will naturally wish to inquire why one of the leading scientific societies already in existence was not adapted to serving as a national academy of sciences. First of all, the conferring of such a signal honor upon any society already in existence would have led to undesirable jealousies and charges of provincialism. The American Philosophical Society had functioned very actively and had contributed in no small measure toward making Philadelphia the scientific capital of the nation in the early decades of the nineteenth century. And the American Academy of Arts and Sciences had likewise furnished real intellectual leadership. But, as was perhaps inevitable with the westward expansion of the nation, both organizations lost much of their position of primacy as the nineteenth century wore on. The collapse of the National Institute was keenly felt by those who thought that Washington ought to be the scientific center of the nation as well as its political capital. While the American Association for the Advancement of Science was serving admirably as a forum for nationwide discussion, even some of its leading members realized that it was not the

---

124 Letter of C. H. Davis, dated February 24, 1863. C. H. Davis, *Life of Charles Henry Davis, Rear-Admiral, 1807-1877* (New York, 1890), 289.
125 *Congressional Globe*, LXII, 37th Congress, 3d Session, 1155.

kind of body, with small but select membership able to assemble frequently, which could be relied upon to furnish the government quickly with scientific advice.[126]

Just why the Smithsonian was not reorganized and pressed into service in a larger capacity has been, and still is, somewhat of a mystery to the author. It seems probable that it was feared that any such new capacity would prove subversive of the intention of the original grant and that it would necessarily, in aiding the government, infringe upon the Smithsonian's primary purpose to devote its attentions to enriching the world's store of knowledge by pure research. Henry, Bache, and Agassiz must have known only too well that the Institution was very adequately endowed to carry on the tasks it was already trying to perform for pure science, without being called upon to bear the costs of whatever Congress or the Executive might request in the way of utilitarian investigations. Henry, it will be remembered, had waged a long campaign against all such tendencies to alter the character of the Smithsonian. And now, the proposition that the immediate needs of the government called for an academy of sciences to give it aid and counsel, while doubtless sincere and perhaps the crux of the whole matter, had nevertheless a political angle and would serve as a talking point in case of debate in Congress and as a bait to secure votes for the project. All this could scarcely be reconciled with the conception of the Smithsonian trust that was held by Henry and his governing associates. At any rate, the men who were most prominent in the establishment of the National Academy of Sciences were either officers of the Smithsonian or were well acquainted with its activities, and it may be assumed that they acted as they did for good and sufficient reasons.

Yet, while the primary circumstance which called forth the creation of the National Academy of Sciences was the desire for a body to give advice to the Federal Government, there was also firmly implanted in the minds of the founders, of course, the urge to promote abstract scientific research. This was well expressed in a statement of the aims of the Academy which read, "The objects of this association are principally to advance abstract science, and to examine, investigate, and

126 A. D. Bache, it will be recalled, had advocated a separate and new national academy of sciences on the ground that it "would supply a place not occupied by existing institutions," which the American Association for the Advancement of Science was "from its temporary and voluntary character, not able to supply." A. D. Bache, *op. cit.*, li.

experiment upon subjects upon which information is desired by the government."[127]

The first meeting of the National Academy of Sciences, held in New York on April 22, 1863, was devoted to problems of organization, the members arranging themselves into two classes: (*a*) mathematics and physics and (*b*) natural history. At the first scientific session, held in Washington from January 4 to 9, 1864, a program of scientific papers was presented by Alexander D. Bache, the first president, Joseph Henry, Louis Agassiz, and others. The first *Report* of the Academy was for the year 1863, and its *Annual* was published in 1865.

The fifty incorporators of the National Academy of Sciences included the names of many who ranked high in American science. Especial mention might be made of Louis Agassiz, Alexander D. Bache, James D. Dana, Wolcott Gibbs, Benjamin Gould, James Hall, Joseph Henry, Joseph Leidy, Hubert A. Newton, and Benjamin Silliman, Jr. Taken all in all, the National Academy of Sciences comprised an eminent group of investigators and one which well bore out Joseph Henry's statement that it was

> implied in the organization of such a body that it should be exclusively composed of men distinguished for original research, and that to be chosen one of its members would be considered a high honor, and consequently a stimulus to scientific labor, and that no one would be elected who had not earned the distinction by actual discoveries enlarging the field of human knowledge.[128]

Inasmuch as the belief was expressed that, "The want of an institution, by which the scientific strength of the country may be brought from time to time to the aid of the government in guiding action by the knowledge of scientific principles and experiments, has long been felt by patriotic and scientific men of the United States,"[129] it is worth while to inquire into the role of the Academy as the scientific adviser of the government[130] in the first few years after its creation.

[127] National Academy of Sciences, *Report of the National Academy of Sciences for the Year 1867* (Washington, 1868), *Senate Miscellaneous Documents*, no. 106, 40th Congress, 2d Session.

[128] *Report of the National Academy of Sciences for the Year 1867* (Washington, 1868), 1.

[129] A. D. Bache in *Report of the National Academy of Sciences for the Year 1863* (Washington, 1864), 1.

[130] "Section 3. *And be it further enacted,* That . . . the Academy shall, whenever called upon by any department of the Government, investigate, examine, experiment and report upon any subject of science or art. . . . ." Act of Congress

No less than five committees were appointed at the request of the Government within a month after the organization of the Academy, and several others came into existence before the close of the War.[131] The Committee on Weights, Measures, and Coinage was appointed on May 4, 1863, at the solicitation of Salmon P. Chase, secretary of the treasury, to consider the question of the "uniformity of weights, measures and coins . . . in relation to domestic and international commerce."[132] This committee included Joseph Henry, Benjamin Silliman, Jr., William Chauvenet, and Alexander D. Bache. Its first definite report, not submitted until January 27, 1866, favored the metric system and suggested that it be legalized by Congress. Appropriate legislation by that body followed a few months later. This is a clear case where the policy of the national government was determined in accordance with the sought advice of the National Academy of Sciences. A committee on Protecting the Bottoms of Iron Vessels against Corrosion, appointed in May of 1863, reported the following year that it could make no recommendations although it was willing to conduct experiments if funds were forthcoming therefor, which was not the case. The truth seems to be that the scientific information needed to solve the problem was not at that time in existence, nor were the proper chemical preparations found until about fifty years later.[133]

A Committee on Magnetic Deviation in Iron Ships, appointed on May 20, 1863, including Alexander D. Bache, Charles H. Davis, Joseph Henry, Wolcott Gibbs, and Benjamin Peirce, wrestled with the problem of shielding the compass from the disturbing action of masses of iron inevitable in the construction of the "iron-clads." The committee adopted the method devised by the English astronomer Airy for counteracting the disturbing influences by means of placing bar magnets in suitable locations and it supervised the correction of the compass on twenty-seven vessels, including monitors and gunboats. A Committee on Wind and Current Charts and Sailing Directions was appointed on May 25, 1863, at the request of the Navy Department to

Establishing a National Academy of Sciences, *Congressional* Globe, LXXII, 37th Congress, 3d Session, 1501.

131 This subject is fully treated in F. W. True, edit., *A History of the First Half-Century of the National Academy of Sciences, 1863-1913* (Washington, 1913), ch. IV, "The Academy as the Scientific Adviser of the Government."

132 *Ibid.*, 206.

133 Henry Williams, "Anticorrosive and Antifouling Paints for Ships' Bottoms; Experience of the U. S. Navy," *Engineering News*, LXVI (1911), 136; and by the same author, "Modern Painting Methods in the Navy," *Scientific American* Supplement, LXXIV (1912), 354.

consider whether Matthew Maury's "Wind and Current Charts" and his "Sailing Directions" should continue to be published at government expense. The opinion of the committee was unfavorable both as to the scientific and the practical merits of the publications,[134] although previously hundreds of thousands of copies had been printed. Since Maury was a Southerner and had withdrawn from the service of the Federal Government, the charge immediately went forth that the decision of the committee was an unfair one. This seems not at all to have been the truth, nor was it in any way influenced by the fact that Maury had cast his lot with the Confederacy. The report recognized the merit of the idea back of the publications and it showed no animus against Maury. The publications were reissued in a revised form several decades later when more scientific data were available.

The Committee on the Question of Tests for the Purity of Whiskey was appointed on January 14, 1864, and, at the request of the acting surgeon general of the army, Benjamin Silliman, Jr., served as its chairman. In a report presented the following year the committee recommended the use of diluted alcohol rather than whisky as a stimulant. A Committee on Experiments on the Expansion of Steam, appointed in 1864, conducted experiments until 1880, but never reported much progress. The Committee on Materials for the Manufacture of Cent Coins, appointed April 11, 1864, at the request of Salmon P. Chase, conducted experiments with aluminum bronze and transmitted a bar of metal to the Director of the Mint. It was not favorably received, and with the return of peace the question of an aluminum bronze currency was dropped. A Committee on the explosion of the United States Steamer *Chenango* was appointed May 2, 1864, to investigate the cause of a frightful boiler explosion resulting in the death of twenty-eight of the ship's crew. An inquest had been held in New York and a jury had been unable to agree in fixing the blame, the majority opinion seeming to imply that the fault lay in the specifications made by the Navy Department. Not being satisfied with this outcome, the Navy sought an impartial investigation by members of the Academy. The report of the Committee of the Academy threw

134 It recommended that it be *"Resolved by the National Academy of Sciences, That in the opinion of this Academy the volumes entitled 'Sailing Directions,' heretofore issued to navigators from the Naval Observatory, and the 'Wind and Current Charts,' which they are designed to illustrate and explain, embrace much which is unsound in philosophy, and little that is practically useful; and that therefore these publications ought no longer be issued in their present forms."* *Report of the National Academy of Sciences for 1863* (Washington, 1864), 112.

the main responsibility on the private constructors rather than on the naval engineers.

In conclusion, while one would not care to maintain that the advice rendered by the Academy through its committees during the war years was fraught with momentous consequences, nevertheless, it is only fair to say that the National Academy of Sciences was a trusted adviser and that its counsel shaped governmental policy in several instances, notably in that of legalizing the metric system in the United States.

The main trends of the period 1790-1865 covered in this chapter may be summarized in a few concluding words. During these years the Northern and Eastern states became thickly dotted with state and local academies of science. The South, meanwhile, still under the baleful influence of slavery, was lagging far behind the North in the matter of scientific endeavor just as it was trailing behind in the onward march of industrial activity. National specialized scientific societies came into permanent existence in the United States for the first time during this period, and these paved the way for the formation of dozens of such organizations which came to be founded in the years between the close of the Civil War and the turn of the century, as we shall see in the next chapter. Integration of the scientific activity of the nation was attempted at the close of the period in the establishment of the National Institution, the Smithsonian Institution, the American Association for the Advancement of Science, and the National Academy of Sciences. American scientific societies had, in general, taken on a highly democratic character and liberally opened their membership to those from every walk and station of life who sought admission to the ranks of those who banded together in the name of science.

Chapter III

THE TRIUMPH OF SPECIALIZATION, 1866-1918

A consideration of the history of American scientific societies during the period from the close of the Civil War to the close of the World War reveals at once three main developments. First, the keynote of the period was the tendency toward specialization; second, there was a slow but sure drift in the direction of national centralization within the specialties; and third, there was an increasing tendency to form strictly technological societies in response to the demands imposed by the industrialization and urbanization of the new "machine age." The factors associated with each of these trends will now be considered in turn.

American scientists in the years prior to the Civil War were, for the most part, generalists who delighted in ranging widely over the whole gamut of science. On the other hand, the men of science who labored in the closing decades of the nineteenth century and the opening ones of the twentieth became increasingly specialists, devoting their time entirely to some particular science or branch of it. Specialization and specialization within specialization became the rule. This led, of course, to greater emphasis upon scientific societies of specialized character, and literally hundreds of such organizations were forthcoming, national, state, and local. The increase of specialization hastened the advance of science to new frontiers. Only a few of the many scientific achievements in the long period between 1866 and 1918 can be mentioned here.

The closing decades of the nineteenth century and the opening ones of the twentieth saw a veritable host of able mathematicians in Europe turning out countless and highly specialized monographs on mathematical topics. Arthur Cayley, J. J. Sylvester, Felix Klein, and Leopold Kronecker were among the most notable contributors to algebra. Francis Galton and Karl Pearson were pioneers in "biomathematics," a new field which they helped to create through the application of statistical methods to biological data. Richard Dedekind, Georg Cantor, and Charles Hermite were leaders among those whose studies of the calculus furnished abstruse analyses in the realm of number theory.

Henri Poincaré, dean of French mathematicians, exerted a wide influence in many fields. In the twentieth century, Lebesgue, Borel, Lorentz, and Einstein were outstanding among those who sought to gain insight into the nature of space by studying the implications of non-Euclidean geometry. American mathematicians, save for a few notable exceptions, lagged behind Europeans in the period covered by this chapter. Benjamin Peirce continued to be the leading mathematician in this country until his death in 1880. He and his son, James M. Peirce, both members of the Harvard faculty, produced many fertile works on higher analysis. Another son of Benjamin Peirce, Charles S. Peirce, a scientist in the service of the Government, made an enviable record as a mathematical contributor. Benjamin Osgood Peirce, a relative of the aforementioned family, and of Harvard, ranged broadly through the physical sciences and was the author of numerous mathematical tracts. From 1877 to 1883 J. J. Sylvester taught at Johns Hopkins and by his writings and his contacts exerted an especially leavening influence upon American mathematical thought. Willard Gibbs of Yale made a number of important mathematical discoveries as by-products of his researches into physics. Simon Newcomb and G. W. Hill were able computers who dealt with celestial mechanics. They found competent successors at the opening of the twentieth century in E. W. Brown of Yale and F. R. Moulton of the University of Chicago. E. H. Moore and Oswald Veblen proved to be analysts of a high order of genius.

In the field of astronomy, two Britishers, Norman Lockyer and William Huggins, filled the closing decades of the nineteenth century with their achievements in solar spectroscopy. Angelo Secchi did comparable work in Italy in the sixties and seventies. Henry Draper, C. A. Young, S. P. Langley, and, in the twentieth century, G. E. Hale pursued similar studies in America. G. H. Darwin, English student of lunar theory, and Simon Newcomb and F. R. Moulton of the United States turned their attention to problems of gravitational astronomy. Henry Draper, E. C. Pickering, and W. H. Pickering in America were instrumental in bringing about the wide application of photography to astronomy. A. J. Cannon, at Harvard, spent the opening decades of the twentieth century in compiling a monumental catalogue of stellar spectra. Larger and better instruments brought new triumphs of observational astronomy. Giovanni Schiaparelli of Italy made studies of Mars in the closing decades of the nineteenth century which aroused much interest and curiosity in America, where Percival Lowell took up similar work and wrote several books on the subject. In 1877 Asaph Hall at Washington discovered two satellites of Mars and a dozen years later E. E. Barnard at the Lick Observatory in California was the first to see the fifth satellite of Jupiter. B. A. Gould in 1879 published an enormous catalogue of plans of the Southern Hemisphere. Lewis Swift and R. W. Brooks of the United States and N. Camille Flammarion of France achieved renown as discoverers of comets.

In the field of physics, European scientists were pushing on to new frontiers. Ernst Mach launched an attack on Newtonian mechanics. R. J. E. Clausius, John Tyndall, and William Thomson rounded out a century of notable advances in the knowledge of thermodynamics.

The study of radiant energy became of increasing interest to physicists in the latter part of the nineteenth century. J. C. Maxwell continued his mathematical investigations of electricity and light. H. R. Hertz, from 1885 to 1889, was engaged in discovering the waves which bear his name and are utilized in wireless telegraphy, brought into practical use by Guglielmo Marconi in 1896. In 1895 W. K. Röntgen discovered the rays which sometimes bear his name but are more commonly known as X-rays. A. H. Becquerel and Pierre Curie and his wife, Marie Curie, a little later were working on radium radiation; Sir Ernest Rutherford in 1899 discovered the "alpha rays" and the "beta rays"; and shortly thereafter Villard announced discovery of the "gamma rays." Built on the work of J. J. Thomson at the close of the century, the conception of the atom was revolutionized in the early years of the twentieth century by Frederick Soddy, Max Planck, H. G. Moseley, A. S. Russell, Niels Bohr, and many others, who developed wave mechanics and the quantum theory. Pieter Zeeman, H. A. Lorentz, and many others, including especially Albert Einstein, who presented his restricted theory of relativity in 1905, were advancing new views about time, space, and matter.

America, too, produced notable physicists. Joseph Henry continued his electrical investigations until his death in 1878. Willard Gibbs, a mathematical physicist of the first order, wrote extensively in the closing decades of the nineteenth century. Johns Hopkins became a center of physics research. Among the students trained there, who became especially distinguished in physics, were E. H. Hall, J. S. Ames, and A. L. Kimball. H. A. Rowland and S. P. Langley at the close of the nineteenth century were ingenious designers of new instruments, as were A. A. Michelson and G. E. Hale in the twentieth. A. A. Michelson and E. W. Morley made accurate determinations of the velocity of light from 1887 to 1906, and Michelson lived to work on the problem for several decades more. R. A. Millikan and A. H. Compton became in the early years of the twentieth century among the recognized American leaders in the study of radiation and atomic theory. R. A. Millikan, C. P. Steinmetz, and M. I. Pupin ranked among the world's foremost students of electricity.

The enunciation of the periodic table by the Russian D. I. Mendeleev in 1869 raised a host of problems for European and American investigators. S. A. Arrhenius, Wilhelm Ostwald, J. H. van't Hoff, and J. A. Le Bel achieved great reputations for their studies in inorganic chemistry during the closing decades of the nineteenth century and the opening ones of the twentieth. Henri Becquerel, Pierre Curie, and Marie Curie were announcing their epochal studies on radium and radioactive substances at the turn of the century; their studies aroused much interest in America. In the field of biological chemistry, Claude Bernard in the fifties and sixties and Louis Pasteur in the closing decades of the century held the center of attention in the European scenes. In the latter decades of the nineteenth century and the opening years of the twentieth, Germany held the lead in industrial chemistry.

In America, Willard Gibbs of Yale, who ranged widely through mathematics, physics, and chemistry, contributed numerous and important scientific papers in the closing decades of the nineteenth century, as did Wolcott Gibbs of Harvard. A different type of chemist was C. M. Lea of Philadelphia, a self-taught man who issued from his private laboratory papers on photochemistry that secured for him international renown in the latter decades of the nineteenth century. T. W. Richards, already teaching at Harvard in the eighteen nineties, and continuing his work there for almost four decades, won the Nobel prize in 1914 for his work on atomic weights.

J. D. Dana and James Hall, veterans of the older geology, continued their stratigraphic work until the nineties. Joseph Leidy, O. C. Marsh, and E. D. Cope were actively engaged in paleontology during the period 1870 to 1890. Louis Agassiz, until his death in 1873, was an active proponent of glacial geology in America. The writings on glacial geology of James Geikie, a Scotch geologist, came to be circulated in America in the closing decades of the nineteenth century and the opening ones of the twentieth, as did also those of the German, Eduard Suess, on historical stratigraphy. T. C. Chamberlin took up the work in the nineties. Perhaps his greatest work was *The Origin of the Earth*, published in 1916, which proposed the substitution of the "planetesimal hypothesis" for the time-honored nebular hypothesis. Further work in this direction by him and his colleague, F. R. Moulton, discredited the older theory. Clarence King spent the last four decades of the nineteenth century in studying the geology of the west; and, in addition to making notable studies of the Cordilleran country, he was instrumental in the organization of the United States Geological Survey, becoming its first head in 1878. C. E. Hutton had a somewhat comparable career in the Utah and Arizona region, and F. V. Hayden did similar work in the Wyoming-Montana area in the years 1860 to 1885. Following the lead of the German scholars, Ferdinand Zirkel and K. H. Rosenbusch, American geologists such as Whitman Cross, R. D. Irving, J. P. Iddings, G. H. Wilson, and G. P. Merrill brought the new science of petrography to America and began using microscopes and chemical analysis extensively in their technique.

The African explorations in the eighteen sixties and seventies of David Livingstone, a Scotch missionary, and Sir H. M. Stanley, an American-British journalist, aroused American interest in that continent. Interest in the polar regions was reawakened in the early years of the twentieth century by the explorations of R. E. Peary, an American who first reached the north pole in 1909, and by the expeditions of Roald Amundsen of Norway to the south pole in 1911 and of R. F. Scott of Great Britain to the same place in 1912.

The closing decades of the nineteenth century found a host of able European writers presenting views upon evolution. Charles Darwin followed his *Origin of Species* with the *Descent of Man* and other writings. Alfred Wallace and Thomas Huxley expounded the new views with facile pens. Herbert Spencer, espousing the evolutionary doctrines, spent almost half a century before his death, in 1903, in elaborating the philosophical implications of evolution and extending his main concepts to many branches of knowledge. Men such as E. L. Youmans and John Fiske were instrumental in popularizing his views in America, and Spencer himself crossed the ocean to enjoy a cordial American reception. Ernst Haeckel of Germany lent the great weight of his authority to the evolutionary views for over half a century after they were first announced. G. J. Mendel, August Weismann, and Hugo De Vries conducted many experiments in genetics which shed light upon the laws of heredity. Pioneer studies in biometrics were made at the close of the nineteenth century and the opening of the twentieth in England, by Sir Francis Galton, Karl Pearson, and William Bateson. It was in Asa Gray, veteran biologist at Harvard, that the Darwinians found their foremost champion in the first rounds of a long struggle against those who clung to beliefs in "special creation."

Outside academic halls, the debate on evolution soon came to convulse press and pulpit and arouse barriers and animosities, the rancor of which had not yet quite subsided decades later, as the reverberations of the controversy arousing strong emotions over the trial of J. T. Scopes at Dayton, Tennessee, taking place in 1925 showed.[1] While the discussion over evolution went merrily, or rather furiously, on, American biologists went steadily forward with their main task of augmenting the information in their specialties. O. C. Marsh, Joseph Leidy, and E. D. Cope form a trio of versatile geniuses whose work in zoology and paleontology will long be remembered. The dean of American zoologists, L. J. R. Agassiz, continued his labors until his death in 1873. The following year closed the career of Jeffries Wyman, his colleague and a well-known anatomist and ethnologist. Alexander Agassiz, continuing in the traditions of his sire, built up the Harvard Museum, lavished his own personal wealth upon it, explored widely, and wrote extensively.

Pursuing his amateur interest in the propagation of fishes, Seth Green entered upon studies which, by his death in 1888, had earned for him the title of father of pisciculture in America. S. F. Baird and G. B. Goode, his assistant at the Smithsonian in the seventies and eighties, were indefatigable cataloguers of zoological specimens. W. K. Brooks, appointed head of the biological department of Johns Hopkins, was known as a stimulating teacher and author as well as for his researches on invertebrates. A. S. Packard and S. H. Scudder filled the closing years of the nineteenth century and the opening ones of the twentieth with their entomological investigations. In the closing decade of the nineteenth and the opening ones of the twentieth, D. S. Jordan was publishing his monumental works on ichthyology, but his interests ranged to many biological branches other than the study of fishes. C. S. Minot did basic work in embryology at the close of the nineteenth century and early years of the twentieth. E. G. Conklin has dealt broadly with embryology and evolution.

Among the leaders in genetics and biometrics in America have been T. H. Morgan, C. B. Davenport, Raymond Pearl, and H. S. Jennings. L. J. Henderson and Jacques Loeb were among the more active of the

---

[1] The history of the conflict over evolution in America is traced in A. D. White, *History of the Warfare of Science with Theology in Christendom* (2 vols., New York, 1896) ; in H. J. Kreider, *The Impact of Evolutionary Thought on America in the Nineteenth Century*, Master's Essay, University of Rochester, 1930; and in B. J. Loewenberg, *The Impact of the Doctrine of Evolution on American Thought, 1859-1900*, Doctoral Dissertation, Harvard University, 1934; B. J. Loewenberg, "The Reaction of American Scientists to Darwinism," *American Historical Review*, XXXVIII (1933) , 687-701. An important section of the country in regard to the conflict is treated in B. J. Loewenberg, "Evolution in New England, 1859-1873," *New England Quarterly*, VIII (1935) , 232-257.

biological chemists in the first decades of the present century. The writings of the nature poet, John Burroughs, and the horticulturist, Luther Burbank, caught the popular fancy at the close of the nineteenth century and still more in the opening decades of the twentieth. In the field of anthropology, Lewis H. Morgan, by the time of his death in 1881, had won a renown which grows through the years. H. F. Osborn and Franz Boas were leaders among the twentieth century anthropologists.

The work of Joseph Lister on antiseptics, at the close of the eighteen sixties and thereafter, opened a new era in the history of surgery. Louis Pasteur, working in the last four decades of the nineteenth century, startled the world by his discoveries in bacteriology, leading to his announcement and elaboration of the germ theory. In the course of his investigations he discovered vaccines for anthrax, rabies, and other mysterious maladies. Robert Koch did comparable work and, notably, in 1882 isolated the tuberculosis bacillus. The discovery of the syphilis parasite by Fritz Schaudinn was followed in 1905 by Paul Ehrlich's discovery of an efficacious remedy. The work of the Russian, I. P. Pavlov, on the "conditioned reflex," in 1890, and that of C. S. Sherrington of Great Britain on nerve connections especially influenced American physiological thought.

America, too, had its great names in the history of medicine. J. M. Sims and O. W. Holmes were prominent as gynecologists, the latter being widely known as a teacher in the Harvard Medical School. W. J. Mayo and C. H. Mayo in the eighties began to build up the reputation of their clinic at Rochester, Minnesota. S. W. Billings wrote on toxicological topics and on nervous diseases. J. S. Billings spent the later decades of the nineteenth century in compiling bibliographical guides to medical literature. E. L. Trudeau's establishment for the treatment of tuberculosis in the Adirondack mountains was the first such sanitarium on the American Continent. The popular imagination was captured in 1900 and the succeeding years by the work of Walter Reed, James Cassall, and Jesse Lazear, who traced yellow fever to the bite of a certain species of mosquito existing especially in the tropics. This enabled W. C. Gorgas to check the ravages of this dreaded disease in Cuba and in Panama during the construction of the Canal.

During the closing decades of the nineteenth century psychology emerged from the realm of philosophy and took a place among the experimental sciences. Wilhelm Wundt established the first psychological laboratory at Leipzig in 1879. G. S. Hall, after studying with

Wundt in Germany, established the first psychological laboratory in America at Johns Hopkins in 1883. William James had been assembling experimental apparatus at Harvard for over a decade, and a laboratory was formally opened there in 1892. By the turn of the century there were more than two dozen such laboratories in the United States. Some of Wundt's students, G. S. Hall of Johns Hopkins University, later president of Clark University, J. McK. Cattell of the University of Pennsylvania and later of Columbia, E. B. Titchener of Cornell, and Hugo Münsterberg, were active among those introducing the new methods into America. E. L. Thorndike became, in the opening years of the twentieth century, one of the foremost pioneers in the study of the psychology of learning. R. M. Yerkes became an outstanding investigator of animal psychology, as did Thorndike also. Bridging the gap between the older and the newer approaches to psychology was William James, whose *Principles of Psychology*, appearing in 1890, had wide influence upon the teaching of psychology of emotions and habits in America. John Dewey, from about 1900 onward, was elaborating his program of educational psychology stressing social adjustment. J. R. Angell, a former student of James', was soon enunciating "structuralism," an interpretation emphasizing biological factors. J. B. Watson, by the second decade of the twentieth century, was completely in revolt against the introspective methods of the older approaches and was himself enunciating "behaviorism," stressing mechanistic response. The work at the close of the nineteenth century of Emil Kraepelin, J. M. Charcot, Pierre Janet, and especially that of Sigmund Freud, with its novel interpretation of the role played in mental life by hidden desires, came increasingly to influence American psychological thought.

Americans in the period 1866-1918 established an enviable record in the matter of making inventions and applying scientific and mechanical principles to the conveniences of material existence. Travel by railroad was rendered more comfortable by George Pullman, who introduced his "palace cars" in 1864, and it was made faster and safer by Thomas S. Hall's invention of block signalling devices in 1867 and by George Westinghouse's airbrake, patented in 1872. In 1866 Cyrus W. Field succeeded in establishing cable communication with Europe. Ten years later, Alexander G. Bell and Elisha Gray simultaneously took out patents for telephones. Edison's quadruplex telegraph of 1874 multiplied the efficiency of the telegraph by making it possible to transmit several messages over a line at the same time. Although major credit for the invention of wireless telegraphy goes to Guglielmo Marconi, of Italo-Irish extraction, the part played by Edison was significant and led to the discovery of the so-called Edison effect, an electrical discharge taking place in vacuum tubes and one utilized in wireless reception. The crowning achievement in the career of Edison was his production of the incandescent electric light in 1879. Turning night into day, this invention was fraught with far-reaching implications, changing the vocational and the avocational life of the American perhaps more than any other. It was Edison also who furnished the phonograph in 1877, about the time several Europeans were groping toward similar instruments, enabling the human voice and sound in general to bridge time as well as distance. Experimentation in the early eighteen eighties by Edison and others with "motion pictures" paved the way for making them a practical and commercial reality some twenty years later.

A new epoch in photography was opened in the eighteen eighties by George Eastman, who replaced the clumsy and expensive "wet" plate with the "dry" plate or celluloid film which was readily inserted in the "kodak," his patent box equipped with an instantaneous shutter, capable of being readily operated by anyone. Amateur and professional photography shot forward by leaps and bounds. Ottmar Mergenthaler produced the linotype in 1885. The business office underwent a metamorphosis at the hands of C. L. Sholes, inventor of the typewriter in 1868, L. E. Waterman, inventor of the fountain pen in 1884, W. S. Burroughs, inventor of a practical and widely used adding machine in 1888, and John R. Gregg, who at the close of the century devised a greatly improved system of stenography based on sound. During the eighteen eighties and nineties Americans discovered how electrical furnaces might be used to produce aluminum, silicon carbide, and calcium carbide and its derivatives. J. W. Hyatt's preparation, known as celluloid, was a scarcely less important chemical contribution. Colonel Albert A. Pope, patentee of the first safety bicycle, began manufacture of it in 1886. The first commercially successful electric railway in America was opened in Richmond, Virginia, in 1887-1888, by Frank J. Sprague, although Edison and other inventors had earlier made successful experiments and needed only time to perfect the generators at their central power houses. Charles P. Steinmetz and Nikola Tesla were pioneers in developing ways of insulating high-tension wires for the transmission of electricity to a distance from places such as Niagara Falls.

As early as 1879 George B. Selden had filed the first patent for an automobile, although it was not until 1895 that he actually drove his vehicle. By that time Elwood Haynes, R. E. Olds, Charles Duryea, and

Henry Ford had begun the commercial manufacture of automobiles. Principal credit for making the motor car an efficient and low-priced vehicle goes to Ford, who put into practice his creed of making a small profit on each sale and turning out cars on such a quantity basis as to insure all the economies of large-scale production. When, on December 17, 1903, Orville and Wilbur Wright made the first successful flight with a heavier-than-air machine, or an airplane as it is called today, driving it with a light gasoline motor, mankind's age-old yearning dream of being able to fly had been realized. Another modern Daedalus was Glenn Curtiss, who devised the hydroplane in 1911. In 1903 Reginald Fessenden succeeded in transmitting the human voice by means of wireless waves, and the era of the radio had begun. It was to receive great impetus from Lee De Forest's invention of the vacuum tube in 1906 and from the Armstrong "feed back" regenerative circuit devised a few years later.

Spectacular as these highlights of invention are, it is quite probable that more real progress was made by a host of unremembered inventors who unobtrusively added increments of improvements to the original inventions. In the Civil War era there were countless unremembered inventors, mostly amateurs, who flourished in the workshops hastily improvised in shed or barn,[2] and who, by adding a little here, a little there, made modern machine tools a possibility. It was to an ever-widening audience of this sort that E. L. Youmans sent his numbers of *Popular Science Monthly*, "the first magazine in the world to report in popular terms the news of laboratory and workshop."[3]

Thomas Edison and Alexander G. Bell, proving to be able industrialists as well as inventors, established large corporations bearing their names. These companies became leaders in industrial research, maintaining staffs of inventors and scientists who expended time and treasure in perfecting and adding to the basic patents. John D. Rockefeller and Andrew Carnegie were among those who lavishly provided for industrial research and then in later years established research foundations for the advancement of pure science. Charles P. Steinmetz, with an expense account but asking no salary, was a mathematical prodigy who spent his life enriching humanity while in the service of the General Electric Company. Among the scientific journals of the

[2] H. H. Saylor, "The Home Workshop of 1872," *Popular Science*, CXL, No. 5 (1942), 194-197.

[3] "Popular Science Introduced a New Type of Journalism," *Popular Science*, Seventieth Anniversary Issue, CXL, No. 5 (1942), 53.

better class spreading the new discoveries of science and invention were the *American Journal of Science* and the *Scientific American*, dating from the previous era, *Popular Science Monthly*, founded under the editorship of E. L. Youmans in 1872, and *Science*, established in 1883, which came under the vigorous editorship of J. McK. Cattell at the close of the nineteenth century.

Not only were the advances in science that resulted from increased specialization and better research facilities preparing the American intellectual and professional classes for larger participation in scientific societies, but certain aspects of American political and economic life worked to accomplish the same end. First and foremost on the political side was the sense of enhanced nationality with which the United States emerged from the throes of the Civil War. This paved the way for national cooperation in science as well as in other spheres of activity. Abandonment of the "radical" program of reconstruction in 1877, and the Nashville meeting of the American Association for the Advancement of Science in the same year, which was well attended by Southerners, were alike heralds of a reunited nation. The Spanish American War and ensuing policies of imperialism stimulated American nationalism. Although the pouring in of millions of immigrants from central and southeastern Europe in the opening decade of the nineteenth century caused the "melting pot" of American nationality to seethe and spill over a bit, the close of the World War found the nation once again in a mood of exultant nationalism.

Economic forces as well were solidifying the nation. Railroad builders like E. H. Harriman, J. J. Hill, and Cornelius Vanderbilt girded the country together with bands of steel. Invisible bonds of union were created by such octopus industrial enterprises as the Standard Oil Company and the Carnegie Steel Corporation. Still less in evidence, but perhaps as real, were the golden chains of credit which bound the financial interests of the country to the offices of J. D. Rockefeller and J. P. Morgan and their associates on Wall Street. Telegraph and telephone lines were the nerve fibers of the new economic life. Labor, too, was organizing on a national scale. The National Labor Union was formed in 1866, the Knights of Labor in 1869, and the American Federation of Labor in 1881. Legislation and Supreme Court decisions gradually increased the power of the federal government over the new interstate enterprises.

Paralleling the trends toward organization on a national scale in the fields of transportation, manufacture, finance, and labor were the attempts of scholars to organize nationally. The movement got under way in the eighteen eighties and has gained momentum with each decade. It was inevitable in a period of increasing specialization that the new national societies should be devoted to specialties. It was also inevitable that, in an era of rapid industrial development, a strong impulse should be given in the direction of establishing societies to deal with the applied aspects of science.

The closing decades of the nineteenth century also witnessed the emergence of cities to a new importance in American life. The period 1866-1918 saw the nation change from being one preponderantly rural and agricultural to one primarily urban and industrial. The part played by the rise of cities in bringing about the growth of scientific societies can scarcely be emphasized enough. In the cities, the accumulation of wealth brought schools, colleges, libraries, and museums. Lured by these superior educational advantages, thither flocked those desirous of engaging in the pursuit of learning. Congregating together, at first informally, groups of scholars and professional men became the nuclei of learned and scientific societies.

With the emergence, then, of the basic forces in modern American life, many factors, intellectual, political, economic, and social, conspired to make scientific societies flourish in this period as in no time previously in American history. We shall turn our attention, first of all, to a study of the most conspicuous trend in the organization of the scientific societies in this period, namely, to a consideration of the rise of the specialized societies.

Mathematics came to play a much larger part in the study of science in the latter half of the nineteenth century than ever before. Physicists, chemists, and engineers began to rely more and more upon precise measurements and to work by formulas rather than by "rule of thumb." It was fitting enough that the first steps to form a national mathematical society should be taken in New York, the nation's greatest city. On November 24, 1888, six members of the department of mathematics at Columbia University formed a society to meet monthly. Professor J. H. Van Amringe, later dean of Columbia College, served as its first president. T. S. Fiske was another of the moving spirits of the new group. It was soon decided to invite the cooperation of all persons professionally interested in the subject of mathematics who lived in or near New York. The group became known as the New York Mathematical Society, proclaiming its object to be "to encourage and maintain an active interest in mathematical science."[4] The *Bulletin* was

---

[4] New York Mathematical Society, *Organization of the New York Mathematical Society, Constitution, By-Laws, and List of Members, June, 1891* (New York, 1891), 14. Bound with New York Mathematical Society, *Bulletin* I (1891).

begun in 1890. A widening of the membership resulted in the adoption of the name, the American Mathematical Society, in 1894, and thereafter the society definitely assumed a national character. *Transactions* were begun in 1900; a library was built up.[5] An active group of mathematicians, including T. S. Fiske, Harold Jacoby, Emory McClintock, M. I. Pupin, and C. P. Steinmetz, soon established the reputation of the society beyond question.

Mathematics clubs were formed in many of the larger American universities in the opening decades of the present century; but these were generally informal groups of students and professors and the meetings were of a seminar nature. Therefore these groups hardly deserve to be considered societies calling for separate or detailed discussion. Mathematics sections were formed in state teachers' associations in a number of states.[6] The American Federation of the Mathematical and the Natural Sciences was formed in 1907, consisting of many associations of teachers of mathematics and the sciences.[7] The mathematics clubs and the mathematics sections were especially active in the mid-West. The Kansas Association of Teachers of College Mathematics in 1915 led soon to the formation of similar societies in near-by states. This mathematical activity centering in the mid-West led to the calling of a convention at Columbus, Ohio, which met on December 30, 1915, and formed the Mathematical Association of America.[8] Among the leaders in the new society were E. R. Hedrick, first president, E. V. Huntington and G. A. Miller, vice-presidents, L. E.

[5] The history of the society is reviewed in detail in T. S. Fiske, "Mathematical Progress in America," *Science*, n. s., XXI (1905) , 209-215; see also Emory McClintock, "The Past and Future of the Society," *American Mathematical Society, Bulletin* I (1895) , 85-94.

[6] The subject of societies for the teaching of mathematics is one which has not yet been explored. The beginnings were made farther back than is commonly supposed. Thus an ephemeral association for the Improvement of Geometrical Teaching is met with in the early seventies, and a College Association of Mathematics Teachers existed in the early nineties.

[7] Included in the federation were: Association of Teachers of Mathematics of the Middle States and Maryland, New York State Science Teachers Association, Central Association of Science and Mathematics Teachers, Association of Teachers of Mathematics of New England, Physics Teachers Association of Washington City, Missouri Society of Teachers of Mathematics and Science, New Jersey State Science Teachers Association, Michigan Schoolmasters Club, New England Association of Chemistry Teachers, New York Physics Club, Indiana Association of Science and Mathematics Teachers, Association of Ohio Teachers of Mathematics and Science. J. D. Thompson, comp., *Handbook of Learned Societies and Institutions: America* (Carnegie Institution, Publications, no. 39, Washington, 1908) , 540.

[8] For the formation and constitution of this society, see "The Mathematical Association of America," *The American Mathematical Monthly*, 1-6.

Dickson, Oswald Veblen, R. C. Archibald, Florian Cajori, and E. H. Moore. This new national mathematical society aimed at the advancement of the teaching side of mathematics, whereas the older organization, the American Mathematical Society, had placed the emphasis on research.[9] Many individuals entered both societies. The Mathematical Association of America promptly took over the sponsorship of the *American Mathematical Monthly*. The societies of teachers of mathematics in the mid-Western states hastened to affiliate with the Mathematical Association of America as sections. The American Institute of Actuaries dates from 1909.

Never a numerous group and always a widely scattered one, America's professional astronomers were a little slower in organizing on a national scale than were some of the other scientists. Yet programs of simultaneous or consecutive observations are especially necessary in astronomy, and the establishment of large observatories at the close of the century, notably the Lick and the Mt. Wilson observatories, paved the way for an increasing degree of cooperation among American astronomers. The Astronomical Society of the Pacific, established in 1889 under the leadership of Edward S. Holden, the first director of the Lick observatory, became active in publishing advanced monographs upon astronomical topics as well as non-technical leaflets for the layman.

As we saw in the preceding chapter, an attempt had been made at Harvard to start an astronomical society of national scope,[10] but the project came to an end a few years later. An informal conference at Harvard College Observatory in 1898 smoothed the way for national organization, and in September of the following year the Astronomical and Astrophysical Society of America was organized by a conference of astronomers and physicists assembled at the Yerkes Observatory, Williams Bay, Wisconsin. "The purpose of this society is the advancement of astronomy, astrophysics and related branches of physics," stated the constitution of the new society.[11] Simon Newcomb and G. E. Hale were among those most prominent in bringing about the new organization. In 1914 the society adopted its present name of American Astronomical Society. The work and meetings of the society are

9 The relations between the two societies are discussed in T. S. Fiske, "Relations between the Association and the Society," *American Mathematical Monthly*, XXIII (1916) , 296-297.

10 Solon Bailey, *History and Work of Harvard Observatory, 1839-1927* (New York and London, 1927) , 31.

11 *Science*, n. s., X (1899) , 786.

faithfully and fully reported in the monthly issues of *Popular Astronomy*. Across the lakes, a Toronto astronomical society formed in 1890, later to become the Royal Astronomical Society of Canada, afforded an example in cooperation for astronomers, as did several successful European astronomical societies. A western association for stellar photography, formed in 1904, was active for a few years. More important was the Solar Union, an international body which was started at a meeting in St. Louis in 1904. The American Meteor Society, organized in 1911 at the Leander McCormick Observatory of the University of Virginia, under the guidance of Charles Olivier, organized enthusiastic amateurs all over the country for the study of meteors. Another specialized society, the American Association of Variable Star Observers, founded in 1911, grew into an international organization with its headquarters at Harvard College Observatory.[12] Several important comets and novae have been discovered by its members. In addition to issuing independent publications, both the American Meteor Society and the American Association of Variable Star Observers have published extensive reports of observations in *Popular Astronomy*.

The trends both in agricultural and in industrial development in the latter half of the nineteeenth century were preparing the way for chemistry to play an increasingly larger role in American life. Agricultural experiment stations, established by the national government or in connection with state universities and agricultural colleges, began to teach "scientific agriculture," emphasizing such things as crop rotation, fertilization of the soil, and the proper feeding of stock. Such emphasis was bound to give an impetus to the study of biological chemistry. Industry, too, began to place a premium on a knowledge of chemistry. Andrew Carnegie was one of the first to employ trained chemists to insure better and more uniform steel products. The expanding oil industry, the meat packing industry, the optical industry soon were demanding chemists. Secondary and collegiate institutions placed chemistry and physics in their curricula alongside the time-honored classical studies, Greek and Latin. The growing fraternity of academic and professional chemists began to demand independent organizations of their own.

At an informal meeting of chemists, presided over by S. A. Lattimore, held in conjunction with the meeting of the American Association for the Advancement of Science at Portland, Maine, in 1873, it was de-

12 J. H. Logan, "The American Association of Variable Star Observers as an Organization," *Popular Astronomy*, XLI (1933) , 474.

cided to ask the governing body for permission to form a subsection for the study of chemistry. Permission was accorded and the subsection became Section C in 1881.[13] However, many chemists felt that more freedom of action could be secured by forming, in addition, an independent national society. Impetus to the formation of such a society was given by a meeting of chemists, national in character, at the Priestley Centennial Celebration held at Northumberland, Pennsylvania, on August 1, 1874. It was there suggested, particularly by Persifor Frazer, that the time was ripe for the formation of an American chemical society. In March of 1876 a call was sent out to bring together those interested in forming such an organization, and on April 6 the first meeting of the American Chemical Society took place with C. F. Chandler presiding.[14]

The American Chemical Society, meeting monthly in New York City, maintained something of the character of a local society for a time. Meanwhile local chemical societies were founded in Cincinnati, Washington, and other cities. In order to bring about more effective cooperation, it was decided to hold the annual meetings of the American Chemical Society in cities other than New York. The first of these was a satisfactory meeting at Newport, Rhode Island, on August 6, 1891. The problem of affiliating the local societies with the American Chemical Society was attacked by delegates from the existing societies who met in conference at Washington in 1891. After several meetings, a new constitution was adopted, and the reorganized American Chemical Society became composed of four local sections, for New York, Washington, Cincinnati, and Rhode Island. By 1941 the number of local sections affiliated with the American Chemical Society had grown to ninety-six.[15] Since 1879, the society has published the *Journal of the American Chemical Society*. It has issued also a number of other journals and monograph series devoted to various specialized branches of chemistry.[16] A chemical professional organization, the Chemists'

[13] Marcus Benjamin, "Organization and Development of the Chemical Section of the American Association for the Advancement of Science," Twenty-Fifth Anniversary of the American Chemical Society (Easton, Pa., 1902) , 86-98.

[14] For an account of this meeting, see *American Chemist*, VI (1876) , 401-406.

[15] These are listed with their dates of formation in Callie Hull, Mildred Paddock, S. J. Cook, and P. A. Howard, comps., *Handbook of Scientific and Technical Societies and Institutions of the United States and Canada* (National Research Council, Bulletin 106, Washington, D. C., 1942) , 34-35.

[16] For the history of the American Chemical Society, see especially C. A. Browne, edit., *A Half Century of Chemistry in America*, 1876-1926, American Chemical Society, *Journal*, XLVIII, no. 8A, 1926. Many other references will be found in my Bibliography.

Club, formed in 1898, with headquarters in New York, but with a membership of national scope, built up an enormous library of chemical literature.

We have seen in the preceding chapter how the Association of American Geologists gradually transformed itself into the American Association for the Advancement of Science. Although Section E of the American Association for the Advancement of Science, devoted to geology, functioned actively, the conviction grew that the geologists ought to have an independent organization of their own. In 1881, at the Cincinnati meeting of the American Association for the Advancement of Science, a committee was appointed to take steps for the formation of a new society.[17] Little was accomplished, however, for some years, but in 1888 the committee issued a call in the June number of the *American Geologist,* a monthly magazine begun in January of that year, urging immediate action.[18] In answer to this call, a group of geologists assembled at Cleveland and unanimously resolved to found the American Geological Society. In December of the same year, final arrangements were concluded and officers were elected at a meeting in Ithaca, New York. James Hall, the venerable Nestor of American geologists, was accorded the honor of being the first president of the society. The more arduous duties devolved upon J. J. Stevenson, first secretary of the society. Alexander Winchell, J. J. Stevenson, and O. C. Marsh were especially active members in the early years of the society, as well as H. L. Fairchild, who became secretary in 1890 and served for many years. In 1889, the society revised its constitution, adopted its present name, the Geological Society of America, and defined its aim as "the promotion of the science of Geology in North America."[19] The Cordilleran Section of the society was organized in 1899 for the convenience of members residing in the Far West. The society has recently published an able and complete history of its activities.[20] The Seismological Society of America was founded in 1906. The Paleontological Society was formed in 1908. Among the local societies were the New York Mineralogical Club, which was founded in 1886, the Philadelphia Mineralogical Society in 1892, and the Geological

[17] For the early steps leading to the founding of the society, see Alexander Winchell, "Organization of the Geological Society of America," Geological Society of America, *Bulletin* I (1889) , 1-6.

[18] "A Note on The Proposed Geological Society," *American Geologist,* I (1888) , 395. Signed by N. H. Winchell and C. H. Hitchcock.

[19] Geological Society of America, *Bulletin* I (1889) , 571.

[20] H. L. Fairchild, *The Geological Society of America, 1888-1930* (New York, 1932) .

Society of Washington (D. C.) in 1893. The Petrologists' Club, also in Washington, was founded in 1910.

The New England Meteorological Society was formed in 1884 at Boston for the study of "atmospheric phenomena in New England."[21] The society was particularly active until 1892, when much of its work was transferred to the New England Weather Service formed at that time under the direction of the National Weather Bureau. The Pennsylvania Meteorological Society, organized in 1892,[22] evidently had an ephemeral existence.

Although state weather services had been established in New York and Pennsylvania as early as 1825 and 1837, respectively, it was not until after the establishment of the National Weather Service in 1870 that the movement really gained momentum. The formation of the Iowa State Weather Service in 1875 was but the first of many such bodies that sprang up all over the country with amazing rapidity until, by the turn of the century,[23] there were about forty-five such services to be found in the states and territories. These had become loosely affiliated into what was sometimes known as the American Association of State Weather Services, and their activities were coordinated through the National Weather Bureau.

The American Climatological Association, established in 1884, having for its object, as stated in its constitution, "the study of Climatology and Hydrology and of Diseases of the Respiratory and Circulatory Organs,"[24] concerned itself with the biological and medical aspects of meteorology in its effects upon the human organism. Increasingly, the society became absorbed with the problem of combating tuberculosis in the days when the "great white plague" was not as yet at all under control.

Although the American Geographical Society had functioned actively from its start in 1851, it was felt that there would be also room for a more popular geographical society. Accordingly, a group of men including its first president, G. G. Hubbard, H. G. Ogden, A. W. Greely, C. H. Merriam, and G. B. Goode formed the National Geographic Society, "for the increase and diffusion of geographic knowledge,"[25] in 1888. In the same year it began to publish the *National*

[21] *Bulletin of the New England Meteorological Society* (1884), (1).

[22] *The American Meteorological Journal*, XII (1895-1896), 201.

[23] *Maryland Weather Service*, I (1899), 360.

[24] Constitution printed in annual *Transactions of the American Climatological Association*.

[25] National Geographic Society, *National Geographic Magazine*, I (1888), no. 1, i.

Geographic Magazine.[26] The society has published many monographs and maps, and it has helped finance a number of exploring expeditions, including Peary's notable expedition to the North Pole. Membership in the National Geographic Society grew rapidly from 200 in 1888 to about 1,300,000 in 1930.

The Association of American Geographers was established in 1904. Membership in this organization of professional geographers was restricted to those who had done original research work in geography. These societies for exploration were formed in this period: the Peary Arctic Club in 1899 and the Records of Past Exploration Society in 1901. Of the regional and local geographical societies founded in this period, the following may be mentioned: the Geographical Society of California and the Geographical Society of Philadelphia in 1891, the Geographic Society of Chicago in 1898, and the Geographical Society of Baltimore in 1902.

Explorers, too, were organizing their own societies. The Appalachian Mountain Club was founded in 1876, the Sierra Club in 1892, and Mazamas, a society for exploring the mountains of the Pacific Northwest, in 1894. The American Alpine Club, formed at Washington in 1903, aims at the exploration of the highest mountain peaks on earth and at the exploration of the Arctic and the Antarctic regions. The Explorers Club, formed at New York in the following year, has built up a splendid library of books on travel and exploration. The Associated Mountaineering Clubs of America, founded by LeRoy Jeffers in 1915, now goes under the name of Associated Outdoor Clubs of America. The federated clubs are interested in nature study and the development of state and national parks and forest preserves. The Pathfinders of America dates from 1914.

Physics, like chemistry and mathematics, became in the latter half of the nineteenth century one of the pillars of the new industrial and technological revolution that was transforming the United States. Recognizing the importance of determining fundamental constants, Wolcott Gibbs, H. A. Newton, F. A. P. Barnard, and others met at Columbia University in 1873 and organized the American Metrological Society. A constitution, drafted by F. A. P. Barnard, was adopted

[26] An announcement of this magazine read: "It will contain memoirs, essays, notes, correspondence, reviews, etc., relating to Geographic matters. As it is not intended to be simply the organ of the Society, its pages will be open to all persons interested in Geography, in the hope that it may become a channel of inter-communication, stimulate geographic investigation and prove an acceptable medium for the publication of results." *National Geographic Magazine* I (1888), no. 1, i.

which stated: "The primary object of this Association shall be to originate measures, or to aid in promoting measures elsewhere originating, designed to improve the system of Weights, Measures and Moneys, at present existing among men, and to bring the same into relations of simple commensurability with each other."[27] The society also sought to bring about the general adoption of the metric system,[28] and proposed to memorialize congress, state legislatures, and boards of education to this end.[29]

In response to a circular, physicists representing seventeen institutions assembled at Columbia University on May 30, 1899, and organized the American Physical Society. H. A. Rowland was elected president. Prominent also among the group of organizers were A. A. Michelson, the vice-president, A. G. Webster, J. S. Ames, B. O. Peirce, W. F. Magie, E. L. Nichols, and M. I. Pupin. The Constitution adopted stated the objects of the society to be "the advancement and diffusion of the knowledge of physics."[30] The society promptly began to issue a *Bulletin,* and in 1893 began the *Physical Review.* To these, within recent years, the *Physical Review Supplement,* later changed to *Reviews of Modern Physics,* has been added.

The growth of specialization within the field of physics brought with it new societies. Thus the Optical Society of America and the Metric Association were founded in 1916, and the American Institute of Weights and Measures in 1917. The American Radio Relay League, founded in 1914 and made up of amateurs and professionals, was

[27] American Metrological Society, *Proceedings,* I (1873-1878) , 5. The constitution goes on to say: "An object secondary to this will be to secure the universal adoption of common units of measure for the expression of quantities which require to be stated in presenting the results of physical observation or investigation, and for which the ordinary systems of metrology do not provide; such as the divisions of the barometer, thermometer and densimeter; the amount of work done by machines; the amount of mechanical energy, active or potential, of bodies at given temperatures, or generated by combustion or otherwise; the quantity and intensity of electrodynamic currents; the aggregate or efficient power of prime movers; the accelerative force of gravity; the pressure of steam and of the atmosphere; and other matters analogous to these. The Association will endeavor also to secure uniformity of usage in regard to standard points of reference, or to those physical conditions to which observations must be reduced for purposes of comparison; especially the temperature and pressure to which are referred the specific gravities of bodies, and the zero of longitude on the earth." *Ibid.,* 6.

[28] *Ibid.*

[29] *Ibid.*

[30] For an account of the formation of the American Physical Society, see *Science,* n. s., IX (1899) , 784.

especially active in publishing discoveries in a rapidly expanding field. By 1930 the society had almost 20,000 members.

The rise of machine-age civilization with its revolution in manufacturing, transportation, communication, and urban construction called forth new general and specialized engineering societies. The American Institute of Mining and Metallurgical Engineers was established in 1871, the Architectural League of New York, the American Society of Mechanical Engineers, and the American Order of Steam Engineers in 1880, the American Institute of Electrical Engineers in 1884, the American Society of Naval Engineers in 1888, the Society for the Promotion of Engineering Education and the Society of Naval Architects and Marine Engineers in 1893, the American Society of Heating and Ventilating Engineers in 1895, the American Railway Engineering Association in 1899, the American Road Builders' Association in 1902, the American Society of Refrigerating Engineers, the Society of Automotive Engineers, and the United Engineering Society in 1904, the Illuminating Engineering Society in 1906, the New York Society of Architects and the American Institute of Chemical Engineers in 1908, the American Institute of Consulting Engineers in 1910, the Radio Club of America in 1911, the Institute of Radio Engineers in 1912, the American Association of Engineers and the National Advisory Committee for Aeronautics in 1915, the Society of Motion Picture Engineers, the Engineering Foundation, and the National Industrial Conference Board in 1916, and the Society of Industrial Engineers and the Society of Terminal Engineers in 1917.[31] The influence of these societies in

31 Numerous regional, state, and local engineering societies, too, came into existence. Though primarily professional in character, they were not without their influence upon the broader aspects of engineering science and ought not to be wholly neglected in a study of this sort. Among the most active regional societies were the following: the Western Society of Engineers in 1869, the Western Society of Civil Engineers in 1880, the Engineering Association of the South in 1889, and the Pacific Northwest Society of Engineers in 1902. Among the more active of the local engineering societies founded during this period were the following: the Engineers Club of St. Louis, founded in 1868, the Engineers Society of Western Pennsylvania and the Michigan Engineering Society in 1880, the Indiana Engineering Society in 1881, the University of Michigan Engineering Society in 1882, the Engineers' Club of Minneapolis in 1883, the Engineers' Society of St. Paul in the same year, the Connecticut Society of Civil Engineers in 1884, the Illinois Society of Engineers in 1886, the Montana Society of Engineers in 1887, the Engineers' Club of Cincinnati in 1888, the Iowa Engineering Society in 1889, the Engineers Society of the University of Minnesota in 1893, the Engineers Society of Western New York and the Providence Engineering Society in 1894, the University of Illinois Association of Engineering Societies in 1895, the Brooklyn Engineers Club

stimulating engineering research and in raising the standards of engineering education can scarcely be overestimated.[32]

Aviation enthusiasts were meanwhile sponsoring numerous clubs to promote the realization of the age-old desire of man to fly. In 1910 there came into existence the Aero Club of Illinois, fostered by Octave Chanute. By the following year, the International Aviation Association had been formed. By 1912 the American Federation of Aviators had come into existence. In 1915 the federal government had taken sufficient cognizance of the progress of aviation to set up the National Advisory Committee for Aeronautics. The law creating this body provided for it to "supervise and direct scientific study of the problems of flight with a view to their practical solution and also to direct and conduct research and experiments in aeronautics." Representatives from both army and navy were included among the membership of this committee, which has had a long and active history.

The science, or rather the art, of glider piloting was developed in the United States in the 1890's by such men as Octave Chanute and A. M. Herring. The Mohawk Aerial Navigation Company (1894), under the leadership of Charles P. Steinmetz, is alleged to have been the first glider club in the world.[32a] In 1895 J. Means of Boston organized the Boston Aeronautical Society. Similar groups soon sprang up in Chicago and other large cities.

The new industrial trends, placing a premium on the knowledge of technological experts, inevitably gave a stimulus to the formation of numerous technological societies. Most of these lie outside the scope of this study, but a few of those whose work or publications were con-

---

in 1896, the Nebraska Engineering Society and the Rochester Engineering Society in 1897, the Louisiana Engineering Society in 1898, the Tufts College Engineering Society in 1899, the University of Nebraska Engineering Society in 1900, the Memphis Engineering Society in 1901, the Municipal Engineers of the City of New York in 1903, the United Engineering Trustees in 1904, the Alabama Society of Engineers in 1905, the Utah Society of Engineers in 1907, the Engineering Society of Wisconsin in 1909, the Vermont Society of Engineers and the Engineering Society of Buffalo in 1912.

[32] "The most important factors in promoting the advance of the engineering profession and in disseminating and rendering available to the world the valuable experience and data accumulated by engineers in the practice of their profession, are the professional associations of national engineering societies." J. N. Lieb, "The Organization and Administration of National Engineering Societies," Presidential Address, American Institute of Electrical Engineers, 1905, *Science*, n. s., XXII (1905), 65.

[32a] N. H. Randers-Pehrson, "History of Motorless Flight," in L. B. Barringer, *Flight without Power* (New York, 1940), 3.

cerned to some extent with the broader aspects of science will be given passing mention here. The New York Electrical Society was formed in 1881, the Technical Society of the Pacific and the Deutsch-Amerikanischer Techniker-Verband were founded in 1884, the Society of Chemical Industry (American Section) in 1894, the American Foundrymen's Association in 1896, the American Mining Congress in 1898, the American Society for Testing Materials in 1898, the American Ceramic Society in 1899, the Electrochemical Society and the National Petroleum Association in 1902, the Technology Club of Syracuse and Affiliated Societies in 1903, the Western Association of Technical Chemists and Metallurgists in 1904, the American Iron and Steel Institute in 1908, the American Institute of Refrigeration in 1910, the Northwestern Mining Association in 1911, the American Automobile Association in 1912, and the American Gas Association and the Zinc Institute in 1918.

Of sufficient value to industrial research to merit passing mention were the following: the National Federation of Textiles (1872), the American Paper and Pulp Association (1878), the National Association of Manufacturers (1895), the Steel Founders Society of America (1902), the American Power Boat Association (1903), the National Association of Engineering and Boat Manufacturers (1904), the Lithographers National Association (1906), the National Canners Association (1907), the American Federation of Aviators (1912), the American Institute of Graphic Arts and the National Alliance of Art and Industry (1914), and the Technical Association of the Pulp and Paper Industry (1915).

One aspect of the nation's technological progress in the early years of the twentieth century that ought not to escape notice is the advent of vocational associations, aimed principally at encouraging industrial education in the public school systems of the land. Earliest of these to attain national scope was the National Society for the Promotion of Industrial Education. The N.S.P.I.E., as it was commonly called, came into being November 16, 1906, at Cooper Union in New York City. Henry S. Pritchett, of the Massachusetts Institute of Technology, became its first president at this meeting, which was addressed by such prominent persons as Nicholas Murray Butler, Jane Addams, and Frederick P. Fish, president of the American Telephone and Telegraph Company. Among the objects mentioned in the constitution was that the society should "bring to public attention the importance of industrial education as a factor in the industrial development of the United States; . . . and . . . promote the establishment of institutions of industrial training." The society became very active in promoting subsidiary state vocational societies, state aid for vocational curricula, vocational training for women, and state vocational surveys.[33]

The National Society for the Promotion of Industrial Education had some 1700 members a decade later when it sponsored passage of the National Vocational Education Act in February, 1917. A year later the society changed its name to National Society for Vocational Education and was headed by a vigorous president in the person of David Snedden. Within a few years a dozen state vocational organizations and several national and regional vocational associations were associated with it. Meanwhile, the Vocational Education Association of the Middle West (1914-1915) had sprung up at the University of Chicago.[34] The National Society for Vocational Education and the Vocational Education Association of the Middle West coalesced in 1925 to form the American Vocational Association,[35] with some fifteen affiliated associations.[36, 37]

A number of industrial firms, for example, the Westinghouse Machine Company, East Pittsburgh, Pennsylvania, and the General Electric Company, Schenectady, New York, were pioneering before 1900 in schools for training apprentices. This movement culminated in 1913 in a meeting at New York University at which the National Association of Corporation Schools was formed, one of the aims of which, frankly utilitarian, was "to have the courses in established educational institutions modified to meet more fully the needs of industry."[38] Among the purely trade associations to concern themselves with technological education were the National Founders' Association, National Association of Wall-Paper Manufacturers, and the International Typographical Union.

The writings of Asa Gray, Louis Agassiz, O. C. Marsh, E. D. Cope,

---

33 F. T. Struck, "The National Society for the Promotion of Industrial Education," *Foundations of Industrial Education* (John Wiley and Sons, New York, 1930) , 94-115. Quotation by permission.

34 F. T. Struck, "The American Vocational Association and Related Organizations," *op. cit.* (New York, 1930) , 117-129.

35 Editorial, *Manual Training Magazine* (April, 1922) , "The Vocational Education Association of the Middle West."

36 American Vocational Association, *News Bulletin*, I, no. 4 (1926) , 12-13.

37 See Z. M. Smith, "Affiliated State Associations," American Vocational Association, *News Bulletin*, III, no. 3 (1928) .

38 National Association of Corporation Schools, *Seventh Annual Proceedings*, 5.

W. K. Brooks, E. B. Wilson, C. O. Whitman, and others led to a vast increase in American biological knowledge in the latter half of the nineteenth century. As we have seen earlier in the chapter, the evolution controversy was another factor in arousing interest in the biological studies. Arboretums, zoological parks, and aquariums in Boston, New York, Philadelphia, Washington, Chicago, St. Louis, San Francisco, and other large cities afforded more tangible evidence of the variety and evolution of living things. The coming of scientific agriculture meant the laying of emphasis upon the economic aspects of biology. National and state experiment stations and bureaus were demanding trained biologists. The two major sub-fields of biology—zoology and botany—became further subdivided. Bacteriology emerged as almost a coordinate entity beside zoology and botany. Taken all in all, there were numerous forces paving the way for the rapid development of the biological societies which occurred.

The American Society of Naturalists began in 1883 as the Society of Naturalists of the Eastern United States. The Delaware Valley (Pennsylvania) Naturalists Union was founded in 1894. In 1902 the Naturalists of the Central States (organized in 1901) amalgamated with it. With a membership limited to persons engaged professionally in some branch of natural history, the society is made up of teachers, investigators, and museum administrators. A number of the more specialized and technical biological societies are affiliated with it. The American Nature Study Society, organized in 1908, aims at the promotion of nature study in the schools and the promotion of non-technical elementary-school nature education. Nature clubs of interested laymen have flourished in many American cities. Similar school clubs have sprung up in large numbers throughout the land since the turn of the century. The Association for Maintaining the American Women's Table at the Zoological Station, organized in 1898, became in 1917 the Association to Aid Scientific Research by Women. For the thirty-five years of its existence the society was maintained by annual subscriptions of fifty dollars each, and its funds were devoted to maintaining a table for the use of women at the Zoological Station at Naples. The society was able to recognize and encourage women's research achievement by an award known as the Ella Richards prize. Madame Curie and Dr. Annie J. Cannon of Harvard University have been among the recipients of this honor. The society held its final meeting on April 30, 1932, and passed a resolution which was highly significant of the improved conditions for research work which had become available to women within a generation and which rendered the continuation of the society no longer necessary.[39]

The American Association of Museums (1906) aims to promote knowledge relating to museums through its annual meetings, regional conferences, and through its publications. The American Society of Biological Chemists was formed in 1906. The New England Federation of Natural History Societies was founded in 1906 and was composed of some twenty-five affiliated organizations by 1930. The Phi Sigma Society was formed in 1915 as an honorary professional society of the guild of biologists. The American Genetic Association was organized in 1913. Among the more active of the general local societies were the Young Naturalists' Society (Minneapolis, 1875), the Linnaean Society of New York in 1878, the Biological Society of Washington (D. C.) in 1880, the Long Island Biological Association in 1890, and the Biological Club of Ohio State University in 1891. The Agassiz Association[40] (1875), with headquarters at Pittsfield, Massachusetts, developed into a nature-study organization primarily, although not exclusively, with a membership among the young which had grown to 12,000 in 1902.

The formation of the New York Zoological Society in 1895 and the New England Zoological Club in 1899 paved the way for the establishment of the American Society of Zoologists in 1903. As entomology became a more and more specialized branch of zoology, new societies for the study of it sprang up. The American Association of Economic Entomologists was begun in 1889, and the Entomological Society of America in 1906. Among the local groups were the Brooklyn Entomological Society in 1872, the Cambridge Entomological Club in 1874, the Entomological Society of Washington (D. C.) in 1884, the New York Entomological Society in 1892, the Entomological Society of Albany in 1899, and the Pacific Coast Entomological Society in 1901. Ornithology, too, became increasingly set apart from other branches of zoology[41] and came to have specialized societies devoted to it. Thus

[39] "*Whereas*, the objects for which this Association has worked for thirty-five years have been achieved, since women are given opportunities to engage in Scientific Research on an equality with men, and to gain recognition for their achievements, be it

"*Resolved*, that this association cease to exist after the adjournment of this meeting." H. J. Crawford, "The Association to Aid Scientific Research by Women," *Science*, n. s., LXXVI (1932), 493. This article contains the salient facts in the history of the society.

[40] Guyot, *Memoir of Louis Agassiz* (Princeton, 1883).

[41] For the developments in ornithology, see *Fifty Years of American Ornithology, 1883-1933* (Lancaster, Pa., 1933).

the American Ornithologists' Union was founded in 1883. The local societies included the Nuttall Ornithologists Club of Cambridge, founded in 1873, the Wilson Ornithological Club (Oberlin, Ohio) and the Delaware Valley Ornithological Club (Philadelphia) in 1890, the Cooper Ornithological Club (Pasadena, California) in 1893, the Michigan Ornithological Club in 1894, the Maine Ornithological Society in 1895, the Massachusetts Audubon Society in 1896, the Illinois Audubon Society in 1897, the Nebraska Ornithologists' Union in 1899, and the Vermont Bird Club in 1900. Toward the close of the century, lovers of bird and animal life began to manifest increasing concern lest many species be utterly exterminated by hunters seeking to cater to clothiers and milliners. As the game conservation movement gained headway, societies began to be formed.

One of the leaders in the conservation movement was G. B. Grinnell, who announced the formation of the first Audubon Society in 1886.[42] The Audubon movement spread rapidly. A federation of state Audubon societies was established in 1900-1901 under the name of National Committee of Audubon Societies of America, which in 1905 became the National Association of Audubon Societies for the Protection of Wild Birds and Animals.[43] The association has issued numerous publications, among them illustrated leaflets which have reached many high school bird- and nature-study clubs. By 1913 there were forty-three Audubon societies as well as similar societies in the District of Columbia and Hawaii.[44] Numerous local affiliated societies have been formed as well. In 1940 the name was changed to the National Audubon Society. The National Aquarium Society dates from 1892. The American Game Protective and Propagation Association was formed in 1911 with J. B. Burnham as president.[45]

Eager searchers of the world of bacterial life were establishing societies of their own. The American Microscopical Society was formed in 1878, and the Society of American Bacteriologists in 1900. Among the local societies were the San Francisco Microscopical Society in 1870, the New York Microscopical Society in 1877, and the Rochester Microscopical Society in 1879.

The writings of L. H. Morgan, L. F. Ward, and A. F. A. Bandelier, in the latter half of the nineteenth century, and of H. F. Osborn, Franz

---

[42] "George Bird Grinnell," *Bird Lore*, XIV (1912), 77.

[43] For the history of the Audubon movement, see *Bird Lore*, VII (1905), 45, 56, and XIV (1912), 77-80.

[44] I base my count on *Index to Bird Lore*, I-XV (New York, 1916), 2.

[45] "American Game Protective and Propagation Association," *Bird Lore*, XIV (1912), 76-77.

Boas, Aleš Hrdlička, and others in the twentieth were instrumental, among other things, in arousing increasing interest in primitive man and the comparative study of aboriginal cultures. The Archaeological Institute, founded at Boston in 1879, consisted of about thirty affiliated societies by 1942.[46] The American Folklore Society, established in 1888 at Cambridge, Massachusetts, specializes in the study of the folklore of the Amerinds.[47] The Anthropological Society of Washington (D. C.), founded in 1879, was especially active and paved the way for the incorporation of the American Anthropological Association in 1902. W J McGee, Franz Boas, and Aleš Hrdlička were among the leader in organizing the new society.[48] In collaboration with the Anthropological Society of Washington and the American Ethnological Society (which had recently experienced a revival), the newly formed American Anthropological Society took over the editorship of the *American Anthropologist*. The Iowa Anthropological Association dates from 1903, and the Alabama Anthropological Association from 1909.

Among the organizations devoted to the scientific study of mankind were a number of humanistic and social science societies. A full treatment of these lies outside the scope of this book. Mention may be made of a few of the more representative of them: The American Philological Association formed in 1869, the American Library Association (1876), the Modern Language Association of America (1883), the American Economic Association (1885), the American Academy of Political and Social Science (1889), the American Asiatic Association (1898), the National Institute of Social Sciences (1899), the American Schools of Oriental Research (1900), the American Political Science Association (1903), the American Academy of Arts and Letters (1904), the Bibliographical Society of America (1904), the American Sociological Society (1905), and the American Association of Museums (1906). Much basic work along similar lines was done by various statewide learned bodies. At this point only one will be mentioned by way of illustration, the Ohio State Archaeological and Historical Society (1885).

Thousands of Americans received mental stimulation from the new

---

[46] The branch societies are listed in Callie Hull, Mildred Paddock, S. J. Cook, and P. A. Howard, comps., *A Handbook of Scientific and Technical Societies and Institutions of the United States and Canada* (National Research Council, *Bulletin* 106, Washington, 1942), 101-102.

[47] In the next few years branches were formed in Boston, Cincinnati, Baltimore, Philadelphia, New Orleans, New York, and Montreal.

[48] The early history of the organization is carefully traced in "The American Anthropological Association," *American Anthropologist*, V (1903), 178-192.

scholarship through the somewhat attenuated popularizations furnished by the Chautauqua Literary and Scientific Circle (1878) and its numerous local subsidiaries, and through the efforts of the International Lyceum Association (1903) and its affiliates.

The botanical section (Section G) of the American Association for the Advancement of Science held its first meeting in 1892. At that time a committee was appointed to consider the formation of an independent national botanical society, and in the following year charter members were elected at the meeting of the American Association for the Advancement of Science at Madison, Wisconsin. The next year, 1894, formal organization took place under the name of Botanical Society of America. In 1906, the Society for Plant Morphology and Physiology (organized at Ithaca in 1897) and the American Mycological Society (organized at St. Louis in 1903) united with it to form the Botanical Society of America, under which name it exists today. A number of random regional, state, and local botanical societies sprang into existence during this period also.[49] The Torrey Botanical Club (New York, 1867) was very active in the closing decades of the nineteenth century.[50] The Philadelphia Botanical Club (1891) publishes the well-known annual, *Bartonia.*

Specialized botanical societies included the American Fern Society founded in 1893, the Sullivant Moss Society in 1898,[51] the American Rose Society in 1899, the American Gladiolus Society in 1910, and the American Plant Propagators Association in 1918. Nurserymen formed a number of professional, national, and regional organizations that lent encouragement to scientific plant breeding.

Persons interested in the conservation of flowers were organizing too. Thus the New England Wild Flower Preservation Society was founded in 1900, and two years later a national group known as the Wild Flower Preservation Society came into existence.

Agriculture in the United States, no less than industry, underwent a revolution in the closing decades of the nineteenth century. State agricultural colleges and experiment stations, springing up under the pro-

[49] Among the more active were the Botanical Society of Western Pennsylvania in 1886, the Vermont Botanical Club, the New England Botanical Club, and the Josselyn Botanical Society of Maine in 1895, the Botanical Society of Pennsylvania (University of Pennsylvania) in 1897, the Botanical Society of Washington (D. C.), in 1901, the Botanists of the Central States in 1905.

[50] See "The Seventy-fifth Anniversary Celebration of the Torrey Botanical Club," *Science*, n. s., VC (1942), 498.

[51] Local groups for the study of fungi included the Boston Mycological Club, formed in 1895, and the Wisconsin Mycological Club in 1903.

visions of the Morrill Acts of 1862 and 1890, as well as the introduction of secondary schools of agriculture in a number of states in the eighteen nineties, made agricultural education available as never before.[52] Agricultural libraries, too, were an educational influence.[53] Government chemists,[54] meteorologists,[55] entomologists,[56] and engineers[57] were engaged more than ever in an attempt to work out the scientific principles of agriculture. New inventions of farm machinery,[58] selected breeding of plants and animals,[59] and warfare on animal and plant disease[60] changed the methods of farming more in the nineteenth century than they had changed from the days of the Egyptians.

Such developments could not fail to stimulate the growth of agricultural societies. The Association of Official Agricultural Chemists was founded in 1884, the Association of American Agricultural Colleges and Experiment Stations in 1887, the American Society for Horticultural Science in 1903, the American Society of Agricultural Engineers and the American Society of Agronomy in 1907, and the Association of Official Seed Analysts of North America in 1908. New state and count-

---

[52] For the growth of agricultural education, see, especially, A. C. True, "Agricultural Education in the United States," *Yearbook of the United States Department of Agriculture, 1899* (Washington, 1900) , 157-190, and, by the same author, "Agricultural Experiment Stations in the United States," *ibid.*, 513-548.

[53] The rise and influence of agricultural libraries are traced in C. H. Greathouse, "Development of Agricultural Libraries," *ibid.*, 491-512.

[54] See H. W. Wiley, "The Relation of Chemistry to the Progress of Agriculture," *ibid.*, 201-258, and Milton Whitney, "Soil Investigations in the United States," *ibid.*, 335-346.

[55] See F. H. Bigelow, "Work of the Meteorologist for the Benefit of Agriculture, Commerce and Navigation," *ibid.*, 71-92.

[56] See L. O. Howard, "Progress in Economic Entomology in the United States," *ibid.*, 135-156; and T. S. Palmer, "A Review of Economic Ornithology in the United States," *ibid.*, 259-292.

[57] See Elwood Mead, "Rise and Future of Irrigation in the United States," 591-612. The relation of the growth of transportation facilities to agriculture is explained in M. O. Eldridge, "Progress of Road Building in the United States," *ibid.*, 367-380; and in Angus Sinclair, "Development of Transportation in the United States," *ibid.*, 643-663.

[58] See G. K. Holmes, "Progress of Agriculture in the United States," *ibid.*, 307-334.

[59] See John Clay, "Work of the Breeder in Improving Live Stock," *ibid.*, 627-642; H. J. Webber and E. A. Bessey, "Progress of Plant Breeding in the United States," *ibid.*, 465-490; and I. Lamson-Scribner, "Progress of Economic and Scientific Agrostology," *ibid.*, 347-366. Lists of animal and plant breeders associations are to be found *ibid.*, appendix, 688-708.

[60] See B. T. Galloway, "Progress in the Treatment of Plant Diseases in the United States," *ibid.*, 191-200; and D. E. Salmon, "Some Examples of the Development of Knowledge concerning Animal Diseases," *ibid.*, 93-134.

less local agricultural societies came into being during the period, many of them being of a transitory nature.[61] Among the new state horticultural societies were the Nebraska State Horticultural Society in 1863, the Kansas State Horticultural Society, founded in 1869, the New Jersey State Horticultural Society in 1875, the South Dakota State Horticultural Society in 1890, and the Horticultural Society of New York in 1902.

Forestry organizations also began to spring up. Among them were the American Forestry Association, founded in 1882, the Pennsylvania Forestry Association in 1886, Woodmen of the World in 1890, the Connecticut Forest and Park Association in 1895, the Massachusetts Forestry Association in 1898, the North Carolina Forestry Association in 1911, the Massachusetts Tree Wardens and Foresters Association in 1913, and the New York State Forestry Association in 1913.[62]

The medical profession which, as we have seen, was the first to organize, continued to push ahead with its professional and research organizations. The influence of new medical schools,[63] and especially of the first graduate schools of medicine at Johns Hopkins, Harvard, Yale, Columbia, and Chicago, helped not only to make a more enlightened medical profession but also exerted an influence in bringing together men who would later keep up professional contacts and cooperate with each other in medical societies. While the American Medical Association has continued to hold the leadership among the general national medical societies, it has by no means held a monopoly of the field. The National Eclectic Medical Association dates from 1870, the American Public Health Association from 1872, the Conference of State and Provincial Health Authorities of North America from 1884, the Association of American Physicians from 1886, the Congress of American Physicians and Surgeons from 1887, the Association of American Medical Colleges from 1890, the National Medical Association from 1895, the Medical Women's National Association from 1913, and the Inter-State Postgraduate Medical Association of North America from 1916.

61 Lists of these occur in the volumes of the *Yearbook of the United States Department of Agriculture.*

62 The rise of forestry in the United States is traced in Gifford Pinchot, "Progress of Forestry in the United States," *Yearbook of the U. S. Department of Agriculture,* 1899, 293-306.

63 "Between 1810 and 1840 twenty-six new schools were founded; from 1840 to 1876, forty seven . . . no less than one hundred and fourteen were founded between 1873 and 1890." H. E. Sigerist, *American Medicine* (New York, 1934). By permission of the publishers.

Fourteen new state medical societies were established in this period, as well as regional, and hundreds of local professional organizations of physicians.[64] Ultimately, most of them became federated with the American Medical Association. Because they worked continually for higher standards of medical ethics and for more thorough medical education, by the close of the period much of the quackery and ignorance common in American medical practice prior to the Civil War had disappeared.

Although numerous sections were formed within the American Medical Association, it soon became evident that medical "specialists" felt they ought also to form organizations composed of those with interests in restricted departments of medicine. A long list of such national organizations includes: the American Ophthalmological Society, established in 1864; the American Otological Society in 1868; the American Neurological Association in 1875; the American Dermatological Association and the American Gynecological Society in 1876; the American Surgical Association in 1880; the American Physiological Society in 1887; the American Pediatric Society in 1888; the Association of Life Insurance Medical Directors of America in 1889; the Child Study Association of America and the American Physical Therapy Association in 1890; the Association of Military Surgeons of the United States in 1891; the American Laryngological, Rhinological and Otological Society in 1895; the American Academy of Ophthalmology and Otolaryngology in 1896; the American Gastro-Enterological Association in 1897; the American Proctologic Society in 1899; the Society of American Bacteriologists, the American Association of Pathologists and Bacteriologists, the American Roentgen Ray Society, and the American Physical Education Association in 1900; the American Urological Association and the American Society of Vertebrate Pathologists in 1902; the American Society of Tropical Medicine in 1903; the National Tuberculosis Association in 1904; the American Sanatorium Association in

[64] The state societies were formed as follows: the West Virginia State Medical Association in 1867, the Nebraska State Medical Association in 1868, the Kansas State Medical Association in 1870 (?), the Colorado State Medical Society in 1871, the Oregon State Medical Society, and the Washington State Medical Association in 1874, the Louisiana State Medical Society in 1878, the Montana Medical Association in 1879, the New Mexico Medical Society in 1881, the South Dakota State Medical Association in 1882, the Arizona State Medical Association in 1892, the Minnesota State Medical Association, incorporated in 1892, the Utah State Medical Association in 1895, the Wyoming State Medical Society in 1902. In addition, the Tri-State Medical Association of Mississippi, Arkansas and Tennessee was founded in 1883, and the Tri-State Medical Association of the Carolinas and Virginia in 1898.

1905; the American Association for Cancer Research in 1907; the American Society for Clinical Investigation in 1908; the American Association of Clinical Research in 1909; the Clinical Orthopedic Society in 1912; the American College of Surgeons, the American Social Hygiene Association, the American Society for Experimental Pathology, and the American Society for the Control of Cancer in 1913; the American Academy of Periodontology and the American Association of Immunologists in 1914; the National Society for the Prevention of Blindness, the Central States Pediatric Society, the American College of Physicians, and the Radiological Society of North America in 1915; the American Association of Industrial Physicians and Surgeons and the American Radium Society in 1916; the American Association for Thoracic Surgery, the American Occupational Therapy Association, and the Association for the Study of Internal Secretions in 1917; and the Eugenics Research Association in 1918. Numerous regional, state, and local specialized medical societies, too, came into existence.[65]

It was during this period that osteopathy emerged as an offshoot of medicine under the leadership of Dr. Andrew T. Still, a Kansas physician, who had achieved considerable skill in the matter of neurological adjustments through manipulation. He enunciated his theory as early as 1874. The American Osteopathic Association was founded in 1897. A number of state and district societies have been established by members of the profession.

Dentists, like the physicians, were gaining in the matter of organization. Numerous state and local professional associations were estab-

[65] A few of the more active research groups may be mentioned: the New York Dermatological Society founded in 1869; the New York Neurological Society in 1872; the Chicago Pathological Society in 1878, the New York Surgical Society in 1879; the Philadelphia Neurological Society in 1884; the Southern Surgical Association in 1887; the Brooklyn Gynecological Society in 1890; the Boston Society of Medical Science in 1896; the New England Ophthalmological Society in 1897; the Charaka Club (New York City), dealing with the historical, literary, and artistic phases of medicine in 1898; the Chicago Urological Society, the Los Angeles Clinical and Pathological Society, and the Pacific Coast Oto-Ophthalmological Society in 1903; the Physiological Society of Philadelphia and the University of Pennsylvania Society of Normal and Pathological Physiology in 1904; the Harvey Society (New York City) in 1905, for the diffusion of medical knowledge by lectures, and the Philadelphia Roentgen Ray Society in the same year; the Chicago Tuberculosis Institute in 1906, the Minnesota Pathological Society in 1912, and the New England Pediatric Society in the same year; the Boston Surgical Society in 1914; the Memphis Society of Ophthalmology and Otolaryngology, the Chicago Society of Internal Medicine, and the Minnesota Dermatological Society in 1915; the New England Surgical Society in 1916.

lished.[66] A purely research group was the Harriet Newell Lowell Society for Dental Research, formed in Boston in 1912. Americans took part in the establishing of an International Dental Federation in 1912.

Pharmacologists, too, were meanwhile organizing numerous state and local professional societies.[67] A national wholesale druggists' association was formed in 1876. The National Association of Boards of Pharmacy was organized in 1904. The American Society for Pharmacology and Experimental Therapeutics was established in 1908. Typical of the professional optometric associations was the New York State Optometric Association (1895).

Among the organizations dealing with the hospital care and medical and nursing treatment of patients were the American Hospital Association (1898) and the National League of Nursing Education (1893). The Women's Medical Association of New York City (1900) was followed by similar societies of women physicians. The movement culminated in the organization of the American Medical Women's Association in 1915. Indicative of the beginning of a new era in the history of preventive medicine in America was the National Organization for Public Health Nursing (1912).

The emergence of the experimental approach to psychology and the growth of an active group of "scientific" psychologists working in some twenty new American psychological laboratories and largely deriving their inspiration from German masters led to the formation of the American Psychological Association.[68] It was G. S. Hall who issued the call for an organization meeting at Clark University on July 8, 1892, in response to which G. S. Fullerton, Joseph Jastrow, William James, G. T. Ladd, J. McK. Cattell, and J. M. Baldwin assembled. G. S. Hall

[66] Many of these are listed in *Polk's Dental Register* (13th ed., Chicago, 1925), "Dental Associations and Societies," 39-58.

[67] The progress in the formation of state pharmaceutical associations may be indicated as follows: the South Carolina Pharmaceutical Association was founded in 1872, the New Hampshire Pharmaceutical Association in 1874, the Pennsylvania Pharmaceutical Association in 1877, the Kentucky Pharmaceutical Association in 1878, the Missouri Pharmaceutical Association, the New York State Pharmaceutical Association, the Texas Pharmaceutical Association, and the Ohio State Pharmaceutical Association in 1879, the Illinois Pharmaceutical Association, the North Carolina Pharmaceutical Association, and the Wisconsin Pharmaceutical Association in 1880, the Virginia Pharmaceutical Association and the Louisiana State Pharmaceutical Association in 1882, the Maryland Pharmaceutical Association and the Michigan State Pharmaceutical Association in 1883, the Alabama Pharmaceutical Association in 1884, and the South Dakota Pharmaceutical Association in 1899.

[68] R. C. Davis, "American Psychology, 1800-1885," *Psychological Review*, XLIII (1936), 471-493.

became the first president. At a meeting held later in the year during the Christmas holidays the society began active operations with a program of papers. In 1900 the Council of the society provided that members of the American Psychological Association might form local sections, and promptly such groups were formed in New York, Chicago, and Cambridge. Sections in clinical, educational, and industrial psychology date from the nineteen twenties.

The earlier failure of attempts to form a philosophical section led to the formation of the American Philosophical Association as a separate organization in 1900. A regional Eastern Branch of the American Psychological Association was established in 1937 and a mid-Western branch somewhat earlier. The society began to publish the *Psychological Review* in 1894, the *Psychological Monographs* the following year, the *Psychological Bulletin* in 1904, the *Journal of Abnormal and Social Psychology* in 1906, and the *Journal of Experimental Psychology* in 1916.[69] Independent organizations dealing with psychology in part were the American Association for the Study of the Feebleminded, founded in 1876, the American Society for Psychical Research in 1885, and the Southern Society for Philosophy and Psychology, founded in 1904. A group calling themselves "The Experimentalists" met first at Cornell University in 1904 under the leadership of E. B. Titchener. After Titchener's death, the group reorganized in 1929, calling themselves the Society of Experimental Psychologists.

As therapeutics of the mind came to receive increasing attention and as psychoanalytic technique led to improved clinical care of the mentally ill and emphasized measures for the prevention of major personality disasters, societies for the treatment of mental disease became increasingly common. The New England Society of Psychiatry was founded in 1875, the Boston Society of Psychiatry and Neurology in 1880, and the New York Psychiatric Society in 1904. Stimulus to the mental hygiene movement was given by C. W. Beers, whose efforts led to the formation of the Connecticut Society for Mental Hygiene in 1908, the first of many such societies.[70] His work led also to the establishment of a National Committee for Mental Hygiene in 1909, which in 1917 began to publish *Mental Hygiene*. The American Psychoan-

---

[69] The history of the American Psychological Association is ably and fully traced in S. W. Fernberger, "The American Psychological Association; A Historical Summary, 1892-1930," *Psychological Bulletin*, XXIX (1932), 1-89. .

[70] The earliest of these were the New York State Committee for Mental Hygiene and the Illinois Society for Mental Hygiene in 1910, and the Massachusetts Society for Mental Hygiene and the Mental Hygiene Society of Maryland in 1913.

alytic Association was formed in 1911, and the American Psychopathological Association the following year.

The specialization movement rendered it all the more imperative that there should be societies of general scope to keep the workers in the various branches of science in contact with each other. The following state academies of science were established in this period: the Kansas Academy of Science, founded in 1868, the Wisconsin Academy of Sciences, Arts and Letters in 1870, the Minnesota Academy of Science in 1873, the Athenée Louisianais in 1876, the Colorado Scientific Society in 1882, the Indiana Academy of Science in 1885, the Iowa Academy of Science in 1887, the Alabama Industrial and Scientific Society in 1890, the Nebraska Academy of Sciences in 1891, the Ohio Academy of Science in the same year, the Michigan Academy of Science, Arts and Letters in 1894, the Montana Academy of Sciences, Arts and Letters in 1901, the North Carolina Academy of Science in 1902, the North Dakota Academy of Sciences in 1908, the Oklahoma Academy of Sciences in 1910, and the Tennessee Academy of Science in 1912.

Local societies of a general scope which were founded in this period include the Davenport Academy of Sciences and the Kent Institute (Grand Rapids, Michigan), founded in 1867; the Cincinnati Society of Natural History in 1870; the Philosophical Society of Washington (D. C.) in 1871; the San Diego Society of Natural History in 1874; the Boston Scientific Society and the Santa Barbara Society of Natural History in 1876; the Meriden Scientific Association in 1880; the Rochester Academy of Science in 1881; the Elisha Mitchell Scientific Society (Chapel Hill, North Carolina) and the Newport Natural History Society (Rhode Island) in 1883; the Lackawanna Institute of History and Science (Scranton, Pennsylvania), the Hartford Scientific Society and the Sioux City Academy of Science and Letters in 1885; the Denison Scientific Association (Denison University, Granville, Ohio) in 1887; the Academy of Science and Art of Pittsburgh in 1890; the Southern California Academy of Sciences in 1891; the Onondaga Academy of Science (Syracuse, New York) in 1896; the Pasadena Academy of Science in 1897; the Manchester Institute of Arts and Sciences (Manchester, New Hampshire) in 1898; the Washington Academy of Sciences (Washington, D. C.) in 1898; the Bridgeport Scientific and Historical Society in 1899; the Warren Academy of Science in 1903; the Scientific Society of San Antonio in 1904; the Staten Island Association of Arts and Sciences in 1905; and the Knox Academy of Arts and Sciences (Rockland, Maine) in 1913. By 1918 twenty-six states of the union had statewide academies of science, and, in addition, two others had one or more local academies.

An interesting general investigation was undertaken by E. G. Dexter, who made a statistical study of the founding of scientific and learned societies in the United States from 1800 to 1900. For purposes of analysis he classified the societies roughly under the following headings: (*a*) general scientific societies, (*b*) historical societies and those of allied interests, (*c*) societies of natural history and the biological sciences, (*d*) associations for the study of special subjects, (*e*) professional associations, and (*f*) teachers' associations. He then compiled the following table showing the number of existing societies under these heads and the number of societies in each class organized in each decade.[71] The table is not complete for the last decade.

	a	b	c	d	e	f	Total
Before 1800	3	1	..	..	..	..	4
1800-	..	1	..	..	1	..	2
1810-1819	1	1	1	..	1	..	4
1820-1829	3	6	1	..	5	..	15
1830-1839	1	7	2	2	..	..	12
1840-1849	2	8	1	1	6	3	21
1850-1859	5	18	4	..	4	3	34
1860-1869	7	18	6	4	4	9	48
1870-1879	15	28	13	7	13	4	80
1880-1889	18	43	19	21	20	9	130
1890-1900	8	22	6	5	4	12	57
Total	63	153	53	40	58	40	407

The American Philosophical Society, the American Academy of Arts and Sciences, the Smithsonian Institution, the American Association for the Advancement of Science, and the National Academy of Sciences all entered upon careers of larger usefulness in the period 1866-1918. Moreover, they were so far successful in adapting their integrative efforts to the needs of the new era of specialization that, outside of the establishment of Sigma Xi, an honorary scientific fraternity, no concerted efforts were made to establish any other general national scientific organization until in 1916, when the National Academy of Sciences took the initiative in creating a new coordinating agency, the

71 E. G. Dexter, *History of Education in the United States* (New York and London, 1904), 552. By permission of The Macmillan Company, publishers.

National Research Council, as a measure of preparedness and national defense.

Throughout the period the American Philosophical Society issued ever-thickening volumes of *Proceedings* and *Transactions* which constitute one of the main repositories of American scientific knowledge available to the historian of the epoch.[72] In 1880 the American Philosophical Society held a celebration in commemoration of the hundredth anniversary of its incorporation.[73] One of the speakers was D. C. Gilman, president of the then recently founded Johns Hopkins University, whose subject was "The Alliance of Universities and the Learned Societies,"[74] contrasting the opposition of university and learned society in the days of Leibnitz with their complementary functions in his own time.

> I take it that the prime purpose of the university is education, its secondary object is research; while the converse is true of the academy, which should always make its major task investigation, and its minor instruction. The best university will include among its professors those who can advance the sciences to which they are devoted, and among the associates of an academy there will always be those who are capable and ready to diffuse among men the knowledge discovered. The university will develop the talents of youth, the academy will task the powers of full grown men. Universities plant seed; academies reap fruit.[75]

In 1889 the centennial anniversary of the occupation of the Hall of the society was made the occasion of another celebration.[76]

The American Academy of Arts and Sciences continued to issue its *Memoirs* and *Proceedings*. It also built up a splendid library of serials of other learned bodies. In 1912 its library was housed in a new building presented to the academy by the heirs of Alexander Agassiz, a sometime president of the academy. Increasingly, as the period wore on, the papers transmitted to the academy came from professors working in the new laboratories of Harvard, Yale, and other colleges, or

[72] The society published a *General Index to the Proceedings of the American Philosophical Society . . . , Volumes 1-50* (1838-1911) (Philadelphia, 1912), which is a most useful compendium for the study of the papers transmitted to the American Philosophical Society.

[73] The American Philosophical Society, *Proceedings at the Dinner Commemorative of the Centennial Anniversary of the Incorporation of the Society Held at the St. George Hotel, Philadelphia, March 15, 1880* (Philadelphia, 1880).

[74] *Ibid.*, 31-35.

[75] *Ibid.*, 32-33.

[76] American Philosophical Society, *Proceedings Commemorative of the Centennial Anniversary of the First Occupation of the Hall of the Society, November 21, 1889* (Philadelphia, 1890).

from professional scientists in government or industrial employ, rather than from amateurs. Mathematics, chemistry, and physics came more and more into prominence in the papers presented to the society as these sciences became more sharply differentiated and received greater emphasis in the latter decades of the nineteenth century and thereafter. The academy held its centennial celebration in 1880, and a few years later a substantial *Centennial Volume* in commemoration of it was issued,[77] containing addresses by Oliver Wendell Holmes and Asa Gray, and an important paper, "Stellar Photography," by Edward C. Pickering. In connection with the centennial celebration a publication fund was inaugurated totaling over $35,000 at the outset, to which Charles Francis Adams, president of the society, H. I. Bowditch, and Alexander Agassiz each contributed $5,000.

The Smithsonian Institution went forward with its program of advancing knowledge during this period. Joseph Henry continued to serve as its secretary until his death in 1878, rounding out thirty-two years in office and furnishing an unbroken continuity of policy. His successor, S. F. Baird, who served as secretary until 1887, was the moving spirit in the establishment of the United States Fish Commission. He was also one of the originators of the marine biological station at Woods Hole, Massachusetts. His work at the Smithsonian was especially concerned with building up the National Museum,[78] and with organizing the system of international exchange of scientific publications. Intimately associated with him was G. B. Goode, appointed assistant secretary in 1887, who, in addition to other labors, prepared a monumental history of the Smithsonian Institution just before his untimely and much regretted death in 1897.[79]

The third secretary was S. P. Langley, who served from 1887 to 1906. Called from his studies of astrophysics at the University of Pittsburgh, he organized the Astrophysical Observatory of the Smithsonian Institution in 1890, and its most important branch, the Mt. Wilson Observatory in California in 1905. He was rather more widely known to an unsympathetic public as the designer of airplanes, and he encountered much ridicule when his man-carrying machine was unsuccessfully launched and crashed into the Potomac in 1903. The plane,

[77] American Academy of Arts and Sciences, *Memoirs,* XI, pt. 1 (1882).

[78] Impetus had been given to the museum side of the Smithsonian in 1876, when collections were secured from the exhibitions of foreign governments at the close of the Centennial Exposition at Philadelphia.

[79] G. B. Goode, edit., *The Smithsonian Institution, 1846-1896; the History of Its First Half Century* (Washington, 1897).

reconditioned, was flown successfully eleven years later by Glenn Curtiss. There are those, therefore, who believe that Langley should be given either full or at least partial credit for the invention of the airplane. C. D. Walcott, noted as a paleontologist, who became secretary in 1907 and served until 1927, was instrumental in the establishment of the Carnegie Institution of Washington. In 1915 he became chairman of the National Advisory Committee for Aeronautics.

The *Smithsonian Contributions to Knowledge,* published from 1848 to 1916 and designed to embrace only additions to knowledge based on original research, were supplemented by the *Smithsonian Miscellaneous Collections* (1862-   ) containing papers often scarcely less technical. *Annual Reports of the Board of Regents* have been issued since 1846, giving a continuous record not only of the work of the Smithsonian itself but also of the progress in the different fields of knowledge and reprinting many papers from learned and scientific journals. In addition, numerous publications were issued by the separate divisions: the Astrophysical Observatory, the Bureau of American Ethnology, and the United States National Museum. The publications of the Smithsonian have been distributed to scientific societies throughout the world, either gratis or at nominal prices. In return for exchanges, the Institution acquired an enormous library of the volumes of American and foreign scientific societies.[80]

Meetings of the American Association for the Advancement of Science were suspended during the Civil War,[81] and those in the years immediately thereafter drew only a small attendance. Indeed, not

---

[80] For the history of the Smithsonian in this period, see G. B. Goode, edit., *The Smithsonian Institution, 1846-1896; the History of Its First Half Century* (Washington, 1897) ; and by the same author, *An Account of the Smithsonian Institution: Its Origin, History, Objects and Achievements* (Washington, 1895) ; H. C. Bolton, "The Smithsonian Institution: Its Origin, Growth and Activities," *Popular Science Monthly,* XLVIII (1896) , 289-303, 449-464; W. J. Rhees, comp., *The Smithsonian Institution: Documents Relative to Its Origin and History, 1835-1892* (2 vols., Washington, 1901) ; and by the same author, *The Smithsonian Institution: Journals of the Board of Regents, Reports of Committees, Statistics, etc., Smithsonian Miscellaneous Collections,* no. 329, XVIII (1880) ; "The Smithsonian Institution," *Smithsonian Report for 1906,* 97-102; A. H. Clark, "Administration and Activities of the Smithsonian Institution," *Smithsonian Report of 1916,* 137-145, a good historical summary; "The Smithsonian Institution," *Smithsonian Report for 1925,* 575-590; *Conference on the Future of the Smithsonian Institution* (Washington, 1927) .

[81] See "The Association from 1861 to 1870," American Academy of Arts and Sciences, *Bulletin* I, no. 4 (1942) , 29-30.

until the Nashville meeting in 1877 did the association resume its truly national scope.[82] Meanwhile, it had been incorporated by Congress in 1874.

The history of the American Association for the Advancement of Science may be divided into two periods. For the first half century of its existence it served and preserved practically unaltered its unitary nature. Since about 1900, however, the actual technical work has been largely carried on by "affiliated" and "associated" organizations, the "association being the correlating and unifying body and the recognized representative and spokesman."[83] The conditions which brought about this change were fundamental ones and indicate a shift in the direction of specialization on the part of American scientific organizations. Until 1882 the scientific work at the meetings of the American Association for the Advancement of Science was conducted by two general sections presided over by vice-presidents, and whatever subdivisions happened to be needed were under chairmen. At the Montreal meeting in that year the American Association for the Advancement of Science divided itself into nine lettered sections; by 1915 the number had grown to twelve.[84]

But the specialization movement was too strong for the scientists to be content merely with forming specialized divisions *within* the American Association for the Advancement of Science. The founding of independent specialized societies did not abate but rather increased, and "by 1890 the multiplication of special technical societies had become a recognized potential danger to the association."[85] In the following year, seven specialized societies were meeting with the American Association for the Advancement of Science. At the Springfield meeting of 1895 it was voted that a "committee of eleven consisting of the President, Permanent Secretary and one member from each section be appointed . . . to enunciate the policy of this Association; that this Committee is empowered to confer . . . with affiliated scientific so-

[82] H. L. Fairchild, "History of the American Association for the Advancement of Science," *Science*, n. s., LIX (1924) , 386.

[83] *Ibid.*

[84] Section A, Mathematics and Astronomy; Section B, Physics; Section C, Chemistry; Section D, Engineering; Section E, Geology and Geography; Section F, Zoology; Section G, Botany; Section H, Anthropology and Psychology; Section I, Social and Economic Science; Section K, Physiology and Experimental Medicine; Section L, Education; and Section M, Agriculture. American Association for the Advancement of Science, *Proceedings*, XL (1915) , 51-52.

[85] Fairchild, *loc. cit.*, 368.

cieties to secure their full and hearty cooperation in the efforts of this Association. . . . "[86]

It gradually became apparent that, if the specialized societies could be brought into cordial relationship with the sections and their meetings, these societies would become a means of reinforcing and vitalizing the American Association for the Advancement of Science. Accordingly, the "Affiliated Societies" were recognized in the Constitution of 1899,[87] and in 1901 they were given representation on the Council.[88] By 1910 thirty societies met in convocation with the Association, and by 1916 the number had risen to forty-three.[89] The affiliated societies, while entirely retaining their autonomous character, secured a larger degree of cooperation through their contacts with the Association.[90] It is significant to note that, whereas membership in the American Association for the Advancement of Science, which had risen from 461 in 1866 to about 2,000 in 1885, had actually declined slightly by 1900, thereafter rose very rapidly, reaching over 4,000 by 1905, about 8,000 by 1910, and over 20,000 in 1920.[91] Apparently, the new arrangements for giving greater scope to specialization very definitely increased the appeal of the American Association for the Advancement of Science.

Geography, as well as specialization, proved a source of tribulation to the American Association for the Advancement of Science. Cost and distance made it quite impractical for many western members to attend the meetings.[92] At the San Francisco meeting of 1915 the twelve specialized societies on the Pacific Coast[93] were brought into harmonious re-

[86] American Association for the Advancement of Science, *Proceedings*, XLIV (1895) , 392.

[87] *Ibid.*, XLVIII (1899) , 26.

[88] *Ibid.*, L (1901) , 30.

[89] H. L. Fairchild, *loc. cit.*, 368.

[90] The relationship was well summed up somewhat later as follows: "The Association has no control over the affiliated societies and academies, but becomes in effect an association of these societies, enabling them to cooperate in all directions where union is desirable." The American Association for the Advancement of Science, *Science*, n. s., XLIV (1919) , 113.

[91] American Association for the Advancement of Science, *Summarized Proceedings, 1915-1921* (Washington, 1921) , a graph on the back cover which shows the growth of membership in the American Association for the Advancement of Science from 1848 to 1920, prepared by B. E. Livingston, permanent secretary of the American Association for the Advancement of Science.

[92] See also "The Pacific Association of Scientific Societies," *Science*, n. s., XLI (1915) , 637-638.

[93] The twelve constituent societies of the Pacific Association of Scientific Societies

lationship with the American Association for the Advancement of Science, and in the following year the Pacific Division of the Association held its first meeting in San Diego.

It was also the hope of the American Association for the Advancement of Science that a widespread network of local branch associations might be organized. Affiliation of local societies with the American Association for the Advancement of Science was forcefully advocated in an address before that body as far back as the close of the nineteenth century:

> Another line of desirable influence would be in relation to local societies. The association from the outset sought and accomplished the great advantage of bringing together scattered and isolated workers in science throughout the land. This social and personal intercourse has been and still is one of the strongest and best elements in annual gatherings. But the local societies throughout the country are still in much the same isolation as the individual workers were fifty years ago; and some system of communication and cooperation among them would be a strength and a stimulus to all. Why might not the association bring about some method of intercourse or federation among these bodies, that would prove of great interest and value?[94]

In 1913 it was

> *Resolved,* That the council of the American Association for the Advancement of Science authorizes the establishment of local branches of the association in places where the members are prepared to conduct branches which will forward the objects of the association.
> *Resolved,* That the standing committee on organization and membership be instructed to promote the establishment of such local branches.[95]

If a program of this sort had been carried out, it would have affiliated most of the local academies of the country. It has not proved feasible

---

(first organized in 1910) to adopt the new arrangement were the Biological Society of the Pacific Coast, the Pacific Coast Paleontological Society, the Cordilleran Section of the Geological Society of America, the Seismological Society of America, the Astronomical Society of the Pacific, the Technical Society of the Pacific Coast, the Cooper Ornithological Club, the California Academy of Sciences, Puget Sound Section of the American Chemical Society, the Pacific Slope Association of Economic Entomologists, the San Francisco Society of the Archaeological Institute of America, and the San Francisco Section of the American Mathematical Society. "Pacific Association of Scientific Societies," *Science,* n. s., XLI (1915), 526.

[94] D. S. Martin, "The First Half Century of the American Association," *Popular Science Monthly,* LIII (1898), 834.

[95] American Association for the Advancement of Science, *Proceedings,* LXV (1913), 464.

to date to put it into practice, although an active unit was organized at Pennsylvania State College.

The American Association for the Advancement of Science held special meetings in Washington, 1916, in connection with the Second Pan-American Scientific Congress. The Congress was described at the time as "the largest official international gathering which has ever assembled in the history of the national capital and, at the same time, the largest Pan-American official gathering which has ever been called together in any capital of the Western Hemisphere."[96] The American Association for the Advancement of Science put on a splendid program of entertainment and speeches, featuring especially an address by Dr. W. W. Campbell, director of the Lick Observatory.

The American Association for the Advancement of Science has adopted the custom of opening its sessions to the public, and that without charge. Toward the close of the period, attendance at meetings often numbered over a thousand, especially when they happened to be held in one of the large seaboard cities. Low dues of two dollars and later of three dollars have never presented any real barrier to securing membership.

The American Association for the Advancement of Science has issued *Proceedings* since its organization in 1848. A volume of *Memoirs* appeared in 1875. *Science,* a weekly periodical under the editorship of J. McK. Cattell, became the official organ of the Association in 1901. In its pages appear the affairs of the Association, current scientific news, notable addresses and papers, and a continuous record of the activities of numerous American and foreign scientific organizations.

The founders of the National Academy of Sciences, as we have seen in the preceding chapter, had two main objectives in view for it: first, that it should serve the government in an advisory capacity; and, second, that it should be a body composed only of eminent men of science who had enlarged the field of human knowledge by abstract research. Though the academy has achieved notable success in regard to the latter of these aims, it has never fully realized the other fond hope of its founders.[97] Yet it has achieved something in this direction. In all, no less than forty-four committees on behalf of the government

96 L. O. Howard. "The American Association for the Advancement of Science Special Meeting." *Science*, n. s., XLIII (1916) , 247.

97 For the history of the National Academy of Sciences, see F. W. True, edit., *A History of the First Half-Century of the National Academy of Sciences, 1863-1913* (Washington, 1913) ; and *The Semi-Centennial Anniversary of the National Academy of Sciences. 1863-1913* (Washington, 1913) .

were appointed by the National Academy of Sciences between the years 1866 and 1908,[98] and twenty-four of their reports were published in the *Annual Reports* of the academy during these years.[99] Most of the committees were appointed at the request of the Treasury and the Navy and War Departments, and their work consisted in the main of making relatively unimportant chemical analyses or of determining physical standards. It must have seemed a bit ridiculous and humiliating to Academicians to be called upon to serve on a "Committee on the Preservation of Paint on Army Knapsacks," or to be requested to analyze smuggled narcotics.

In these later days such duties are more properly assigned to government bureau chemists and physicists. More in accordance with the dignity of the academy was a memorable report of 1896 "On the Inauguration of a Rational Forest Policy for the Forested Lands of the United States." The committee which prepared the report at the request of the Department of the Interior, after a study of the forest policies in Germany, had the satisfaction of seeing its recommendations adopted by President Cleveland; and, although Congressional politics retarded the policies for a time, subsequent administrations have embraced the general principles of the epochal report. In 1905 Gifford Pinchot, one of America's foremost advocates of scientific forestry, wrote:

> The work of the committee of the National Academy of Sciences, while it failed of much that it might have accomplished, nevertheless was the spring from which the present activity in forest matters was derived. The proclamation of the reserves which it recommended drew the attention of the country as nothing else had ever done to the question of forestry.[100]

While on the whole the influence of the National Academy of Sciences on national policy was not very great, nevertheless the advisory committees did undeniably render the government real service; and in some minor instances Congress saw fit to enact their recommendations immediately with appropriate legislation.

Mention might be made of a few of the leading achievements of prominent members of the National Academy of Sciences in this period. Prior to 1900, Simon Newcomb and A. A. Michelson conducted

---

[98] *A History of the First Half-Century of the National Academy of Sciences,* "Committees Appointed by the Academy on Behalf of the Government," 203-205.

[99] For a summary account of the works of these committees, see *A History of the First Half-Century of the National Academy of Sciences, op. cit.,* ch. IV, "The Academy as the Scientific Adviser of the Government."

[100] *Yearbook of the Department of Agriculture, 1899,* 297.

experiments for the purpose of determining more accurately the velocity of light. B. A. Gould compiled enormous catalogues of the stars of the southern hemisphere, spending years of his life in Argentina for the purpose. Henry Draper and S. P. Langley were prominent among those who studied solar physics. H. A. Newton and Simon Newcomb investigated meteor and planetary orbits, respectively. Under the direction of E. C. Pickering of Harvard College Observatory, celestial photography at that institution was employed to build a huge "library" of the sky. Wolcott Gibbs and Benjamin Silliman, Jr., were known for their chemical discoveries. J. D. Dana and James Hall were renowned as geologists. O. C. Marsh and Joseph Leidy ranged broadly over the biological sciences. With the passing of the older galaxy, new names appeared no less distinguished. H. F. Osborn achieved renown as a paleontologist and anthropologist. Franz Boas was a distinguished worker in the latter field also. A. A. Michelson continued his experiments in physics, being aided by E. W. Morley in his studies of "ether drift," the results of which baffled interpretation until they were explained by the theory of relativity. R. A. Millikan was winning recognition for his electrical discoveries in the opening decades of the twentieth century. G. E. Hale, solar physicist at the Yerkes and Mt. Wilson Observatories, in addition to his scientific discoveries was an active leader in organizing international astronomical cooperation. C. G. Abbot of the Smithsonian Institution was a designer of ingenious apparatus for the study of radiation.

Among the most notable presidencies of the National Academy of Sciences were those of Joseph Henry, who became president in 1868 and served for a decade, of O. C. Marsh, who served almost continuously either as acting president or president from 1878 to 1895, and of Wolcott Gibbs, who took up office in 1895 and served until 1900.

The principal publications of the academy in the period were the *Memoirs,* 1866- , the *Biographical Memoirs,* 1877- , which appeared irregularly, the *Annual Reports,* 1863- , the *Annual,* 1863-1886, the *Proceedings,* 1863-1894, and the *Proceedings,* new series, 1915- . These, like the publications of the Smithsonian Institution, were made up of papers of an advanced sort; and they constitute an important intellectual legacy of the period.

Acquiring the Alexander Dallas Bache Fund in 1867, the academy started upon a career of custodianship for administering numerous funds and prizes, including notably the Henry Draper Fund, the Wolcott Gibbs Fund, the Marsh Fund, the Joseph Henry Fund, and the Benjamin A. Gould Fund.

In 1900 the National Academy of Sciences sent delegates to the Wiesbaden Conference, which organized the International Association of Academies, the most important of the international scientific bodies prior to the organization of the International Research Council during the World War.

We have seen how the American Association for the Advancement of Science underwent a modification of its structure and divisional arrangement under the impact of the new forces of specialization. Similarly, it was but natural that the National Academy of Sciences should undergo some corresponding change in its internal organization. The original constitution, it will be recalled, provided for the members to arrange themselves into two classes: (*a*) mathematics and physics and (*b*) natural history, with further subdivisions of five sections in each class. This arrangement remained in force until 1872, when the classes were abolished. In 1885, 1890, 1892, and again in 1895 the question of divisions came up for discussion,[101] but no action was taken until 1899, when the constitution was amended so as to establish six committees: (1) mathematics and astronomy; (2) physics and engineering; (3) chemistry; (4) geology and paleontology; (5) biology; and (6) anthropology.

In 1911 this scheme was amended to replace the committees on biology and anthropology with four committees: (*a*) botany, (*b*) zoology and animal morphology, (*c*) physiology and pathology, and (*d*) anthropology and psychology.[102] In 1914 the committee on astronomy and mathematics was subdivided, making separate committees for each branch.[103] In the *Report of the National Academy of Science for the Year 1915* the committees were for the first time designated as sections. Other scientific committees were created from time to time. Thus, in 1918, for example, a committee on weights and measures was created and another on solar research, the latter including among its membership G. E. Hale, W. W. Campbell, and A. A. Michelson.

It was in 1916 that the National Academy of Sciences took its greatest step in the direction of organizing the scientific and the technical agencies of the nation by bringing into existence the National Research Council, which in the course of the next few years perfected an elaborate plan of divisions to give direction to research in the various fields of science.[104]

101 *A History of the National Academy of Sciences, 1863-1913*, 68-70.
102 *Report of the National Academy of Sciences for the Year 1911*, 14.
103 *Report of the National Academy of Sciences for the Year 1914*, 20.
104 *Report of the National Academy of Sciences for the Year 1918*, 110.

The eighteen-eighties witnessed the birth of an important general scientific organization different from those just discussed. Designed to encourage students on the threshold of a scientific career by making scientific work seem more attractive to them, to aid in sustaining their morale, and to promote their social and intellectual relationships, Sigma Xi, launched about half a century ago, has been a potent influence in stimulating research.

In the spring of 1886 a student geological society was organized at Cornell University under the inspiration of Professor H. S. Williams, who was soon planning a broader organization to be known as the Society· of Modern Scientists. Simultaneously, Frank Van Kleck, an instructor in engineering at Cornell, and W. A. Day, a senior at Cornell, had plans for forming a scientific society. In the fall, seven other students of engineering were drawn into the movement, and the name selected for the society was Sigma Xi.[105] Professor Williams was soon made acquainted with Sigma Xi. Becoming its first president, he delivered a memorable presidential address on June 15, 1887, entitled "The Ideal Modern Scholarship." He presented the purpose of the society, stressing investigation and specialization as ideals for the scientific worker and urging facility in the use of the French and German languages.[106] Branch chapters were soon organized at other colleges and universities. This honorary intercollegiate society has performed signal service in encouraging young productive research workers during the formative period in their careers, and has served well the fraternal needs of the more mature possessors of the coveted keys of Sigma Xi.[107]

[105] The first and undated constitution of the society, probably published in 1887, contained these words:

"Friendship in Science. While those whose heart and soul is in their work are coping with the great problems of Nature, let them remember that the ties of friendship cannot be investigated, but only felt. Let them join heart and hand, forming a brotherhood in Science and Engineering, thus promoting and encouraging by those strong, personal attachments of friendship, the highest and the truest advances in the scientific field. To lend aid and encouragement to those newer brothers, who likewise laboring in the same sphere are aspiring to honored positions. And in collegiate halls to award an honor, which to scientific recipients shall signify, 'Come up higher.'

"Therefore, with these ends and objects, the signers of this paper, do hereby agree to lend their efforts to the establishment of an organization to be publicly known as the SOCIETY OF THE SIGMA XI." Reprinted in Sigma Xi, *Sigma Xi Quarter Century Record and History, 1886-1911* (Urbana-Champaign [1913?]) , 4.

[106] Extracts from this address are printed in *ibid.*, 7-8.

[107] The aims of Sigma Xi are well summarized in its constitution of 1911:

"The object of this Society shall be to encourage original investigation in science, pure and applied, by meeting for the discussion of scientific subjects; by the publica-

Gamma Alpha Graduate Scientific Fraternity was formed in 1908 by the union of the Society of Gamma Alpha, which had been organized in 1889 at Cornell University, and the Alpha Delta Epsilon Scientific Fraternity, which had been organized in 1904 at Johns Hopkins University. Gamma Alpha aims to promote the fraternal spirit among graduate scientific workers in universities and research centers.

By the close of the period the Smithsonian Institution by no means stood alone as the only research institution in the United States. Others had come into being through the grants of wealthy individuals. The earliest of the notable new institutions was the Carnegie Institution of Washington, founded by Andrew Carnegie in 1902 and endowed by him to the extent of $22,000,000 by 1911.[108] Its aims were set forth in articles of incorporation as the following:

> That the objects of the corporation shall be to encourage, in the broadest and most liberal manner, investigation, research, and discovery, and the application of knowledge to the improvement of mankind; and in particular—
> To conduct, endow, and assist investigation in any department of science, literature or art, and to this end to cooperate with governments, universities, colleges, technical schools, learned societies, and individuals.[109]

The second of the great foundations was the Rockefeller Institute for Medical Research, incorporated in 1901. The primary work of the foundation, the advancement of knowledge, was administered by its president, assisted by an International Health Division and by four directors, for the natural sciences, the medical sciences, the social sciences, and the humanities, respectively. The medical laboratories, organized in 1905, were placed under the direction of Abraham Flexner. A hospital was opened in 1910 and a department of animal pathology in 1916.[110]

The third and largest of the new research institutions is the Rocke-

---

tion of such scientific matter as may be deemed desirable; by establishing fraternal relations among investigators in the scientific centers; and in granting the privilege of membership to such students as during their college course have given special promise of future achievement." Reprinted in *ibid.*, 45.

108 For the organization and work of the Carnegie Institution, see *The Carnegie Institution of Washington; Scope and Organization* (Washington, 1911) and the annual *Year-books.*

109 Carnegie Institution of Washington, *Yearbook* no. 1, 1902 (Washington, 1903), vii, Articles of Incorporation of Carnegie Institution of Washington.

110 The activities of the Rockefeller Institute for Medical Research are discussed in annual editions of *The Rockefeller Institute for Medical Research* (New York, 1911   ).

feller Foundation,[111] chartered under the laws of New York in 1913, with its object stated to be "to promote the well-being of mankind throughout the world."[112] By 1918 the gifts to the foundation had reached the gigantic sum of $120,765,856.[113]

The new research institutions were able to cooperate with the scientific societies in many ways, and served as a bond between the societies.

Even as the exigencies of the Civil War had led to the formation of the National Academy of Sciences, it was another conflict, the World War, which brought about the creation of the National Research Council for the fuller utilization of the scientific talent of the nation. With the threat of unrestricted German submarine warfare a growing menace to America's peace, the National Academy of Sciences, realizing the unpreparedness of the country for conflict and knowing the assistance which the scientific bodies of the belligerent powers were affording their governments, voted unanimously on April 19, 1916, to offer its services to the harassed president of the United States in the interests of national preparedness.[114] Upon President Wilson's acceptance of this offer, the Academy appointed a committee, including among its membership G. E. Hale (chairman), Simon Flexner, and R. A. Millikan, to devise a scheme of organization. The report which the committee submitted recommended the formation of a National Research Council to bring into cooperation governmental, industrial, and other research agencies.[115]

Thereupon the academy, assisted by consultations with the so-called Committee of One Hundred[116] of the American Association for the Advancement of Science, proceeded to set up the National Research Council. A military committee of the National Research Council was

[111] For the history and work of the Rockefeller Foundation, see its *Annual Reports* and its annual *Reviews*.

[112] Rockefeller Foundation, *Annual Report*, 1913-1914 (New York, 1914), 7. The charter continues, "It shall be within the purposes of said corporation to use as means to that end research, publication, the establishment and maintenance of charitable, benevolent, religious, missionary and public educational activities, agencies and institutions, and the aid of any such activities, agencies and institutions already established and any other means and agencies which from time to time shall seem expedient to its members or trustees." *Ibid.*, 7.

[113] Rockefeller Foundation, *Annual Report*, 1918 (New York, 1918), "Treasurer's Report," 339.

[114] *Report of the National Academy of Sciences for the Year 1916*, 32.

[115] National Research Council, "Preliminary Report of the Organizing Committee to the President of the Academy," National Academy of Sciences, *Proceedings*, II (1916), 507-510.

[116] National Academy of Sciences, *Report for the Year 1916*, 33.

soon working with the Army and Navy chiefs and the Council of National Defense in the prosecution of the war. A research information committee at Washington was established through the National Research Council with branch offices in London and Paris. Its scientific attaches worked in close cooperation with the offices of the Military and Naval Intelligence and had as their function "the securing, classifying, and disseminating of scientific, technical, and industrial research information, especially relating to war problems, and the interchange of such information between the allies in Europe and the United States.[117]

The utility of the Council as a preparedness and war-time measure having been more than amply demonstrated, men of vision began to seek to perpetuate it and make it a permanent member of the national scientific agencies. Accordingly, President Wilson issued an executive order on May 11, 1918,[118] in accordance with which the Council was perpetuated, "having for its essential purpose the promotion of scientific research and of the application and dissemination of scientific knowledge for the benefit of the national strength and well-being.[119]

---

[117] Report of the National Research Council [for 1917], in *Report of the National Academy of Sciences for the Year 1917* (Washington, 1918) , 48. The functions of the foreign branches, it was further explained, were:

"To serve as an agency at the immediate service of the commander in chief of the military and naval forces in Europe for the collection and analysis of scientific and technical research information and as an auxiliary to such direct military and naval agencies as may be in use for the purpose.

"To serve as centers of distribution to the American Expeditionary Forces in France and to the American naval forces in European waters of scientific and technical research information originating in the United States and transmitted through the research information committee in Washington.

"To serve as centers of distribution to our allies in Europe of scientific, technical, and industrial research information originating in the United States and transmitted through the research information committee in Washington." *Ibid.*, 49.

[118] National Research Council, *Organization and Members, 1922-1923* (Washington, 1922) , 6.

[119] *Ibid.* Its aims were more fully defined as follows:

"1. In general, to stimulate research in the mathematical, physical, and biological sciences, and in the application of these sciences to engineering, agriculture, medicine, and other useful arts, with the object of increasing knowledge, of strengthening the national defense, and of contributing in other ways to the public good.

"2. To survey the larger possibilities of science, to formulate comprehensive projects of research, and to develop effective means of utilizing the scientific and technical resources of the country for dealing with these projects.

"3. To promote cooperation in research, at home and abroad, in order to secure concentration of effort, minimize duplication, and stimulate progress; but in all

The cooperation of the scientific and technical bureaus of the government was pledged. Financial aid was furnished by the government during the war. The main activities of the National Research Council fall within the period treated in the next chapter, and they are fully discussed there.

A survey of the period 1866-1918 reveals the triumph of the ideal of specialization among the scientists. Numerous national and local specialized societies came into existence. By 1918 almost every state in the Union possessed a state academy of science, and a large number of municipal academies of science had been founded as well. Sigma Xi, an honorary scientific society, had come into being during this period, affording a stimulus to young scientists upon the threshold of their career, and offering a bond of fellowship for them in later life. The National Research Council came into existence at the very close of the period to coordinate the scientific activity of the nation confronted with entrance into the World War. It so amply demonstrated its utility that it was perpetuated, and it has become permanent and one of the most valuable scientific research agencies in America.

---

cooperative undertakings to give encouragement to individual initiative, as fundamentally important to the advancement of science.

"4. To serve as a means of bringing American and foreign investigators into active cooperation with the scientific and technical services of the War and Navy Departments and with those of the civil branches of the Government.

"5. To direct the attention of scientific and technical investigators to the present importance of military and industrial problems in connection with the war, and to aid in the solution of these problems by organizing specific researches.

"6. To gather and collate scientific and technical information at home and abroad, in cooperation with governmental and other agencies and to render such information available to duly accredited persons." *Ibid.*, 6.

Chapter IV

AMERICAN SCIENTIFIC SOCIETIES AND
WORLD SCIENCE, 1919-1944

The principal developments in American scientific organization that have taken place in the quarter of a century since the close of World War I are discussed in this chapter. There are certain salient trends which serve to distinguish post-war scientific organization from that which preceded it. First of all, American scientists, both as individuals and through associations, have come to play a much larger role in international scientific organization. Second, the National Research Council and other councils not only have been engaged in coordinating scientific endeavor but also have come to assume directive capacity and have initiated research projects. These councils, heedful on the whole of warnings not to narrow scientific inquiry too exclusively to materialistic ends, have year after year secured and allocated funds for scientific purposes. They have cooperated with national, state, and local scientific societies, maintained relations with governmental bureaus, universities, and research foundations, and in countless ways have promoted scientific knowledge. Their singular effectiveness has indubitably won them an enduring place in the American scheme of scientific organization. Finally, the maturing of American science has been reflected in a new awakening to historical perspective, finding expression in the formation of such organizations as the History of Science Society and the Philosophy of Science Association.

In addition, this recent period partakes in some degree of the characteristics of all three of the periods that we have studied previously. Like the natural philosophers of the era prior to 1789, its scholars have speculated boldly upon the general principles of nature and have founded societies emphasizing the underlying unity of all the sciences. Like the men of the period of national growth, 1800-1865, contemporary scientists have sought, and to a considerable extent realized, the functioning of a genuinely national academy of sciences, a tolerably well-endowed Smithsonian Institution, an energetic American Association for the Advancement of Science, and they have continued to found state and local academies of science. Then, too, specialization, industrialization, and national affiliation within the specialties have been carried even farther in the years since World War I than in the epoch 1865-1918. Expansion and consolidation in the direction of time-honored traditions have continued to take place. The closing period of our study finds scientific societies firmly and apparently permanently entrenched among American intellectual institutions.

Before turning to our study of scientific societies in contemporary American life, it will be well to review some of the high lights of scientific endeavor in the period since the close of World War I. Only a few achievements can be selected from the work of the galaxy of investigators here and abroad, and any selection must be, in addition, somewhat arbitrary and lacking in historical perspective.

Lending stimulus to inquiry concerning the philosophical bases of science are the writings of Alfred N. Whitehead and Bertrand Russell. Among the foremost exponents of the new history of science movement in America have been George Sarton, Florian Cajori, Charles H. Haskins, Lynn Thorndike, F. E. Brasch, L. J. Henderson, R. C. Archibald, J. K. Wright, Alexander Pogo, David E. Smith, and Benjamin Ginzburg.

During the years since the close of World War I, progress in astronomy on this side of the Atlantic has been centered in observatories equipped with the world's largest telescopes, produced by American mechanical genius, and made possible through American munificence.[1] A few of the leading astronomical achievements here and abroad may be mentioned briefly. Harlow Shapley, utilizing what is known as the "period luminosity law," arrived at new conclusions as to the immensity of the galactic system and the distance of the extra-galactic nebulae, or "island universes." George E. Hale, utilizing the spectrohelioscope which he perfected, made intensive investigations of solar structure. A. A. Michelson measured stellar diameters with his newly invented interferometer; S. A. Mitchell and Donald Menzel headed numerous eclipse expeditions. Frank Schlesinger measured and catalogued stellar parallaxes. F. R. Moulton and Ernest W. Brown devoted their energies

---

[1] A reflecting telescope with a 100-inch mirror was established at Mt. Wilson, an 85-inch one commenced for the University of Michigan, an 82-inch mirror for the University of Texas, a 74-inch one for the University of Toronto, a 72-inch one for the Dominion Astrophysical Observatory, and telescopes with 60-inch mirrors or larger at Harvard, Mt. Wilson, and Ohio Wesleyan. A disk for a 200-inch mirror was poured in 1933 for an observatory to be located on Mt. Palomar and operated by the California Institute of Technology and the Mt. Wilson Observatory.

to abstruse mathematical discussions of planetary and lunar theory, respectively. Robert G. Aitken catalogued binary systems. Charles Olivier and H. H. Nininger specialized in the study of comets and meteors. H. N. Russell analyzed the composition of stellar structure, and W. W. Campbell studied stellar motions. Annie J. Cannon classified the spectra of innumerable stars. An able group of European scientists, including Sir Arthur Eddington, Sir James Jeans, Abbé Lemaître, W. de Sitter, J. C. Kapteyn, Albert Einstein, and Tullio Levi-Civita, wrestled valiantly with large problems of the origin, evolution, and destiny of cosmology. Planetariums, opened in Chicago, Philadelphia, Los Angeles, and New York, gave a special stimulus to popular interest in astronomy in the United States.

In the field of mathematics, Leonard E. Dickson of the University of Chicago brought out a monumental work on the theory of numbers. Edward V. Huntington of Harvard delved deeply into postulate theory; David E. Smith of Columbia and Florian Cajori of the University of California made exceptionally able studies in the history of mathematics. Albert Einstein, renowned European mathematician and physicist, who had propounded the special theory of relativity in 1905 and the general theory in 1917, and produced his unified field theory in 1931, came to America in 1933 and became a teacher of mathematics in the Institute for Advanced Study organized at Princeton. Charles P. Steinmetz and others in America had already been acquainting American mathematicians with the subject of relativity. Hermann Heyl and Tullio Levi-Civita continued to contribute to the subject in Europe.

Physicists and chemists, both here and abroad, found themselves in the post-war era drawn into ever closer contact with each other in a concerted effort to discover more about the fundamental structure of matter. R. A. Millikan was a Nobel prize winner in 1923 for his feat of isolating the electron and measuring its charge. Albert Einstein won the same honor in 1921 for his work on relativity. Niels Bohr of Denmark received the coveted award in 1922 for his work on atomic structure. R. A. Millikan and A. H. Compton and their disciples carried out extensive and worldwide studies of cosmic radiation. Professor V. F. Hess, who first announced the discovery of cosmic radiation in 1912, was awarded the Nobel prize in 1936. Sharing the prize with him was C. D. Anderson, discoverer of the positron, or positive electron. The "neutron" had been found in the preceding year by James Chadwick of England. By 1936 some 250 isotopes had been discovered, one of the most interesting being the hydrogen isotope deuterium, or "heavy water," discovered in 1932 by Professor Harold C. Urey of Columbia University. Irving Langmuir won the Nobel award in 1932 for his work on the chemistry of surface reactions. Announcement of the principles of "indeterminancy" by Werner Heisenberg of Germany seemed to destroy the possibility of predicting simultaneously both the mass and position of an electron, and its implications left mechanistic philosophers with large difficulties to reconcile. A new device seeming to hold great possibilities for the scientist was the invention by Dr. V. K. Zworykin of the iconoscope, a mosaic photoelectric cell of high sensitivity.

Explorers were busily engaged in conquering the few remaining unknown land areas. Donald MacMillan, Lincoln Ellsworth, Charles Lindbergh, and Richard Byrd conducted Arctic explorations. Byrd was the first man to fly not only over the North Pole but over the South Pole as well. Several American expeditions explored the desert wastes of Tibet and Gobi. A. W. Stevens and O. A. Anderson, exploring upward, ascended over thirteeen miles into the stratosphere; William Beebe explored downward into the bathosphere or ocean depths.

American geologists, in addition to coping with countless special problems, found themselves increasingly drawn into cooperation with the physicists and the astronomers in trying to solve cosmic problems. Analysis of the rate of disintegration of radium indicated the age of the lithosphere to be perhaps two billion years, and the total age of the earth as a planet to be several billion years more. Isostasy and the Wegnerian hypothesis of drifting continents also came in for much study. In the field of meteorology, C. G. Abbot, secretary of the Smithsonian Institution, correlated solar radiation and climate, thereby opening new possibilities of long-range weather forecasting.

The researches of the biologist Thomas H. Morgan and his associates concerning the fruit fly, *Drosophila,* gave support to the gene theory of heredity. Alexis Carrel and Charles Lindbergh developed a new device, the perfusion pump (artificial heart), for keeping organs alive outside the natural body. Henry F. Osborn, authority on primitive man and paleontology, was one of the most prolific writers on science that America has yet produced. Franz Boas contributed studies of anthropology and native American linguistics, and Aleš Hrdlička also wrote widely on anthropological topics.

The workers to achieve renown in the field of medicine were of course numerous. Two Canadians, Banting and Best, in 1922 discovered the efficacy of insulin in the treatment of diabetes. Three years earlier, Edward C. Kendall, at the clinic made famous by the Mayo

Brothers, detected "thyroxin," the active principle of the thyroid gland; and a host of other discoveries relating to endocrinology soon followed. George Whipple, George Minot, and William P. Murphy won the Nobel prize in 1934 for studies of pernicious anemia. Henry Sherman of Columbia University, Elmer V. McCollum, and others made epochal studies showing the relation of vitamins to growth and health. Walter B. Cannon and George W. Crile studied bodily changes accompanying emotions. Harvey Cushing performed notable work in brain surgery. In medicine there was an increasing tendency for discoveries to be made by groups of scientists working in collaboration in large institutions, such as the Rockefeller Institute for Medical Research, the Mayo Clinic, and the Johns Hopkins University. World War I had a profound influence on the development of surgery, sanitation, and medicine, and on psychiatry as well.

While the post-war period found American psychologists in tolerable agreement as to the fundamental laboratory facts of their studies, they disagreed radically about interpretations, so that a great variety of conflicting "schools" of thought arose. John B. Watson was the most active protagonist of behaviorism, denying the existence of instincts and leaning towards a mechanistic interpretation of psychological activity. William McDougall, on the other hand, stressed instincts and emotions, purposive drives, and hormic action. Kurt Koffka and Wolfgang Köhler lectured in America on "gestalt psychology," which stressed "figure and background." Edward B. Titchener, Madison Bentley, and others, expounded "structuralism," stressing physiological structure. Freud had numerous disciples in America. Still others expressed different views or turned to eclecticism. Psychology became rapidly subdivided into subfields. Edward B. Titchener and E. G. Boring were leading experimentalists. Leonard T. Troland attempted a philosophical interpretation of the mind. Rudolf Pintner, L. M. Terman, E. L. Thorndike, H. C. Link, and many others busied themselves with devising intelligence, character, personality, and vocational ability tests. Floyd Allport, Hulsey Cason, and others, interested in social psychology, studied responses to social situations. Frank B. Watson and many others specialized in child psychology. Robert Gesell studied the genetic aspects of psychology. R. M. Yerkes became an authority in the field of animal psychology, H. H. Goddard investigated the mentally deficient, and James McK. Cattell turned to the more cheerful task of studying men of genius.

In recent years scientific discoveries have been broadcast to the nation as never before. Leading newspapers, such, for example, as the

New York Times, have come to lure specially trained science editors and news correspondents. The cinema and the radio report the "latest" in science. The Century of Progress Exhibition held in Chicago in 1933 brought forth panegyrics of praise of the developments in the pure and applied sciences, although social critics might be asking *quo vadis* and "whither mankind?"

From World War I the victorious United States emerged with a heightened sense of national consciousness. After a few years of troubled readjustment it entered upon a period of unparalleled prosperity. Scientific societies reaped the benefits of better times and gained in membership and donations. The lean years following the economic debacle of 1929 proved a testing time for the societies as well as for many other institutions. Feeling the pinch of poverty, unemployed industrial scientists and technicians, or professional men with curtailed incomes, frequently came to view as a luxury affiliation with scientific organizations which a few years previously would have been deemed a necessity. Even the deepening of the depression, however, found new scientific organizations coming into existence.

In every department of science new specialized societies have come into existence in the years since the close of World War I. In the field of mathematics a new body for the study of statistics, the Econometric Society, was formed in 1930.

Three new national astronomical organizations were formed: the Amateur Astronomers Association in 1927, the Society for Research on Meteorites and the American Amateur Astronomical Association in 1933. Local astronomical societies began to spring up in many places. Among the more active of these were the Bond Astronomical Club (Cambridge, Massachusetts), 1924, the Junior Astronomy Club of the American Museum of Natural History, 1929, the Baltimore Astronomical Society in 1931, the Rochester Astronomy Club (Rochester, New York), 1932, the Sky Scrapers (Brown University), 1932, the Indianapolis Astronomers Association, 1933, the Mid-West Meteor Society and the Amateur Astronomical Society at New Orleans in 1934, and the Barnard Astronomical Association of Tennessee in 1935. Amateur astronomers and others, inspired by R. W. Porter, enthusiastic telescope constructionist, and A. G. Ingalls, an editor on the staff of the *Scientific American,* were, at the same time, forming groups of telescope makers at Springfield (Vermont), Boston, Chicago, Cincinnati, Indianapolis, Syracuse, Buffalo, Rochester, Watertown (Wisconsin), Little Rock (Arkansas), Los Angeles, Tacoma (Washington), East Bay (California), Dallas (Texas), and other places. An organization known as Astrolab, established in 1932, aimed at coordinating the efforts of the amateur astronomers and telescope builders. Under the inspiration of C. A. Federer, Jr., an Amateur Astronomers League of America seems at last to have come into being. It is a federation of astronomical societies, and, by February of 1942, at least ten such bodies had ratified its constitution. Amateur telescope makers are being called upon increasingly to utilize their skill in lens grinding for the manufacture of war optical apparatus in World War II. Before this situation arose, however, the amateur telescope makers of America had produced an estimated 25,000 telescopes of the standard six-inch mirror size or larger.[1a]

In the field of mathematics, the last quarter of a century has seen an increasing amount of emphasis on the application of statistics not only to the physical sciences but to the biological and social sciences as well. The Econometric Society (1930) and the Institute of Mathematical Statistics (1935) were reflections of this trend. The National Council of Teachers of Mathematics dates from 1920. By 1940 more than a century had elapsed since the founding of the first statistical society in America. By that time, too, Raymond C. Archibald was able to bring out *A Semi-centennial History of the American Mathematical Society, 1888-1938*. As was the case with astronomy, numerous school clubs were springing up throughout the country to hold informal colloquies on mathematical problems.

Among the principal new national chemical societies of the last two decades are: American Institute of Chemists and the National Colloid Symposium, both founded in 1923, the Chlorine Institute (1924), the Association of American Soap and Glycerine Producers (1926), the Association of Consulting Chemists and Chemical Engineers (1928), the Wood Chemical Institute (1929), the Italian-American Chemical Society (1935), and the Metropolitan Microchemical Society (New York City, 1936). Although this list may not be inclusive, it is certainly representative; and that it is not much longer is indicative of the degree of adaptability and extensibility of the already existing American Chemical Society, whose sections and branches could be made to serve most of the needs of American chemists. One branch of applied chemistry perhaps deserves special mention in this period, namely, ceramics. Representative societies in this field include such state associations as the Ohio Ceramic Industries Association (1924) and the Ceramic Association of New York (1933). A fraternal organization,

[1a] *Scientific American,* CLXV (1941), 300.

Keramos (1932), was the result of merging previously existing similar groups in Ohio and Illinois that were started decades earlier. The Institute of Ceramic Engineers, an organization of national scope, was formed in 1938 as an affiliate of the American Ceramic Society (1899).

Geographical societies coming into existence in the last two decades have included: the Society of Woman Geographers (1925), the Pacific Geographic Society (1929), the Texas Geographic Society (1933), and the New York State Geographical Association and the Association of Pacific Coast Geographers, both founded in 1935. The American Society of Photogrammetry (1934) took the lead in aerial photography, which not only has had a revolutionary effect on mankind's knowledge of such hitherto uncharted regions as Greenland, Alaska, and Antarctica, but it has proved indispensable to warfare as well. The American Polar Society, formed in 1934 "to act as a clearing house of polar information, to be of aid to members of polar expeditions, and to spread the knowledge of the polar regions,"[2] was particularly interested in the regions of Antarctica made known by the Byrd expeditions.

In the interim between World War I and World War II American geologists were not only aggressive in pushing their global quest for oil resources but they were also in the van of the domestic organizers of professional research associations. Some of these associations were national, even international, in scope; others were regional and local. The societies reveal, too, that the sub-fields of geology were emerging rather sharply. In passing, an enumeration is made here of some of the societies, selected from among the above-mentioned categories: American Geophysical Union (a League of Nations affiliate), the Mineralogical Society of America, the American Petroleum Institute, and the Shreveport (Louisiana) Geological Society, all formed in 1919; the Society of Economic Geologists and the Kansas Geological Society (1920), the Branner Geological Club of Southern California (1921), the Rocky Mountain Association of Petroleum Geologists and the Ardmore (Oklahoma) Geological Society (1922), the Houston (Texas) Geological Society (1924), the New York State Geological Association (1925), the Society of Economic Paleontologists and Mineralogists, the American Shore and Beach Preservation Association, the North Texas Geological Society, and the West Texas Geological Society (all in 1926), the International Commission on Glaciers (1927), the Shawnee (Oklahoma) Geological Society (1928), the South Texas Geological Society (1929), the Appalachian Geological Society, the Society of

2 *Objects of the Society* (Leaflet American Polar Society), 1940.

Exploration Geophysicists, the Gypsum Association, and the Woods Hole Oceanographic Institution (1930), the Gemological Institute of America and the East Texas Geological Society (1931), the Philadelphia Geological Society and the American Gem Society (1934), the Oceanographic Society of the Pacific (1935), the South Louisiana Geological Society (1936), the Carolina Geological Society and the Michigan Geological Society (1937), the Association of College Geology Teachers (1938), the National Speleological Society (1939), and the Crystallographic Society (Massachusetts Institute of Technology–Harvard, 1941).

The American Meteorological Society and the International Commission on Maritime Meteorology were formed in 1919 and the International Committee for the Study of Clouds two years later. The American Meteorological Society[3] was organized at St. Louis on December 29, 1919, for "the advancement and diffusion of knowledge of meteorology, including climatology and the development of its application to public health, agriculture, engineering, transportation and inland waterways, navigation of the air and oceans, and other forms of industry and commerce,"[4] as its constitution stated. It marked "the beginning of a movement not only to push forward investigations of weather and climatic conditions, but also to widen the valuable application of the knowledge already at hand."[5] Within a few months, while half of its membership of nearly six hundred were amateur or professional meteorologists, the rest included "aeronautical engineers, architects, astronomers, automobile salesmen, aviators, business men, civil engineers, . . . ecologists, explorers, farmers, fruit growers, geographers, geologists, hydraulic engineers, lawyers, mariners, merchants, nature lovers, newspaper men, officials in rail and auto-transportation, telephone, telegraph, light, and fuel companies, physicians, physicists, shippers, stockmen, students and teachers."[5] In a field like meteorology the collecting of regional data is particularly essential. Hence, it proved highly advantageous to form numerous subdivisions, and, by 1943, no less than twenty local seminars and branches were affiliated with the parent organization.

The years since World War I have seen the field of physics split into various subdivisions, with new specialized societies coming into existence. The American Standards Association was formed in 1917

3 *Bulletin of the American Meteorological Society,* I (1920), 1.
4 *Ibid.,* XXX (1940), cover.
5 *Ibid.* I (1920), 1.

and the Association of Scientific Apparatus Makers of the United States in 1919, the Horological Institute of America in 1921, the International Amateur Radio Union in 1925, the Association for Correlating Thermal Research in 1928, the Society of Rheology and the Acoustical Society of America in 1929, the American Association of Physics Teachers and the American Rocket Society in 1930, the Inter-society Color Council in 1931, the American Society of Biophysics and Cosmology, the Association of Color Research, and the Photographic Society of America in 1936. The second World War has placed increasing demands upon both the professional and amateur photographer. Especially has aerial photography been undergoing rapid change. In 1931 the American Institute of Physics was established by joint action of the American Physical Society, the Optical Society of America, the Acoustical Society of America, and the Society of Rheology for the purpose of better coordinating research in physics.[6] The Institute publishes *Review of Scientific Instruments*.

Every year has brought the formation of a large number of radio clubs and associations. These have recently come into prominence in the war effort for their training programs and experimental research.[7]

The Institute of Standards and the Chicago Society for Measurement and Control both date from 1940. The rapid rise of electronics naturally called into being several new and as yet young but very active societies. Thus the American Society for X-ray and Electron Diffraction was organized in 1941[7a] and the Electron Microscope Society of America in 1943.[7b] On July 7, 1943, had come the announcement that the Council of the American Physical Society had authorized a "Division of Electron and Ion Optics.[7c] Soon the work of these bodies, as well as that of the American Institute of Physics, was to play a major rôle in the war effort.[7d]

---

[6] The announcement of the Institute states: "This new Institute will be governed by a board composed of representatives of the four societies named. Through it these Societies aim better to serve Physics and Physicists. They desire to establish closer relations with other national and local groups of research workers, teachers and students in the field of Physics. They will study the financial and other problems of the physics journals. Finally, they offer the services of the Institute to other Societies, to the Public and to the Press." *Scientific Monthly*, XXXV (1932), iv.

[7] "The American Radio Relay League War Training Program," *QST*, XXXVI, No. 5 (1942), 64.

[7a] *Science*, n. s., XCIV (1941), 16.

[7b] *Journal of Applied Physics*, XIV (1943), 566-567.

[7c] *Ibid.*, 406.

[7d] H. A. Barton, the American Institute of Physics, *Science*, n. s., XCIV (1944),

As was shown in the preceding chapters, the engineers were among the first professional men to organize strongly. This tendency has, of course, been continued in recent decades, with increasing emphasis on specialization and new fields. The following have been selected as indicative of the newer trends: the Society of American Military Engineers was organized in 1919, the American Engineering Council representing some thirty-six engineering societies in 1920, the State Conference of Sanitary Engineers and the American Society of Agricultural Engineers also in 1920, the Newcomen Society of England, American Branch in 1922, the Russian Association of Engineers in 1923, the American Society of Safety Engineers and the American Society of Bakery Engineers in 1924, the Institute of Traffic Engineers and the American Society of Danish Engineers in 1930, the Engineers' Council for Professional Development in 1932, the National Society of Professional Engineers in 1934, the Society for the Advancement of Management in 1936, the American Society of Measurement and Control in 1939, and the Television Engineers Institute of America in 1940.

Among the more active of the regional, state, and local engineering societies founded in this period were: the Engineering Society of Western Massachusetts in 1919, the New Jersey Society of Professional Engineers, the Little Rock (Arkansas) Engineers Club, the Raleigh (North Carolina) Engineers Club, and the Engineering Council of Utah in 1920, the Engineers Club of Hampton Roads (Virginia), the Engineers Club of the Lehigh Valley (Pennsylvania), and the Minnesota Federation of Architectural and Engineering Societies in 1921, the Engineering Societies of New England, the East Bay (California) Engineers' Club, and the Engineers Club of Omaha in 1922, the Hartford Engineers Club in 1923, the Montclair (New Jersey) Society of Engineers, the Engineers Club of Plainfield (New Jersey), the Arkansas Engineers Club, and the St. Louis Institute of Consulting Engineers in 1924, the Massachusetts State Engineers Association, the New York State Society of Professional Engineers, and the New York State Association of Municipal Engineers in 1925, the New York Society of Model Engineers in 1926, the Lansing Engineers Club in 1927, the Worcester Engineering Society in 1928, the South Carolina Society of Engineers, the Engineers' Club of Birmingham (Alabama), the Pro-

172-176; and P. E. Klopsteg, "The Work of the War Policy Committee of the American Institute of Physics," *Review of Scientific Instruments*, XIV (1943), 236-241, for a discussion of the relations of the American Institute with the Office of Scientific Research and Development.

fessional Engineers of Oregon, and the Kalamazoo (Michigan) Engineering Society in 1929, the Structural Engineers Association of California in 1930, the San Francisco Engineering Council in 1932, the Pennsylvania Society of Professional Engineers in 1933, the Kentucky Society of Professional Engineers and the Air Conditioning, Heating and Ventilating Engineers of Baltimore in 1934, the Engineers Club of Tulsa (Oklahoma) in 1935, the District of Columbia Council of Engineering and Architectural Societies in 1936, the Illinois Engineering Council and the Fort Wayne (Indiana) Engineers Club in 1937, and the Engineers Club of Western North Carolina in 1938.

World War I had given considerable impetus to aviation, and new aeronautical societies were constantly springing up. The Aeronautical Chamber of Commerce of America was organized in 1921, the National Aeronautic Association of U.S.A. came into being in 1922, and the National Aeronautic Association in 1927. The Institute of Aeronautical Sciences (1932), with forty-seven branches throughout the country, has an extensive library of aeronautical publications at its headquarters at Rockefeller Center, New York City.

The Soaring Society of America was formed in 1932. Two years earlier, with soaring societies active in many countries, the ISTUS (International Association for the Study of Motorless Flight) had come into being. In 1940 a directory of almost two hundred glider clubs in the United States alone was compiled by Barringer.[7e] The Academy of Model Aeronautics dates from 1934.

Many of the new technical societies of the period deserve passing mention for their scientific research or applications of science. Any list of these organizations must be exceedingly arbitrary, but most of the following societies were significant in this connection. The Scientific Apparatus Makers of America, the American Welding Society, the American Zinc Institute, the Army Ordnance Association, the American Petroleum Institute, the American Asphalt Institute, the Oregon Technical Council, and the Technical Club of Dallas were all formed in 1919, the American Society for Metals, the North Central Electric Association, and the Electric Steel Founders Research Group in 1920, the American Institute of Steel Construction, the American Specification Institute, the Copper and Brass Research Association, and the Associated Technical Societies of Detroit in 1921, the National Slate Association in 1922, the American Refractories Institute and the New England Council in 1925, the National Technical Association, the

[7e] L. B. Barringer, *Flight without Power* (New York, 1940), 206-211.

National Electrical Manufacturers Association, and the Association of Asphalt Paving Technologists in 1926, the American Construction Council in 1927, the Lead Industries Association and the Non-ferrous Ingot Metal Institute in 1928, the Aluminum Research Institute in 1929, the Textile Foundation and the United States Institute for Textile Research, both in 1930, the American Society of Tool Engineers in 1932, the American Military Institute, the Edison Electric Institute, the National Association of Photo-Lithographers, the National Paint, Varnish and Lacquer Association, and the National Oxygen–Acetylene Association in 1933, the Association of American Railroads and the Electric Railroaders' Association in 1934, the American Council of Commercial Laboratories in 1935, the American Transport Association of America in 1936, the Industrial Research Institute in 1938, the American Coordinating Committee on Corrosion in 1939, the National Inventors Council in 1940, the Automatic Council for War Production in 1941, and the Refrigeration Research Foundation in 1943. The technical societies in World War II, as in World War I, have been called upon heavily to unite their efforts in expediting the war program.

The biologists of the country have never been brought together into one general society, but the Union of American Biological Societies, established in 1921 and composed of 32 societies which elect representatives to a council, meant the establishment of a loose federation to coordinate efforts in the biological sciences.[8] A National Council on Elementary Science was organized in 1920. The American

8 The Union of American Biological Societies is composed of the American Association for the Advancement of Science (Sections F, G, N, and O) , the American Association of Anatomists, the American Association of Economic Entomologists, the American Genetic Association, the American Physiological Society, the American Phytopathological Society, the American Society of Agronomy, the Entomological Society of America, the American Society of Biological Chemists, the American Society for Experimental Pathology, the American Society of Naturalists, the American Society for Horticultural Science, the American Society of Zoologists, the Botanical Society of America, the Ecological Society of America, the Society of American Bacteriologists, the Society of American Foresters, the American Ornithologists' Union, the American Society of Mammalogists, the American Society of Parasitologists, the National Research Council, Division of Biology and Agriculture, the American Society of Ichthyologists and Herpetologists, the American Society of Plant Physiologists, the American Association of Immunologists, the Society for Experimental Biology and Medicine, the American Society of Clinical Pathologists, the American Society for Pharmacology and Experimental Therapeutics, the Genetics Society of America, the Poultry Science Association, the National Association of Biology Teachers, the Wildlife Society, and the American Dairy Science Association.

Nature Association dates from 1922. The Association of Virginia Biologists, formed in 1921, became the Virginia Academy of Science two years later.

The botanists, too, were actively engaged in establishing societies. Those interested in the study of forestry organized the Society of American Foresters in 1920 and the American Tree Association in 1922. The American Society of Plant Physiologists was formed in 1924. Among the leading agricultural bodies were the World Agricultural Society and the Agricultural History Society, both formed in 1919, and the Crop Protection Institute in 1920. Fruit growers and vegetable gardeners continued to form horticultural societies as in the preceding period. According to a list of June, 1935, issued by the United States Bureau of Plant Industry, there were then some 281 horticultural organizations in the United States.[9]

Prominent among the flower garden societies were the Women's National Farm and Garden Association, organized in 1919, the Iris Society in 1920, the Delphinium Society and the American Orchid Society in 1921, the Cactus and Succulent Society in 1929, the Amaryllis Society in 1933, and the American Rock Garden Society in 1934. The National Council of State Garden Club Federations, organized at Washington in 1929, deserves special mention. By 1936, almost 2,000 clubs, having all told a membership of over 100,000 in thirty-six states, were represented in the council.[10]

Several new zoological societies have come into existence in recent years. The American Society of Mammalogists was formed in 1919, the Commission on Standardization of Biological Stains in 1921, the American Society of Parasitologists in 1924. Wildlife preservation societies include the Izaak Walton League of America (1922), the Northeastern Bird-Banding Association (1922), the Federation of the Bird Clubs of New England (1924), and the American Wildlife Institute (1935). Other societies for the study of man were added in this period, namely, the Archaeological Society of Washington, incorporated in 1921, the American Association of Physical Anthropologists in 1928, the Society for American Archaeology in 1934, the Society for Applied Anthropology in 1941, and the Inter-American Society of Anthropology and Geography in 1942.

In the field of medicine, the National Health Council, made up of representatives from public health organizations, was formed in 1920. Two regional societies, the Pacific Northwest Medical Association and

9 *Science*, n. s., LXXXIV (1936), 362.
10 *Ibid.*

the Central-Tri-State Medical Association, were formed in 1922 and in 1924, respectively. The State Society of Neurological Surgeons dates from the same year. The National Tuberculosis Association dates from 1918. The American Association of Oral and Plastic Surgeons and the American Birth Control League were formed in 1921; the American Academy of Physical Therapy, the American Congress of Physical Therapy, the American Society of Clinical Pathologists, and the American Association of Chemical Pathologists in 1922; the Association for the Study of Allergy, the American Association for Medical Progress, an organization in opposition to antivivisectionism, and the American Child Health Association in 1923; the American College of Radiology and the American Heart Association in 1924; the American Eugenics Society in 1926; the Aero Medical Association of the United States in 1928; and the American Federation for Clinical Research in 1941. The formation of the American Society of Medical History in 1924, with Victor Robinson as the moving spirit, is evidence of increasing awareness·of historical perspective on the part of American medical scholars. The American Human Serum Association was formed in 1937 for investigation of the properties of human blood, and preservation and distribution of blood serum. World War II served presently to emphasize the need for such an organization. The American Academy of Applied Dental Science dates from 1919. The National Conference on Pharmaceutical Research was established in 1922, and the American Institute for the History of Pharmacy in 1941.

New studies in psychology and psychiatry brought new organizations. The Association for Research in Nervous and Mental Diseases and the New York State Association of Consulting Psychologists were formed in 1920; the New York Society for Clinical Psychiatry and the Central Neuropsychiatric Association in 1922; the American Orthopsychiatric Association in 1924; and the American Foundation for Mental Hygiene in 1928. The American Association on Mental Deficiency (1933) was the outgrowth of an earlier organization known as the Association for the Study of the Feebleminded. The Psychometric Society was formed in the early part of 1935 by L. L. Thurstone, Paul Horst, J. W. Dunlap, and others interested in the application of statistics to psychology. The society later in the year became affiliated with the American Psychological Association. In 1936 it began to publish a journal, *Psychometrika*. The Society for the Psychological Study of Social Issues, formed in 1936, in 1937 requested affiliation with the American Psychological Association. Of interest to psychologists as well as to philosophers is the Association for Symbolic Logic which held its first meeting September 1, 1936, at Harvard University, and began to publish its journal, *Symbolic Logic,* later in the year.

Led by its valiant crusader, Clifford W. Beers, the mental hygiene movement has progressed rapidly. By 1933 there had been formed "State Societies or Committees for Mental Hygiene . . . in Alabama, California, Connecticut, Delaware, District of Columbia, Illinois, Indiana, Kansas, Kentucky, Louisiana, Maryland, Massachusetts, Michigan, Missouri, New Jersey, New York, Ohio, Oregon, Pennsylvania, Rhode Island, South Carolina, Utah, Washington, and Wisconsin."[11]

We shall turn next to the general societies for the advancement of science. Several state academies of science have come into existence in the years since 1918. The New Hampshire Academy of Science was formed in 1919, the Georgia Academy of Science in 1922, the Virginia Academy of Science in 1923, the Pennsylvania Academy of Science, the West Virginia Academy of Science, and the Alabama Academy of Science in 1924, the Colorado-Wyoming Academy of Science and the Louisiana Academy of Science in 1927, the Texas Academy of Science in 1928, the Minnesota Academy of Science in 1932, the Missouri Academy of Science in 1934, the Florida Academy of Sciences in 1935, the Nevada Academy of Natural Sciences in 1941, and the Oregon Academy of Science in 1943. Somewhat similar was the Northwest Scientific Association organized at Spokane, Washington, in 1923. The present time finds a state academy of science or an equivalent body in practically every state in the Union.

Extreme specialization in the contemporary period has led some scholars to re-emphasize the value of general societies in maintaining cooperation between those working in different fields primarily, but that could gain by cooperation. Commenting on the desirability of such general societies in the present era, G. E. Hale said:

> The progress of research, and the rapid advance of knowledge along particular lines, have naturally resulted in the highly specialized organization of science of the present day. Two centuries ago the Royal Society of London and the Paris Academy of Sciences could easily embrace the whole range of science, and include in their membership essentially all of the able investigators of England and France. The establishment of the Linnaean Society in 1788 marked the beginning of a dispersive movement that has continued ever since. While the multiplication of societies dealing with narrowly limited fields of science is a sign of progress, the complete separation of investigators who might work in cooperation is certainly not desirable. In fact, the increase of specialization, instead of rendering unneces-

11 "The Twenty-Fifth Anniversary of the Founding of the Mental Hygiene Movement," *Mental Hygiene,* XVII (1933) , 568.

sary organizations dealing with science as a whole, has served to emphasize their extensive possibilities. It may be doubted whether there was ever a time in the history of science when such bodies could render greater service.[12]

The American Philosophical Society, still issuing its *Proceedings* and *Transactions,* entered upon its third century in 1927. A celebration was held in that year commemorating the two hundredth anniversary of the founding of the Junto and featuring learned addresses. It was attended by numerous delegates from American and foreign societies and institutions of learning. A volume commemorating the occasion and containing the addresses was later issued.[13] A new series of *Memoirs* was begun in 1935. At another celebration, held in 1943 on the two hundredth anniversary of Franklin's proposal for an intercolonial scientific society, papers on the history of early science in America were featured.

The American Academy of Arts and Sciences, issuing its *Memoirs* and *Proceedings,* in this as in the preceding periods, continued to live up to its ancient and scholarly traditions.

Upon the death of C. D. Walcott in 1927, C. G. Abbot became secretary of the Smithsonian Institution. Marshalling many lines of evidence to confirm the theory of the cyclical variability of solar radiation, he seems to have brought the long-range prediction of climate within the bounds of possibility. Under the direction of Abbot, a new major department of the Smithsonian came into existence, the Division of Radiation and Organisms, for the study of the effects of radiation upon living things. In recent years Smithsonian scientists have had frequent occasion to place their knowledge at the hands of those who seek the exploitation of natural resources.

Thus, in the early twenties, oil geologists turned to J. A. Cushman, authority on Foraminifera, for guidance in detecting oil-bearing sands, with the result that new advances in technique were made, saving the oil companies large sums. Of particular practical value also were the Smithsonian botanical or tropical vegetation studies conducted in the Latin American countries. The Smithsonian Institution in the last few decades has found itself increasingly embarrassed by having to

[12] G. E. Hale, "The International Organization of Scientific Research," *International Conciliation,* no. 154 (1920), 431.

[13] American Philosophical Society, *Proceedings,* LXVI (1927), *The Record of the Celebration of the Two Hundredth Anniversary of the Founding of the American Philosophical Society, Held at Philadelphia for Promoting Useful Knowledge, April 27 to April 30, 1927* (Philadelphia, 1927).

face twentieth century needs and responsibilities with a nineteenth century endowment.[14] In the early nineteen twenties the income from its endowment was but a meager $65,000 and its total endowment of about $1,206,000 (in 1924) was less than the annual income of the Carnegie Institution in the same year.[15] Inadequately endowed and facing rising costs of the war and "prosperity" years of the twenties, the Smithsonian forewent many of its cherished projects. In 1910 it had to cease publishing the Smithsonian *Contributions to Knowledge,* and a few years later the Miscellaneous Collections had to be curtailed in size. Its library of 800,000 items, the foremost scientific library in the country, remained partly uncatalogued and unbound, and its work in preparing the American part of an international catalogue of scientific literature was practically halted. Special mention ought to be made of the widely used *Smithsonian Physical Tables,* revised and enlarged editions of which appeared in 1923 and 1927. In 1927 a conference on the future of the Smithsonian Institution was attended by conferees representing many of the principal research agencies of the nation. The booklet which was prepared in commemoration of that event summarized many of the achievements of the first eighty years of the existence of the Smithsonian.[16] Among other achievements, nine government bureaus had sprung from the work of the Smithsonian. The Smithsonian had built up and distributed ten series of technical publications, it possessed the largest scientific library in the country, it administered the nation's largest museum of over 10,000,000 specimens in addition to having given away to schools and museums over a million more, and it had taken part in some 1,500 expeditions covering the globe.[17] Over 5,000,000 copies and parts of copies of its own publications had been distributed.[18] The Smithsonian Institution, in spite of its inadequate funds, still preserves essentially its original character of being an organization for increasing and diffusing knowledge, and this *per orbem,* as its seal reads, more than ever in the contemporary period.

The American Association for the Advancement of Science adopted a constitution in 1920, replacing one which had been in use for forty-

[14] See especially *The Smithsonian Institution; A Revelation* (pamphlet, Washington, 1926), 18-24, "Twentieth Century Work on a Nineteenth Century Endowment."

[15] "The Smithsonian Institution," *Annual Report of the Board of Regents of Smithsonian Institution for the Year Ending June 30, 1925,* 587.

[16] *Conference on the Future of the Smithsonian Institution* (Washington, 1927).

[17] *Ibid.,* 6.

[18] *Smithsonian Report for 1925,* 580.

six years. The new instrument, in adddition to making the governing council more effective, provided for greater influence in its affairs by the affiliated societies.[19] By November 30, 1939, there were 119 associated organizations, of which 88 were affiliated, 23 being academies of science.[20] The number of sections had then grown to fifteen,[21] the same as in 1937. Membership in the Association increased steadily from about 11,500 in 1920 to nearly 16,000 in 1928. The annual convocation winter meetings and summer meetings of the American Association for the Advancement of Science were more and more widely attended by American men of science in the twenties and thirties as an increasing number of the specialized societies met with it and as a larger number of universities adopted the policy of paying part of the expenses of their professors who took an active part in presenting papers at the meetings.[22]

The outbreak of religious and social persecution and the policies of government censorship in many countries of Europe in the early thirties seemed a matter of grave concern to many American scientists as foreign scholars became refugees in America after harrowing experiences. The American Association for the Advancement of Science took cognizance of the deplorable situation in 1934, when its Council went on record with a memorable declaration of intellectual freedom.

> The American Association for the Advancement of Science feels grave concern over persistent and threatening inroads upon intellectual freedom which have been made in recent times in many parts of the world.
>
> Our liberties have been won through ages of struggle and at enormous

[19] "Constitution of the American Association for the Advancement of Science; Adopted at the Third St. Louis Meeting, December 29, 1919, to January 3, 1920, and in force since January 3, 1920," *Summarized Proceedings of the American Association for the Advancement of Science* for the period from June, 1921, to June, 1925 (Washington, 1925) , 11-17.

[20] *The Organization and Work of the American Association for the Advancement of Science* (pamphlet, Washington, 1929) , I.

[21] They were (*A*) mathematics, (*B*) physics, (*C*) chemistry, (*D*) astronomy, (*E*) geology and geography, (*F*) zoological sciences, (*G*) botanical sciences, (*H*) anthropology, (*I*) psychology, (*K*) social and economic sciences, (*L*) historical and philological sciences, (*M*) engineering, (*N*) medical science, (*O*) agriculture, (*P*) education. *Ibid.,* 37-38.

[22] For the more recent history of the American Association for the Advancement of Science, see the account of its affairs almost weekly recorded in *Science;* and also the *Summarized Proceedings* of the Association, issued quadrennially. H. L. Fairchild. "The History of the American Association for the Advancement of Science," Address at the Seventy-fifth Anniversary meeting of the Association, Cincinnati, December 29, 1923, *Science,* n. s., LIX (1924) , 365-369, 385-390, 410-415 is of some value for the early years of the period covered by this chapter.

cost. If these are lost or seriously impaired there can be no hope of continued progress in science, or justice in government, of international or domestic peace or even of lasting material well-being.

We regard the suppression of independent thought and of its free expression as a major crime against civilization itself. Yet oppression of this sort has been inflicted upon investigators, scholars, teachers and professional men in many ways, whether by government action, administrative coercion, or extra-legal violence. We feel it our duty to denounce all such actions as intolerable forms of tyranny.

There can be no compromise on this issue for even the commonwealth of learning cannot endure "half slave and half free."

By our life and training as scientists and by our heritage as Americans we must stand for freedom.[23]

One of the most forward-looking programs of the American Association for the Advancement of Science concerns the sponsoring of junior academies of science composed of high school students. By 1942, such groups were in existence in almost half the states of the union, under the general supervision of the American Association for the Advancement of Science, state academies of science, and public school systems.

The National Academy of Sciences in the years since 1918 has functioned much more actively than it had for some time previously. An energetic group of relatively young men such as R. C. Tolman, Harlow Shapley, Donald Menzel, A. H. Compton, and J. B. Conant found the *Proceedings,* established as we have seen in 1915, an apt vehicle for bringing their contributions to the attention of the scientific world. Then, too, the National Research Council in turn stimulated the Academy into greater activity and broadened its contacts. Many of the meetings and studies of the Council are reported in the *Proceedings.* The annual reports of the National Research Council have been published in the yearly *Report of the National Academy of Sciences.* The National Academy of Sciences also continues to publish its occasional *Memoirs* and its *Biographical Memoirs.*

In April of 1941 the National Science Fund was established by the National Academy of Sciences. "The Academy receives . . . and applies large or small gifts for all physical and biological sciences. . . . The National Science fund is a philanthropic organization that applies funds given for the advancement of science. It also provides an authoritative advisory group for the convenience of any prospective

23 American Association for the Advancement of Science, *Summarized Proceedings, June, 1929, to January, 1934,* LXXXII–LXXXVI (Washington, 1934) , 53.

donor in science."[24] Since its first bequest from Alexander Dallas Bache in 1867 of $60,000, the National Academy of Sciences and the National Research Council have administered several hundred such gifts.

Within the past two decades the National Academy of Sciences has lost something of its former austerity and has become more conscious of its possibilities as a disseminator as well as an accumulator of knowledge. This has been due in no small measure to *Science Service* (1921), which for almost twenty years shared the building of the National Academy of Sciences in Washington and through its affiliation with that body and the National Research Council did an important piece of journalistic pioneering for science.[25]

The National Research Council was reorganized and placed on a permanent basis by the Academy in 1919[26] under an Executive Board of some forty members. About the same time a gift of $5,000,000 from the Carnegie Corporation was forthcoming to aid the work of the National Academy and the National Research Council. The Council presently organized into three divisions to deal with the various branches of knowledge. The reorganization of the Council also provided for enlisting the cooperation of research men throughout the country by inviting national scientific and technical societies to name

24 *Philanthropy in Science; the National Science Fund of the National Academy of Sciences* (New York, 1941) and A. F. Blakeslee, "Origin and Ideas of the National Science Fund," *Science*, n. s., 94 (1941) , 356-358.

25 Watson, Davis, "The Twentieth Anniversary of Science Service," *Scientific Monthly* (1942) , 291.

26 The scheme adopted was:
  Division of General Relations:
  I.   Government Division
  II. Division of Foreign Relations
  III. Division of States Relations
  IV. Division of Educational Relations
  V.   Division of Research Extension
  VI. Research Information Service
  Division of Science and Technology:
  VII. Division of Physical Science
  VIII. Division of Engineering
  IX.   Division of Chemistry and Chemical Technology
  X.   Division of Geology and Geography
  XI.   Division of Medical Sciences
  XII. Division of Biology and Agriculture
  XIII. Division of Anthropology

"Fourth Annual Report of the National Research Council," *Report of the National Academy of Sciences for the Year 1919* (Washington, 1920) , 104.

representatives to the Council. By 1933 some 79 societies were affiliated with the National Research Council in this fashion.[27] In the same year almost 1200 scientific men were directly related to the Council through its board or committees and there was a large number of other collaborators in various projects.[28]

The National Research Council has administered three series of post-doctoral fellowships in physics, chemistry, and mathematics, in the medical sciences and in the biological sciences, including anthropology, psychology, and basic aspects of forestry and agriculture, the money for these being furnished by the Rockefeller Foundation. By 1933 about 140 of these were available with annual stipends of a minimum of 1600 dollars. Grants in aid for research are also administered. The Council sponsored the publication of the seven-volume *International Critical Tables of Numerical Data of Physics, Chemistry and Technology,* and other works as well. The work of each of the separate divisions of the Council has been vigorous in itself and forms a considerable chapter in scientific history.[29]

The work of the Research Information Service, a committee on the Executive Board, has been especially valuable. It has compiled surveys of the laboratories, research funds, scientific societies, and other research agencies of the United States. The National Research Council, active in the organization of the International Research Council in 1919, has cooperated with it in numerous ways and with the International Council of Scientific Unions, which replaced the National Research Council. It also cooperates with the international unions affiliated with the International Council of Scientific Unions. Since 1919 the National Research Council has issued about one hundred bulletins and more than that many numbers of its *Reprint and Circular Series,* both sets of publications being premier vehicles for issuing advanced technical papers and the results of the Council's scientific surveys. Since 1932 the Council has published the *Transactions* of the American Geophysical Union.

The economic crisis of 1933 led to the issuance of an executive order

27 "A History of the National Research Council, 1919-1933," *Reprint and Circular Series of the National Research Council,* no. 106 (Washington, 1933) , 8.

28 *Ibid.*

29 For the history of the separate divisions of the National Research Council, see its annual *Reports* published in the Reports of the National Academy of Sciences; and "A History of the National Research Council, 1919-1933," *Reprint and Circular Series of the National Research Council,* no. 106 (Washington, 1933) , or the same in *Science,* n. s., LXVII (1933) , 355-360, 500-503, 552-554, 618-620, LXVIII (1933) , 26-29, 93-95, 158-161, 203-206, 254-256.

(no. 6238) by President Franklin D. Roosevelt, establishing a Science Advisory Board to serve for two years "with authority, acting through the machinery and under the jurisdiction of the National Academy of Sciences and the National Research Council, to appoint committees to deal with specific problems in the various departments."[30] This board, consisting of nine, and later (1934) fifteen, members, was created at a time when the scientific services of the government were scattered through forty federal bureaus, eighteen of which were primarily scientific.[31] Acting under the chairmanship of K. T. Compton, the board, which received no compensation for its services, entered into its work of disinterested criticism and coordination of the scattered government agencies as it had been the intention of the New Dealers that it should do. Three main principles guided its operation: (1) to give advice only when requested, (2) to consider general functions, standards, and programs rather than personnel, and (3) to look toward social objectives of science, and to see which of them were a necessary obligation of government and how each bureau might contribute toward them.[32] In September of 1934 the Board issued its first report, suggesting numerous changes and supplying evidence for the need of maintaining the scientific activities of the government amidst a hue and cry for budgetary retrenchment. A second and similar report was issued a year later,[33] and the life of the Board itself was extended until December 1, 1935, at which time it was disbanded. It did, however, pave the way for the National Defense Research Committee a few years later.

The National Academy of Sciences and the National Research Council were instrumental in the organization of the International Research Council. In order to bring about a greater degree of cooperation between the scientific societies of the allied and associated belligerent nations, the Royal Society of London summoned an Inter-Allied Conference on International Scientific Organizations which met at London on October 9, 1918, and was attended by delegates from Belgium, Brazil, France, Great Britain, Italy, Japan, Serbia, and the United States. A plan prepared by the Council of the National Academy of Sciences was presented to the conference by the American

[30] Executive Order 6238, reprinted in *Report of the Science Advisory Board, July 31, 1933, to September 1, 1934* (Washington, 1934), 7.

[31] *Ibid.,* 12.

[32] *Report of the Science Advisory Board, July 31, 1933, to September 1, 1934* (Washington, 1934), 15.

[33] *Second Report of the Science Advisory Board, September 1, 1934, to August 31, 1935* (Washington, 1935).

delegates. It called for the creation of an International Research Council to act as a central clearing house for the activities of such research councils as were already in existence in various nations or such as it might be expected would be formed in the near future. The details of the creation of national research councils were left to the national academies of science or other main bodies representing the scientific men of each country, and each council was to be con-stituted so as to represent federated research agencies. At a second conference held at Paris in November of the same year, an executive committee with representatives from Belgium, France, Great Britain, Italy, and the United States was appointed to work out details and to consider the formation of associated unions. The projects came to fruition the following year, when the International Research Council[34] and the associated unions—the International Astronomical Union,[35] the International Union of Geodesy and Geophysics, and the International Union of Pure and Applied Chemistry[36]—were in-

[34] The objects of the International Research Council were stated at the Brussels meeting to be:

"1. To coordinate international activities in the various branches of science and its applications.

"2. To encourage the formation of international associations or unions needed to advance science.

"3. To guide international scientific activities in fields where no adequate organization exists.

"4. To establish relations with the governments represented in the union for the purpose of interesting them in scientific projects."

G. E. Hale, "The International Organization of Scientific Research," *International Conciliation*, no. 154 (1920), 436.

[35] "Organization Meeting of the American Section of the Proposed International Astronomical Union," *Proceedings of the National Academy of Sciences*, V, (1919), 193-196; and International Research Council, *Constitutive Assembly, Held at Brussels, . . . 1919, Reports of Proceedings*, 160-164.

[36] There have been subsequently organized the International Mathematical Union, the International Union for Scientific Wireless Telegraphy, the International Union for Pure and Applied Physics, the International Union for Biological Sciences, and the International Geographical Union. The utility of these unions can be easily seen if we take but a single instance. The International Astronomical Union unites in a single body the former workers preparing a great chart of the heavens from maps made in many lands, the International Union for Cooperation in Solar Research, and the International Conference of Ephemerides, and centralizes the telegraphic report-ing of astronomical discoveries. Each country in the union organizes a national committee to coordinate national astronomical research and to select delegates to the meetings of the International Astronomical Union. It has been found con-venient to organize the International Union of Geodesy and Geophysics into sub-sections dealing with (1) geodesy, (2) seismology, and (3) meteorology.

augurated[37] at a meeting held in Brussels, July 18-28. Statutes were adopted.[38]

The Division of Foreign Relations of the National Research Council acts for the National Research Council in its relations with the International Research Council, and keeps the government informed on scientific questions. On the membership of the division are representatives of the National Academy of Sciences, the American Association for the Advancement of Science, the American Philosophical Society, · the American Academy of Arts and Sciences, and representatives of important international organizations. The National Research Council adheres to the following bodies associated with the International Research Council: the International Astronomical Union, the International Geodetic and Geophysical Union, the International Mathematical Union, the International Union of Pure and Applied Chemistry, the International Union of Pure and Applied Physics, and the International Union of Scientific Radiotelegraphy.

On May 15, 1922, the Council of the League of Nations appointed the International Committee on Intellectual Cooperation[39] as a section of the Secretariat. It is commonly referred to simply as the Committee on Intellectual Cooperation. This Committee has consisted of such eminent men as Einstein, Lorentz, Henri Bergson, and Gilbert Murray. Almost from the first, Professor Millikan was the American representative. The establishment of national committees to act as intermediaries between the International Committee and the national scientific organizations in the various countries was begun in 1924, and such a committee was formed for the United States.

A subcommittee of the Council on Intellectual Cooperation, known as the Subcommittee on Bibliography, was appointed in 1922, with Professor Jacob R. Schramm of the National Research Council the American member. This Committee assumed, in 1926, supervision over the work of the Scientific Relations Section of the Institute and adopted the title of Subcommittee on Science and Bibliography. The Com-

---

37 W. W. Campbell, "Report of the Meetings of the International Research Council and of the Affiliated Unions Held at Brussels, July 18-28, 1919," *Proceedings of the National Academy of Sciences*, VI (1920), 340-348.

38 Conseil International de Recherches, *Assemblée constitutive, tenue . . . à Bruxelles, du 18 au 28 Juillet, 1919, procès verbaux des séances plénières* (International Research Council, *Constituent Assembly Held at Brussels . . . 1919, Reports of Proceedings*), 1-13.

39 For the work of the Committee on Intellectual Cooperation and its created agencies, see the official reports of the International Committee on Intellectual Cooperation. *Series of League of Nations Publications*, XII, A.

mittee on Intellectual Cooperation has been actively engaged in the performance of much useful work, most of which lies altogether outside the scope of this paper.[40]

Upon appeals of the Committee on Intellectual Cooperation for financial assistance, the French government made an offer of the use of a building in Paris and an annual fund of 2,100,000 francs, to be used for the creation of an International Institute. This generous offer was accepted by the Assembly and Council of the League of Nations in 1924, and the Institute was inaugurated on January 16, 1926, with offices in the Palais Royal.[41]   The Governing Body of the Institute is the Committee on Intellectual Cooperation. This appoints a Committee of Directors, one from each of five nations.[42] The governing body also appoints an administrative director and the higher staff officials. The International Institute serves as the executive and advisory body of the Committee on International Cooperation. The work of the Institute is further subdivided into sections, one of which is known as the Scientific Relations Section. It issues the *Bulletin of International Scientific Relations.*[43]

On September 28, 1927, the Council of the League accepted an offer made by the Italian Government to found and maintain an International Educational Cinematographic Institute. The Governing Body included among its membership C. E. Milliken, Secretary of the Motion Picture Producers and Distributors of America, Inc. We have considered briefly some of the steps taken by the League to constitute bodies competent to deal with science in an international way, and have pointed out that Americans have had a part, though not a very conspicuous part, in its work.

In 1926 Germany, Austria, Hungary, and Bulgaria were added to the list of the countries having a right to join the International Re-

[40] The activities of these bodies are summarized in such works as D. P. Myers, *Handbook of the League of Nations since 1920* (Boston, 1930) , 139-214; and H. R. G. Greaves, *The League Committees and World Order* (London, 1931) , 111-138. See also Bibliography of the present book.

[41] See especially Vernon L. Kellogg, "The League of Nations Committee and Institute of International Cooperation," *Science*, n. s., LXIV (1926) , 291-292; and *The International Institute of Intellectual Cooperation*, a Pamphlet of the League of Nations (Paris, 1926) , 4.

[42] Vernon L. Kellogg of the National Research Council has served as the American member most of the time.

[43] See "The First Six Months of the International Institute of Intellectual Cooperation," *Science*, n. s., LXIV (1926) , 175.

search Council.[44] In 1931 the International Research Council underwent a reorganization altering its statutes in such a way as to give a greater degree of freedom to its member unions[45] and become known as the International Council of Scientific Unions.[46] The International Council of Scientific Unions, meeting at Brussels in 1934, cognizant of economic collapse and grave political unrest in central Europe, a rising wave of nationalism, and much fear of impending war, took occasion to make significant affirmation of its belief in the necessity of world peace, the international character of science, and the need for cooperation on the part of all humanity.[47] Congress has from

[44] *Conseil International de Recherches, Assemblée Générale Extraordinaire . . . à Bruxelles,* 1926, 12.

[45] Cinquième Assemblée Générale Ordinaire du Conseil International de Recherches et Première Assemblée Générale Ordinaire du Conseil International des Unions Scientifiques, Tenue . . . à Bruxelles, le 11 Juillet, 1931, *Procès-verbale de la séance* (International Council of Scientific Unions, Fifth Assembly of the International Research Council and the First Assembly of the International Council of Scientific Unions Held at Brussels, July 11, 1931, *Reports of Proceedings*) , 1-13.

[46] "International Council of Scientific Union, Statutes," International Council of Scientific Unions, *Fifth Assembly of the International Research Council and the First Assembly of the International Council of Scientific Unions . . . ,* Reports of Proceedings, 78-82.

[47] "The International Council of Scientific Unions, being aware of the fact that the present economic and political difficulties have brought humanity face to face with a number of the most complicated and dangerous problems and threaten to erect a system of barriers between various nations, expresses its deep faith that ultimately a way will be found leading towards a more harmonious economic structure, and wishes to stress the importance of maintaining by all means international cooperation in the domain of science under whatever circumstances may present themselves.

"As laid down in its Statutes, the Council recognizes the relations between pure and applied science. There is no doubt that both governments and industrial groups will in an ever increasing degree call upon scientists for elucidation of the manifold complexities and problems which human life and relationships are presenting—problems, the most important of which perhaps are those of finding food, space in which to live and employment for the various peoples spread over the earth. If at the present moment an international organisation devoted to the solution of these problems is still beyond our vision, and organisation constructed according to national systems must be provisionally strengthened for fear of losing hold of economic possibilities, it can be foreseen that the scientists of every country will be drawn more and more into these spheres of national organisation. The Council expresses its confidence that scientists, while giving their aid in meeting the needs of their own nations, will never lose sight of the international character of science as a whole, and will ever continue to keep in working order and to develop the connections necessary for international cooperation, even if severe shocks unhappily might come to threaten economic and political relations.

time to time made appropriations for defraying the expenses of the participation of the National Research Council in the work of the International Council of Scientific Unions.

Americans participated in its organization in 1919 and in succeeding years in the organization of the following international scientific unions which adhered to the International Research Council: the International Astronomical Union, the International Union of Geodesy and Geophysics, the International Union of Chemistry, the International Union of Mathematics, the International Union of Physics, the International Union of Scientific Radio Telegraphy, the International Union of Geography, and the International Union of Biological Sciences.

Since 1918 American scientists very frequently have been called upon to serve on international commissions such as the International Electrotechnical Commission, the International Standardization Commission, the International Commission on Illumination, the International Commission on Terrestrial Magnetism and Atmospheric Electricity, and the International Commission on Maritime Meteorology, to cite only a few random examples. Americans, too, have frequently served on international committees such as the International Committee for the Study of Clouds, the International Committee for Physical Research Congresses, and the International Committee for Mental Hygiene. Nor ought the role of Americans to be overlooked in the formation of such new international scientific societies as the International Society for the Study of Microbiology, the International Society of the History of Medicine, and the International Committee of the History of Science.

The last two decades have seen in America an increasing interest in the history of science. This would seem to be the result of two things. First of all, it is indicative of the maturing of scientific research in America that scholars should be imbued with a deeper sense of historical perspective. Evidence of this is to be found in the fact

---

"In professing its faith in the possibility and the necessity of peace between the world's peoples, the Council points out that the 'brotherhood of scientists' can be an important factor towards the establishment of a desire for mutual understanding and helpfulness in order to overcome the dangers involved in a too exclusive nationalism.

"The Council therefore, in emphasizing the significance of science, both pure and applied, as a common treasure for all humanity, which can only be realized through a free-spirited cooperation of the most diverse elements, is of opinion that scientists of the whole world have a task of working for this understanding, and urges all allied organizations to give constant attention to this task." *Nature*, CXXXIV (1934), 89-90.

that in 1919 a group of scholars interested in the historical and cultural aspects of science, especially active among whom was Lynn Thorndike, organized a section in the American Historical Association, and
in the following years a similar section was formed within the American
Association for the Advancement of Science with W. A. Locy chairman.
Then, too, it received stimulation from a somewhat earlier similar
movement in Europe.

The troubled years of World War I and the post-war years led
to the flight of many scholars, who, taking refuge in America, brought
with them the retrospective habits of thought of an older civilization.
A man of such character was George Sarton, who, in association with
David E. Smith, inspired the formation of the History of Science
Society on January 12, 1924. Devoted to the whole range of science on
the historical side, it forms a valuable connecting link between the
American Association for the Advancement of Science, with which
it is affiliated and meets annually, and the American Historical Association, with which it meets in every alternate year.[48] It is also a constituent member of the American Council of Learned Societies. The
society has furnished funds for continuing the publication of *Isis,* an
international quarterly review which performs the unique service of
reprinting the great classics of science and contains comprehensive
bibliographies of current scientific works and articles on history of
science topics as well. The leaders of the movement felt the need for
publication of articles longer than could be printed in *Isis* and,
accordingly, in 1936 they started a new publication, *Osiris.* The first
volume of this new series was dedicated to David E. Smith, a veteran
historian of science. Edited by George Sarton and others, it contained
articles on the history of mathematics.

Under the leadership of men imbued with the ideas of the fundamental unity of all scientific thought and the need for a genetic approach to the growth of scientific knowledge, the History of Science
Society essays the role of promoting more harmonious cooperation
among students of the natural and social sciences and the humanities
alike. The vitality of the new movement is one of the most striking of
the recent trends. The International Academy of the History of Science
(Académie Internationale d'Histoire des Sciences) , also known as the

---

[48] "It may . . . be observed that the proposed society will be the first, on any large
scale, to afford a common meeting ground for scientists, historians, and philosophers.
Indeed, the study of the history of science seems to provide the only feasible method
for bridging the widening gap between the men of science on the one hand and the
men of letters on the other." *Isis,* VI (1924) , 6.

International Committee of the History of Science (Comité Internationale d'Histoire des Sciences), formed at Paris in 1928, counts George Sarton and Lynn Thorndike among its organizers. In 1933 ten out of a total membership of 150 were Americans, including, in addition to the two men mentioned above, David E. Smith, Charles H. Haskins, and Raymond C. Archibald.[49]

A still more recent venture is the Philosophy of Science Association, dating from July 1, 1932, which aims at bringing closer together "philosophy" in the narrow sense and "science" in an also restricted sense. With a growing membership of outstanding scholars, this society is rapidly coming into prominence. In 1934 it began to issue its organ *Philosophy of Science* in which appear papers on such topics as logic, mathematical and mechanistic interpretations of the universe, or determinism.

The Copernicus Society (1941) was pledged to the publication of an English edition of the *De Revolutionibus Orbium Coelestium* in commemoration of the four hundredth anniversary of that work.

During 1938 there was gradually coalescing a "new kind of scientific society, the first in the United States, aimed to orient scientific developments to their social implications, and give scientists a measure of control over the applications of science";[50] and the outcome was the formal launching of the American Association of Scientific Workers by a group of young scientists attending the annual meeting of the American Association for the Advancement of Science.[51] The American Association of Scientific Workers closely resembles the Association of Scientific Workers, organized in 1918 in Great Britain, with which it cooperates. The American Association of Scientific Workers presently organized semi-autonomous branches in Boston, Cambridge, Chicago, New York, and Philadelphia.[52] The society, dedicated to a study of the relations of science and society, aggressively embarked upon programs dealing with public relations, education, socialized medicine, protection of the consumer, government aid to science, and related problems. The organization has striven in particular to maintain freedom of scientific inquiry, freedom of speech, and it has strongly condemned those foreign governments which it deems to have crushed freedom, driven out independent-thinking scientists, and harnessed

[49] Académie Internationale d'Histoire des Sciences (Comité International d'Histoire des Sciences), *Annuaire*, II (Paris, 1933), 3 et 11-15 *passim*, 33.

[50] *New York Times*, Dec. 31, 1938.

[51] *Science* n. s., XXXVIII (1938), 562.

[52] *Ibid.*, n. s., LXXXIX (1939), 58.

science for war and destruction. Membership in the American Association of Scientific Workers by 1940 had grown to several hundred.

Impetus to a better type of journalism in the reporting and editing of scientific views was given by the formation of the American Society for the Dissemination of Science in 1919, Science Service in 1921, and the National Association of Science Writers in 1934.

The nation's drifting toward entrance into World War I had called the National Research Council into being; likewise, as the United States hovered on the brink of entrance into World War II, a struggle characterized by unprecedented mechanized warfare, thoughtful scientists turned to the creation of new agencies for directing and mobilizing the nation's scientific men and resources for defense. In this program much of the locating of scientific talent was begun in 1940-1941 by the National Roster of Scientific and Specialized Personnel, which, under the direction of Leonard Carmichael, undertook a comprehensive survey of the nation's men of science and their availability for governmental research. This survey made extensive use of the membership lists of the principal scientific societies of the nation.

By order of the Council on National Defense a group called the National Defense Research Committee (NDRC) was formed in June, 1940, for the purpose of supplementing the work of the Army and Navy in developing devices of war. This body was intended to function in an executive capacity since the National Academy of Sciences had already existed as an advisory body since the Civil War, when it was created. The NDRC is a civilian committee of six men operating as part of the temporary emergency governmental machinery. Organized into divisions and directed by such men as Vannevar Bush of the Carnegie Institution of Washington, R. C. Tolman of the California Institute of Technology, F. B. Jewett of the Bell Telephone Laboratories, Roger Adams of Illinois, J. B. Conant of Harvard, K. T. Compton of Massachusetts Institute of Technology, and C. P. Coe, Commissioner of Patents, and representatives of the Army and the Navy, NDRC initiated hundreds of contracts for research with colleges, institutes, and industrial laboratories, spending $10,000,000 in the first year of operation. By the time the United States entered World War II, December 8, 1941, it was estimated that about 75 per cent of the available physicists whose names were starred in *American Men of Science* were engaged in war research. Dr. Conant established in London an office for the interchange of technical information with Britain.

After President Roosevelt's declaration of an unlimited national emergency on May 27, 1941, came an executive order on June 28, 1941, setting up the Office of Scientific Research and Development, of which the earlier National Defense Research Committee was an integral part. This order provided that: "There shall be within the Office for Emergency Management of the Executive Office of the President the Office of Scientific Research and Development."[53] Wide powers were given to the Office thus set up. Its duties included advising the president as to scientific and medical research in the defense program, mobilizing of scientific personnel, collaboration with the War and Navy Departments, and the support of scientific research for national defense. It was further authorized to enter into agreements with the National Academy of Sciences, the National Research Council, and other scientific institutions for studies, experiments, and reports.

The order set up within the office: (1) the National Defense Research Committee (NDRC), one member of which was to be the president of the National Academy of Sciences, and (2) the Committee on Medical Research (CMR). Dr. Vannevar Bush, who had been chairman of the NDRC and was succeeded in this post by Dr. Conant, was placed at the head of the Office of Scientific Research and Development. Prominent scientists and representatives of the Army and Navy are on the OSRD, as well as a representative of the National Advisory Committee for Aeronautics (NACA). The OSRD works in close cooperation with the National Academy of Sciences, the National Research Council, the National Inventors Council, the Division of Medical Sciences of the National Research Council, and many other scientific and technical organizations, especially also with the National Roster of Scientific and Specialized Personnel.[54]

Dr. Karl T. Compton, President of the Massachusetts Institute of Technology, delivered a Pilgrim Trust Lecture, under the auspices of the Royal Society of London, May 20, 1943, entitled "Organization of American Scientists for the War,"[55] in which he discussed the part played by the Federal Bureaus, the Armed Services, and the non-

[53] Executive Order Establishing the Office of Scientific Research and Development in the Executive Office of the President and Defining Its Functions and Duties. Executive Order 8807, June 28, 1941. Franklin D. Roosevelt.

[54] Vannevar Bush, "National Defense Research Committee," *Scientific Monthly,* LI (1940), 284-285; "Federal Office of Scientific Research and Development," *Scientific Monthly,* LIII (1941), 196; and "Science and National Defense," *Science,* n. s., XIVC (1941).

[55] K. T. Compton, "Organization of American Scientists for the War," *Science,* n. s., II C (1943), 71-76, 93-98.

governmental agencies, including the scientific societies, in meeting the scientific problems raised by the war. Then, in the latter part of his address, he passed to the special new scientific agencies that had sprung up to meet the crisis, including the National Roster of Scientific and Specialized Personnel, the Office of Scientific Reseach and Development (OSRD), the National Defense Research Committee (NDRC), the Committee on Medical Research (CMR), the Joint Committee on New Weapons and Equipment (JCNWE), the National Inventors Council, Office of Production Research and Development of the War Production Board, Engineering, Science and Management War Training Program, and the Army and Navy Technical Training Programs.

In explaining to the British the reasons for the new agencies he said:

> In spite of an apparently complete peace-time organization of science in America, it has always been our experience, in the time of great emergency, that it appears advisable to establish temporary new agencies to deal particularly with the emergency. I have frequently tried to analyze the reasons for the establishment of special scientific agencies during times of crises. They are, I think, varied and rather fundamental. One of them is that every great crisis involves conditions so different from the normal situation that the types of organizations which can survive and operate during peace-time are not adequate to meet the emergency. It may be, for example, that the emergency calls for exercise of very extensive administrative functions, such as the supervision of research projects and the disbursement of large governmental funds to a far greater extent than in peace-time. Hence a peace-time body of scientific men organized primarily to exercise advisory functions may not be organized in the manner suited to prompt and efficient executive action. Another reason is the impossibility of always maintaining in the administrative positions of peace-time agencies the personnel who would be most effective for handling important projects in a war emergency. Men who have the proper capabilities are frequently too busy and too active in other directions to be willing to hold positions in a peace-time organization which is relatively inactive. Consequently, when the emergency comes, the only alternatives may be to change the leadership in the existing organizations, a difficult if not impossible process, or to set up new temporary agencies to deal with the emergency.[56]

A somewhat fuller discussion of the new agencies appeared a few months later in the *Journal of Applied Physics.*[57]

By the summer of 1943, however, notwithstanding the multiplicity of new coordinating scientific agencies, there were many who thought that the federal government ought to take a more direct and perhaps

[56] *Ibid.*, 94.

[57] "Mobilization of Scientific Resources," *Journal of Applied Physics*, XIV (1943), 373-405.

quite compulsory hand in bringing together American men of science in the war effort. Thought along these lines crystallized in the so-called Science Mobilization Bill, or Kilgore Bill.[58] This bill (1) attempted to define scientific and technical personnel, (2) would create an Office of Scientific and Technical Mobilization and within it a National Scientific and Technical Board and a National Scientific and Technical Committee, (3) would mobilize scientific personnel (including deferment from Selective Service), and (4) would provide for the national mobilization of scientific facilities.

This bill soon aroused a storm of controversy. Senator Kilgore stoutly defended his measure in the press and at public meetings. The American Association of Scientific Workers sponsored a forum discussion on the merits of the Science Mobilization Bill at which Dr. Kirtley Mather of Harvard, its president, remarked, "This is the first time that the representatives of all the professions have met together to consider their place in the wide perspective of national life. . . ."[59] Opposition to the bill was not long in developing, as is revealed in the pages of *Science* for the years 1943 and 1944. The governing body of the American Association of Chemists condemned it.[60] Harlan T. Stetson called it a "Bill for the Regimentation of Science,"[61] and later declared that the National Research Council was already accomplishing the necessary scientific objectives.[62] The American Institute of Physics expressed opposition,[63] as did the Electrochemical Society.[64] Other writers and societies expressed varying degrees of approval or disapproval of the bill. A seven-page analysis by Vannevar Bush appeared in *Science,* the general tone of which, while moderate, condemned the ways in which the new proposed agencies would conflict with existing ones.[65] Probably the most serious blow dealt the bill in the realm of organized science came, however, from the American Association for the Advancement of Science. "After careful consideration of the purposes and provisions of the Service Mobilization Bill, . . . the American Association for the Advancement of Science, an organization

---

[58] The text of the Kilgore-Patman Bill (S. 702, H.R. 2100) was reprinted in *Science*, n. s., III C (1943), 442-443. Soon more than a dozen articles had appeared in *Science* discussing it.

[59] A. B. Novihoff, "A Wartime Conference," *Scientific Monthly,* LVI (1943), 583.

[60] *Science,* n. s., II C (1943), 213.

[61] *Science,* n. s., III C (1943), 509.

[62] *Science,* n. s., II C (1943), 365.

[63] *Science,* n. s., III C (1943), 482-483.

[64] *Science,* n. s., II C (1943), 461.

[65] Vannevar Bush, "The Kilgore Bill," *Science,* n. s., II C (1943), 571-577.

having nearly 28,000 members (and having 187 associated and affiliated societies with a combined membership of over 500,000 persons whose interests cover broadly all the natural and social sciences) now through its Council of about 250 members chosen from among the leaders of American science, respectfully recommends to the Senate and to the House of Representatives of the United States that the Kilgore Bill (S. 702) be not passed either in its present form or in any other form containing similar provisions."[66] Whether the bill or a similar one would pass was a question still unanswered in the early months of 1944.

In the contemporary period are interwoven the tangled skeins of many developments. The American Philosophical Society and the American Academy of Arts and Sciences are still treasured legacies of the eighteenth century. The nineteenth century heritage is of course much greater. The Smithsonian Institution, the American Association for the Advancement of Science, and the National Academy of Sciences are integral parts of American scientific organization. Many state and local academies of science and scientific societies live on after decades of service and have seen similar sister societies springing up until every state in the union is supplied with at least one and usually with more. New broad societies of national scope afford an opportunity for scholars with historical or philosophical insight to publish their writings. Specialized societies have multiplied *ad infinitum* but have been coordinated through the American Association for the Advancement of Science and the National Research Council.

Finally, Americans were taking an active part in contemporary enterprises of international scientific cooperation until the international situation brought an eclipse of international scientific activity. The United States in World War II finds itself making unprecedented demands on its scientific organizations through such new agencies as the Office of Scientific Research and Development and the National Roster of Scientific and Specialized Personnel. The specialized and technical societies are assuming new burdens in the all-out war effort. The older, more general societies are no less busy. The present war has disrupted international scientific organization, or rather split it into two warring camps, embracing the scientists in the United Nations on the one side and those of the Axis on the other. Of political and military necessity, there can be little present cooperation between them. Some astronomical discoveries are still transmitted over cables, but the transmission of meteorological knowledge or information will

[66] *Science*, n. s., 98 (1943), 137.

be under a tight ban of censorship for the duration of the war. A host of scientific discoveries in this and other fields will be kept secret until the war is over. When the present conflict terminates, as sometime it must, the scientific societies will doubtless be called up to render heavy assistance in reconstructing the world intellectually.

Chapter V

THE INCREASE AND DIFFUSION OF KNOWLEDGE

"Promote, then, as an object of primary importance institutions for the general diffusion of knowledge," George Washington urged his countrymen in his Farewell Address.[1] It would doubtless have pleased full well the Father of his Country to know that American scientific societies, in which he had lively interest, would multiply and become important agencies for the "increase and diffusion of knowledge."[2] This chapter is devoted to a brief analysis of some of the means whereby American scientific societies have augmented and disseminated scientific knowledge. The human side of the scientific societies also deserves some mention.

National scientific gatherings have become such commonplace occurrences that it is hard to realize that they have come into vogue in America only during the past century. These symposia now furnish indispensable forums for the presentation and discussion of scientific papers. The sessions offer the most convenient time for threshing out disagreements and clearing up difficulties. The critical atmosphere which pervades the scientific meeting offers some guarantee against smug complacency or a tendency to be too dogmatic or assertive.[3]

Attractions other than the purely scientific sessions are afforded by the major scientific conventions. Undue prolixity of speakers is often compensated for by the renewal of congenial professional contacts. Flagging spirits may be revived by entertaining excursions to museums, observatories, or industrial plants. It was gradually discovered that the Christmas holidays were a particularly convenient time for holding scientific convocations, and this festive season an-

1 J. D. Richardson, *A Compilation of Messages and Papers of the Presidents* (20 vols., New York, 1922) , I, 212.

2 This phraseology, often quoted or paraphrased, appears in the will of James Smithson, reprinted in G. B. Goode, edit., *The Smithsonian Institution, 1846-1896* (Washington, 1897) , 20.

3 As Laplace aptly remarked over a century ago, "The isolated scholar can give himself over without fear to the spirit of system: he hears only from afar the contradiction which he experiences. But in a learned society, the clashing of systematic opinions soon ends by destroying them; and the desire to convince each other, establishes necessarily between the members the agreement to admit only the results of observation and of calculation." P. S. Laplace, *Précis de l'histoire de l'astronomie* (Paris, 1821) , 99.

nually finds the wives and families of professors bewailing the whole-sale exodus to some distant point where numerous societies are meeting in conjunction with the winter convention of the American Association for the Advancement of Science.

Perhaps even more influential than scientific meetings in molding scientific thought are the numerous publications of the societies. Scientific journals, proceedings, and transactions appearing monthly or oftener are important agencies for reaching the professions, conveying the knowledge of recent scientific advances and the latest scientific news. Memoirs and bulletin series supplement these publications with special papers. They often contain the best extant biographies of prominent deceased scientists—a valuable, even if somewhat neglected source of historiography. The American Association for the Advancement of Science, through its official organ, *Science,* weekly summarizes the advances in science and its columns report the doings of the leading American and foreign scientific societies. *Nature* performs a similar service in Great Britain. In recent years *Science News Letter,* a weekly published by Science Service, formed under the auspices of the National Research Council, the National Academy of Sciences, and the American Association for the Advancement of Science, and subscribed to by many libraries throughout the country, has had a wide popular appeal. It serves as an antidote to the mushroom crop of contemporary pseudo-scientific tabloids. Important abstracts, such as *Chemical Abstracts, Biological Abstracts,* and *Psychological Abstracts,* and comprehensive indices, such as *Engineering Index,* are other types of publications of scientific societies which are invaluable in saving the time of specialists and in directing them to articles which concern their special fields of interest. Transcending both space and time, the publications of scientific societies have a limitless reach and stretch out to generations yet unborn and preserve for them the great classics of science.[4]

---

[4] In summarizing the importance of the publications of learned and scientific societies, Professor W. H. Brewer, president of the Connecticut Academy of Sciences, once said: "The publications of learned societies, under various names and in various ways, furnish by far the most comprehensive literature of science, philosophy, history, and art, that we have. For a time, this was almost the only way of publishing to the world new discoveries. Today it is as pervasive as it is extensive, and as yet no substitute has been found for this means of publishing and disseminating the details by which results have been obtained, even if the bare results might be available through the periodical press or other channels. These publications are an important part of every public library, but by reason of their enormous extent no library can be complete in them. It is only when we attempt to investigate their number in any

Inasmuch as the publications of scientific societies are the textbooks of scholars, they have a very important influence on the training of scientific writers. Rigid requirements are exacted by the better publications in regard to methods of presenting tables, graphs, footnotes, citations, and abbreviations. The growth of scientific journals since 1900 has been accompanied by an unparalleled increase in the production of scientific papers.[5] The journals have also been media for inculcating rising standards of professional ethics and for maintaining group solidarity and *esprit de corps*. The scientific and technical journals, too, have their message for the enterprising experimenter and inventor. Franklin, Bell, Edison, and many others who have enriched the world with new devices have acknowledged that they turned to the scientific journals for help in completing their inventions. Unquestionably, the publications of scientific societies have done much to stimulate cooperation in research, helping scientists and scientific societies to keep in touch with each other. Those who would promote international good will and cooperation see in the wide exchange of publications a means for tearing down barriers of race, creed, and nationality, and for bringing about unity of effort on the part of the scientific fraternity throughout the world.

Early American academies of science began providing laboratory equipment, or "philosophical apparatus" as it was then called; and, although the growth of universities, commercial laboratories, and government bureaus has largely done away with the necessity for societies to furnish such instrumentalities, the experimental apparatus owned by the societies continues to be of considerable value to investigators. Many of the societies have established libraries and built them up through the exchange of publications. Such libraries because of their specialized and technical nature have peculiar value for scientists and technicians. Other societies have established museums or herbariums.[6] Some of these museums have subsequently been absorbed

---

branch of science that we can appreciate the great influence such associations must have had in diffusing learning and information among the mass of the people and in making it available for their industries, their comfort and their intellectual pleasure." W. H. Brewer, "The Debt of the Century to Learned Societies," Connecticut Academy of Arts and Sciences, *Transactions*, XI (1901), pt. I, lii-liii.

[5] "In 1903 there were only about 4,000 Americans who had published research work, whereas now [1932] there are more than 20,000." J. McK. Cattell, "The Distribution of American Men of Science in 1932," *Science*, n. s., LXVII (1933), 264.

[6] Information concerning laboratories, libraries, and museums is given in Callie Hull, Mildred Paddock, S. J. Cook, and P. A. Howard, comps., *Handbook of Scientific and Technical Societies and Institutions of the United States and Canada*, Fourth Edition (National Research Council, *Bulletin* 106, Washington, 1942).

by the municipal or university museums, but others have continued to exist and prosper as separate entities, although the conception of founding a scientific society to carry on a museum is today almost as archaic as the notion of founding a society to furnish "philosophical apparatus."[7]

Wealthy patrons or benefactors have often established prizes, or made grants for the award of medals.[8] A particularly popular feature has often been the conducting of prize essay contests. These competitions afford a way for young and little-known men to win distinction by bringing their work before competent authorities. In recent years fellowships and scholarships have been maintained by a number of leading councils and societies. Societies devoted to archaeology, anthropology, and geography have taken a particular interest in the support of exploring expeditions.

While research institutions have rarely been willing to endow a scientific society and have maintained that universities are better adapted to sponsoring programs of investigation, private individuals, however, have made generous donations to societies both with and without specifications as to their use. Large grants have sometimes unfortunately led to all too human friction as to how the money should be spent. A recent and particularly notable gift was a legacy of approximately four and a quarter million dollars presented by Dr. Richard F. Penrose, Jr., at his death in 1931 to the Geological Society of America, of which he had been a life member.

Intersociety cooperation has come about increasingly. Economy of effort forbids that struggling organizations should perform overlapping work, issue similar publications, and compete in other ways, when societies, by combining efforts and resources, can secure greater administrative effectiveness and enlarged spheres of influence. Essentially the same conditions of competition that contemporaneously led to the federation of churches, the consolidation of charity agencies, and the integration of industry were at work and have caused societies to form

7 A few of the societies which still maintain museums are the Boston Society of Natural History, the Academy of Natural Sciences of Philadelphia, the American Chemical Society, the Geological Society of America, the Brooklyn Institute of Arts and Sciences, the California Academy of Sciences, the Buffalo Society of Natural Sciences, the Cincinnati Society of Natural History, the Essex Institute of Salem, Massachusetts, and the San Diego Society of Natural History. See also American Association of Museums, *Handbook of American Museums* (Washington, 1932).

8 Callie Hull, Mildred Paddock, S. J. Cook, and P. A. Howard, comps., *op. cit.* See "Research Funds" and "Medals and Prizes" in the index.

associations of associations and a wide variety of looser affiliations as well.

Several methods of integrating scientific societies have been developed. Oldest of them is the formation of a "merger" society by the outright union of two already existing societies. This has often meant no more than the swallowing up of a weak or dormant society by one more vigorous, or, again, instead of absorption of one body by another, two organizations have merged on terms of real equality. We have noticed how the American Philosophical Society was formed by the union of two societies, to which a third also was presently added; and we have mentioned in this paper many societies which were the result of union and absorption. Hundreds of other instances could have been mentioned, and practically all of the larger societies are a conglomerate product, many of whose very names bear witness to their hybrid character.[9] Upon the reorganization of the Brooklyn Institute of Arts and Sciences in 1887-1888, a number of independent societies, such as the Brooklyn Microscopical Society, the Brooklyn Entomological Society, and the Linden Camera Club of Brooklyn, became absorbed in the society as departments of that body.

In addition to amalgamations on the "fusion principle," described above as "mergers," there have come about a number of affiliations on the "federation principle," in which each of the societies maintains its own identity but all closely cooperate.

A "horizontal combination" of scientific societies may be defined as an affiliation of societies devoted to the same subject, for example, the American Medical Association, which is a federation of regional, state, and local societies devoted to one particular branch of science, namely medicine. Other similar combinations which have been effected are the National Association of Audubon Societies for the Protection of Wild Birds and Animals,[9a] the American Chemical Society, the American Institute of Mining and Metallurgical Engineers, the American Osteopathic Association, the American Pharmaceutical Association, the American Society of Civil Engineers, the American Society of Naturalists, the Archaeological Institute of America, the United Engineering Society, the Associated Outdoor Clubs of America, the Federation of American Societies for Experimental Biology, the Union

[9] For example, American Philosophical Society Held at Philadelphia for Promoting Useful Knowledge, Academy of Medicine of Cleveland and Cuyahoga County Medical Society, and Society of Automotive Engineers.

[9a] Name changed in 1940 to the National Audubon Society.

of American Biological Societies, and the Congress of American Physicians and Surgeons.

A "vertical combination" of scientific societies may be defined as an affiliation of societies devoted to different subjects. For example, the National Research Council, which is a federation of representatives from societies devoted to different branches of learning. The formation of national research councils composed of representatives from a group of federated societies has become one of the most important ways evolved for bringing about the coordination of scientific societies. This form of loose affiliation secures valuable cooperation, while allowing the societies their own autonomous government.

The "horizontal-vertical" combination of scientific societies is an affiliation in which both societies devoted to the same subject and societies devoted to different subjects are affiliated; for example, the American Association for the Advancement of Science. It is a horizontal combination in that it is an association representing some thirty state academies, devoted to one and the same object, though a broad one, the promotion of science. It is a vertical combination in that it is an association representing societies devoted to different branches of learning such as the American Chemical Society and the American Psychological Association. With the American Association for the Advancement of Science there are officially associated about 150 independent scientific organizations, 100 of which are officially affiliated.

Other types of integration have been tried. Joint meeting of societies in related fields, periodical congresses, and conventions have been resorted to and are becoming increasingly common. Relations between parent and daughter societies often long persist. Finally, there exists an unofficial and intangible, yet strong, bond of association between all the scientific societies—a sense of sisterhood—fostered by constant intercommunication and the exchange of proceedings, and by a belief in an ultimate unity of all science.

In some cities there are special associations of societies, as the Associated Technical Societies of Detroit. There is a remarkable degree of cooperation on the part of many societies existing together within municipalities even when there is no such formal affiliation as that mentioned above. In particular have municipal academies of science served as the coordinators of more specialized local societies. Thus, the Washington Academy of Sciences acts as the head of a large number of affiliated societies of Washington, D. C., which have become federated through it.[10]

State academies of science serve to coordinate the local societies within a state, and the latter bodies often find it convenient to meet with the academies. Regional scientific efforts, in addition to being coordinated by the regional sections which some national scientific societies have established, have also been unified to a degree by a number of specialized regional scientific societies, such as the Audubon Association of the Pacific.[11] It has been a definite policy of those national societies whose membership was large, and where the nature of the field made it particularly desirable to organize local units, to federate such units with the national body. The American Chemical Society, the American Medical Association, and the National Association of Audubon Societies for the Protection of Wild Birds and Animals[11a] furnish good examples. When the local societies did not precede the national, but rather the need for them arose later, the problem was sometimes solved by founding regional sections; for example, the Cordilleran Section of the Geological Society of America was founded in 1899 for the convenience of members residing in the far west. The

[10] These include the Chemical Society of Washington (Washington Section of the American Chemical Society) ; Washington Section, American Institute of Electrical Engineers; Washington Section, American Society of Mechanical Engineers; Anthropological Society of Washington; Archaeological Society of Washington; Botanical Society of Washington; Columbia Historical Society; Entomological Society of Washington; Geological Society of Washington; Helminthological Society of Washington; Medical Society of the District of Columbia; National Geographic Society; Philosophical Society of Washington; Washington Section, Society of American Foresters; Washington Society of Engineers; Washington Branch, Society of American Bacteriologists; Washington Post, Society of American Military Engineers.

An account of the early constituent societies and of the founding of the Washington Academy of Sciences may be found in Washington Academy of Sciences, *Proceedings*, I (1899) , 1-14.

[11] Other examples are the Astronomical Society of the Pacific, the Central Tristate Medical Society, the Inter-state Post Graduate Medical Association of North America, the New England Botanical Club, the New England Dermatological Society, the New England Federation of Natural History Societies, the New England Ophthalmological Society, the New England Otological and Laryngological Society, the New England Pediatric Society, the New England Society of Psychiatry, the New England Surgical Society, the North Central Electric Association, the Northwest Scientific Association, the Southern Society for Philosophy and Psychology, the Southern Surgical Association, Mid-South Post Graduate Medical Assembly and the Tri-state Medical Association of the Carolinas and Virginia.

[11a] Name changed in 1940 to National Audubon Society.

combination movement, which has been described in this chapter, has progressed much farther than in many countries. It is commonly realized that tremendous progress in this direction has been achieved in recent decades, adding greatly to the usefulness of the societies.

Cooperation with other agencies of research plays a considerable part in the activities of practically all the scientific societies. Particularly intimate have been the relations between the universities and the scientific societies of the United States. Indeed, some of the societies grew directly out of informal clubs within university faculties. College professors, professionally devoted to the pursuit of learning and able to contribute a solid body of knowledge, give a stability and permanence to the membership of the societies such as has been furnished by no other group. Not only have the universities been chiefly responsible for preparing a personnel capable of directing the societies, but they have aided in many other ways. University buildings equipped with apparatus for demonstration purposes are the favorite places for holding scientific gatherings, and the academic atmosphere of these quarters, congenial as it is to scientists, lends congeniality to intercollegiate conferences. The authorities of many universities, in order that their faculties may have the opportunity to make regular contacts with other men in the same line of teaching and research, have passed regulations to the effect that their universities will once each year pay the transportation expenses of each member of professorial rank to the meeting of a scientific society or convention.

The university professors are relied upon to furnish many of the articles appearing in the scientific publications. Moral pressure to undertake research, brought to bear upon the faculty members of many universities, is in part responsible for this. Almost invariably the task of editing a scientific journal has devolved upon a group of academic men. Reports of the work done at university laboratories and museums are often published in journals of the societies, while occasionally the reverse is true and publications financed by universities have been heavily indebted to scientific societies.

Frequently, the library of a scientific society is deposited in perpetuity in some large library,[12] often that of a university, upon agreement of the latter always to keep the collection accessible to the members of the society. Similarly, the museums of scientific societies

[12] See, for example, the "Memorandum of Agreement between the Case Library of Cleveland, Ohio, and the Geological Society of America, Relating to the Custody of the Geological Society Library," Geological Society of America, *Bulletin*, VI (1895), 427.

are frequently deposited with the museum of some university. The greater resources of a university often result ultimately in the total absorption of a society's original collection by gift to the university upon provision for its maintenance.

The leading student of the history of seventeenth century scientific societies, Martha Ornstein, found no small amount of antagonism existing between the aims of the universities as then constituted and devoted to theology and the arts and the newly founded scientific societies, which were established by men rather disdainful of the hidebound universities.[13] There was no such struggle of aims and purposes between the scientific societies and the universities in America. There was friction between the classicists and those who desired to put science into a more prominent place in the curriculum of the universities, and there were occasional flares of opposition to the scientific spirit by some theologians; but, from the first, the scientists in the universities made common cause with those in the scientific societies. Indeed, the two groups were very much identical in America. The universities and the scientific societies are not today antagonistic organizations; rather they are supplementary, and, as shown above, their activities are intertwined in numerous ways. Scientific societies bring together professors of the same subject in different colleges. Scientific gatherings are the classes where the professors are schooled, and the scientific journals are their textbooks. The relations between the two types of bodies for the promotion of learning must ever be intimate. Professional organizations such as the Engineers' Council for Professional Development, the American Medical Association, and the American Chemical Society, in the interest of maintaining standards, have undertaken the inspection and accrediting of educational curricula in their respective fields.

Since American scientific societies are mostly private bodies consisting of voluntary associations of students of science united together for the promotion of science at their own expense, they have had, in general, few formal relations with the state and federal governments. A few societies, however, deserve mention because of their governmental connections. The National Academy of Sciences has received financial aid from the national government and it has, in turn, furnished the government with advice on technical matters from time to time. The rapid development of government scientific bureaus and

13 Martha Ornstein, *The Role of Scientific Societies in the Seventeenth Century* (The University of Chicago Press, Chicago, 1928), Part III, "The Learned Societies and the Universities," *passim*.

the independent position of American scholars in relation to the government have made it impossible for the National Academy of Sciences to assume a role of importance as an aid to the government analogous to the great government-patronized academies of Europe. The National Research Council worked in close cooperation with the technical bureaus of the government for the solution of war problems. The Smithsonian Institution has always been a valuable connecting link between the national government and American and foreign societies.[14] There is usually no antagonism between the government technicians and the members of scientific societies, and, indeed, the two groups merge. As individuals, thousands of members of scientific societies have aided the government bureaus in the capacity of employees or by giving their advice to the government scientists. This has been facilitated by the fact that many of the national scientific societies have their headquarters in Washington.

Some state academies of science have received a measure of state financial aid, and some have been furnished with meeting rooms in state capitols or other government buildings. In return many members of state academies have given freely of their time and services in the solution of the scientific problems of the state. However, there is little that might be called state control of scientific societies; that would be foreign to the American traditional policy of *laissez faire* in regard to learned societies. On numerous occasions presidents of the United States, state governors, and other high officials have complimented American scientific societies on their unselfish cooperation in helping to solve the nation's scientific problems and on their general worth in raising the educational level.

Younger than university and scientific society, more thoroughly devoted to the "practical" than either, the research foundation or institute stands forth as a human institution of great unknown significance for mankind. America, where men have gathered wealth faster than anywhere else, has been the place where endowments have built up these institutions as nowhere else in the world. Their relation to the scientific societies has been only partially worked out as yet.

---

[14] "The most important service by far which the Smithsonian Institution has rendered to the nation from year to year since 1846—intangible, but none the less appreciable—has been its constant cooperation with the Government, public institutions, and individuals in every enterprise, scientific or educational, which needed its advice, support, or aid from its resources." G. B. Goode, *"The Origin of the National Scientific and Educational Institutions of the United States."* In *Annual Report of the Smithsonian Institution, 1897*, II, 320.

Research foundations and institutes have generally deemed it more advisable to prosecute scientific research with their own laboratory facilities than to confer "grants in aid" on universities or to call on a diffuse organization like a scientific society to study a particular scientific problem. The National Research Council, however, has administered a number of research funds. An interesting recent development is the foundation of the Physical Institute, mentioned in the preceding section, by a number of societies devoted to branches of physics. Many of the leading men in such organizations as the Smithsonian Institution, the Carnegie Institution of Washington, and the Rockefeller Institute have contributed much to scientific societies, and these institutions have cooperated with scientific societies in many ways.

The industrial research scientist is ever pushing harder on the heels of the university pure science investigator as a discoverer. Large corporations like the Standard Oil Company, the General Electric Company, Edison Electric, Du Pont, U. S. Steel, General Motors, Ford Motors, and the Eastman Kodak Company maintain research laboratories where well-trained scientists make discoveries primarily for the sake of industry, but which often turn out to be of general scientific importance as well. Some corporations have cooperated in the founding of technical societies like the American Manganese Producers Association and the American Drug Manufacturers Association. Other corporations send representatives to societies which are less exclusively scientific trade associations. Engineering societies have rendered much aid to industrial research; in return, industry has contributed generously in membership and money to the support of engineering and technical societies. Valuable as such an alliance of industry and pure science is, occasional fear has been expressed lest utilitarian interests capture the general science societies and subvert research to their own ends. A note of warning has been sounded that a proper balance must be kept between "science for the sake of science" and a narrow industrial outlook. What must be preserved is enough freedom of action on the part of research men to enable them to make discoveries which may have ultimate, even if not immediate, value.

In the preceding sections, the scientific societies were considered in relation to advanced scientific workers. In this section, the attempts of the societies to reach a public which is only casually interested in science are discussed.

The American public has been permitted and encouraged to partake of the benefits conferred by the societies to a degree which amazes and delights Continental visitors. Meetings are usually open to the public, the publications of the societies may be readily purchased, responsible citizens are generously allowed the use of libraries, prizes are offered for general competition, and all persons are invited to visit the museums, herbaria, and laboratories maintained by the societies.

The societies have rendered valuable service to the public in reporting scientific events and discoveries. The American Association for the Advancement of Science deserves special mention because of its Press Service and its publication, *Science*. The leading scientific societies have come to cooperate increasingly with press syndicates and radio networks for presenting science to the public in interesting and non-technical language. Medical associations, in particular, are to be commended for sponsoring regular features of health education programs. With the facilities of communication ever becoming greater, the societies are continually engaging the attention of a wider public audience.

A result of the coming of highly specialized research has been that it has become increasingly difficult for the amateur, lacking in training and equipment as he often is, to make really worth-while contributions to science. Yet, even in this age of great telescopes, cyclotrons, and electronic microscopes, leaders in many fields of science are summoning the public for assistance and would count it a real loss to get no response. Granting that synthesis and interpretation must be left to the specialists, there still is said to exist plenty of opportunity for amateurs in the way of collecting specimens and making observations for which the expert has not the time. Thus amateurs can hunt for comets, or record observations of variable stars or meteors, or collect minerals of botanical or zoological specimens. Amateur mathematicians have, in America as elsewhere, been in the vanguard of those making real contributions to learning. Franklin and other early founders of American scientific societies, living in an age of far less specialization, had these things in mind, and the tradition has fortunately persisted. What a local scientific society might accomplish was well expressed by Joseph Henry, secretary of the Smithsonian Institution:

> Such an association is an important organization for the advancement of adult education and the diffusion of interesting and useful knowledge throughout a neighborhood. The society must, however, be under the care of a few enthusiastic and industrious persons; it should adopt the policy of awakening and sustaining the greatest number of persons possible in its operations, and for this purpose the meetings must be rendered attractive; care should be taken to provide a series of short communications on various subjects, on which remarks should be invited after they have

been read; all should be pressed into the service, and each solicited to contribute something, the object being to make the special knowledge of *each* the knowledge of *all*.

. . . The importance of an establishment of this kind should not be confined to the mere *diffusion* of knowledge. It should endeavor to *advance* science, by co-operating with other societies in the institution and encouragement of original research. Thus it can make collections of the flora and fauna, of the fossils, rocks, minerals, etc., of a given region, of which the location of the society is the center, and thereby contribute essentially to the knowledge of the general natural history of the continent. It can also make exploration of ancient remains and collect and preserve specimens of the stone-age, which still exist in many parts of our country, and to which so much interest is at present attached. Further, it can induce its members to make records of meteorological phenomena, many of which, of great interest, can be made without instruments such as the times of the beginning and ending of storms, the direction of the wind, the first and last frost, the time of sowing and harvesting, the appearance and disappearance of birds of certain kinds, the time of the blossoming and ripening of various fruits, etc., and, as soon as the means of the establishment will afford, a series of meteorological observations should be entered upon with a perfect set of instruments.[15]

In evaluating the work of the scientific societies and institutions, one of the most competent judges, J. McK. Cattell, has said:

The educational activity of a nation is not confined to its schools. Societies, journals, museums, laboratories, and other institutions devoted to the advancement and diffusion of knowledge are an important part of the educational system of the United States. These agencies are on the one hand for the use of those who teach, and thus represent the most advanced educational work. On the other hand they extend the range of education widely among the people.[16]

The remainder of this chapter is devoted to a few general considerations of the relation of American men of science to the scientific societies. Early in the nineteenth century the men who held the newly created professorships of science came to supplant amateurs as the presiding geniuses of the scientific societies. Such a turn of affairs was further hastened by the trend toward specialization that gained momentum throughout the century and placed the amateur at a serious disadvantage.

Since, in America, membership in scientific societies is generally on

15 Joseph Henry, "On the Organization of Local Scientific Societies. A Letter from Professor Henry, Secretary of the Smithsonian Institution." (Printed in *Annual Report of the Smithsonian Institution for 1875*, 217-219.)

16 J. McK. Cattell, "Scientific Societies and Associations," N. M. Butler, edit., *Education in the United States* (Albany, 1900), II, 865.

the basis of professional interest rather than on a purely honorary one, the American scientist is left free to form such affiliations as may best suit him at a given time. The fact that the cost of membership in American scientific societies is relatively low has also been a factor in enabling the impecunious scholar to participate in organizations. Nor should the role of the colleges and universities in preparing their graduates for entrance into the scientific societies be minimized. Professional men other than those associated with universities form a second principal group of those interested in the scientific societies.

Physicians have for several centuries been accustomed to band together to guard the ethics of their profession, and this is truer of the present time than ever before.[17] As to how far a particular organized body may go in attempting to speak for the profession as a whole is a matter yet to be decided. The question as to whether a voluntary professional society may be construed by the courts as a "combination in restraint of trade" is one that is now to the fore in several legal battles, with the American Medical Association under fire in its attempts to restrict "socialized medicine," either in the interests of public safety and professional ethics or in its own selfish interests, depending on how one looks at the matter.

Engineers, chemists, industrial technicians, and research workers of corporations, institutes, and foundations have come increasingly to figure in the societies. This class has turned more and more to the professional societies as the guardians of the economic interests in a

---

[17] It was, of course, practically necessary for a physician to join a local medical society. These societies kept lists of members of the profession in good standing, mercilessly fought imposters and incompetency, worked for higher standards of medical education, and influenced state legislation.

"The medical profession is in some respects the most powerful labour union in the world. Not only does it exercise complete control over training for medical practice and enforce an ethical code which no member of the profession may violate with impunity, but it also enjoys a legal monoply of work within its field and special legal protection for its members in the performance of their duties. It fixes its own rate of wages, either by establishing uniform minimum rates applicable within the territory of a local association, or by standing solidly behind the individual practitioner who establishes rates for himself. It entirely controls the subsidiary profession of nursing, it has established close working relations, sometimes equivalent to a monopoly, with the profession of pharmacy and the administration of public health laws, and it exercises a large and in certain directions a controlling influence in the business of manufacturing drugs, medical preparations, surgical instruments, and hospital supplies. Throughout the profession the spirit of solidarity is high, and membership in the profession is an honour everywhere recognized." William MacDonald, *The Intellectual Worker and His Work* (New York, 1924), 248. By permission of The Macmillan Company, publishers.

society in which they have to compete with organized capital and labor.[18]

The feminine element played no very large part in the history of American scientific societies until the twentieth century. The American Association for the Advancement of Science has permitted almost from the outset the election of women as members:

> At the first meeting, 1848, a resolution was voted that members might introduce ladies at the sessions. After this gracious concession to feminine curiosity, women are not recognized in the records until 1858. At the 12th meeting, in Baltimore, the standing committee proposed some changes in the rules, one of which was to admit ladies as associate members. The only subsequent mention of the proposed amendments is a sentence in volume 13, page 364, which reads, "Resolved, that no action is necessary in regard to the motion to admit ladies as members, inasmuch as two ladies have already been admitted." Examining the former lists of members it is found that in 1850 two women were elected, or at least named on the roll, Maria Mitchell, Nantucket, Massachusetts, and Margaretta Morris, Germantown, Pennsylvania. Either the two women had been *very quiet for eight years* or else the standing committee was not alert.[19]

Women have been admitted to other scientific societies from time to time, and membership in most of them is now open to women. In a few instances women have created societies of their own, for example, the Society of Woman Geographers, the Women's Medical Society of New York State, and there was established the Association to Aid Scientific Research by Women, designed specifically to aid financially women in scientific research work. American women, however, seem yet to be far behind the men in scientific achievement. Of the two hundred and fifty names starred in the fifth edition of the *Biographical Directory of American Men of Science* (1932) for distinguished scientific achievement, there were only three women. In the total listing of about twenty-two thousand names in this work, only about seven hundred and twenty-five women were included, and of these, 22 per cent were working in the field of psychology alone.[20] Contemporaneously, in the National Academy of Sciences, with a membership of about two hundred and fifty, there are only two women. The ladies of the nineteenth century could point to educational dis-

18 See *ibid.*, especially ch. VI, "The Learned Professions," and ch. XI "The Organization of Intellectual Workers," and ch. XII "Typical Organizations of Intellectual Workers."

19 H. L. Fairchild, "The History of the American Association for the Advancement of Science," *Science*, n. s., LIX (1924), 511.

20 J. McK. Cattell, "The Distribution of American Men of Science in 1932," *Science*, n. s., LXXVII (1933), 267.

crimination against them, or, to put it more mildly, they did not possess equal opportunities until toward the close of the century. The reasons for such great contemporary disparity seem somewhat obscure to the author. It may be explained in part by the fact that many women researchers have had the results of their studies issued under the name of some research institution or have contributed to the writings of some man, who, as director of research investigations, has secured publication of the studies, which henceforth have become associated with his name rather than with that of the assistant who actually did the detailed work. Hence the extent of their contributions has not always been fully apparent.

Several groups in the United States have found it expedient to organize scientific societies along "racial" lines. Thus, Der Naturhistorische Verein von Wisconsin was formed in 1857, the American Society of Swedish Engineers was organized in 1888, the Swedish Engineers' Society in 1908, and the Russian Association of Engineers in 1923.

Frequently scientists have found it desirable to hold membership in several societies. This has become increasingly common as the progressive differentiation of the fields of science has led to the formation of more and more highly specialized scientific societies. A typical man of science might belong to the American Association for the Advancement of Science, to the main national society in his field, to a specialized society dealing with his particular interest within that field, to state and local professional societies. He might be active in state and local municipal academies of science as well. It is not impossible to find men who have become members of a dozen or more learned and scientific societies. While it is impracticable to measure statistically this multiple membership for the country as a whole at any particular epoch, its prevalence may readily be discovered by perusal of any directory of American men of science. The results of one such investigation of multiple membership in the scientific societies of Washington, D. C., at the close of the last century are not without interest and are summarized in the tables on pp. 189 and 190.

There is probably no better way to make up a list of the principal scientists of the United States than by consulting the membership lists of the principal national scientific societies. Compiling his monumental editions of *American Men of Science*, J. McK. Cattell surveyed the membership of leading scientific societies. Several sets of figures compiled by him for showing the distribution of American scientific workers in various fields throw some light on membership in scientific

MEMBERSHIP IN THE JOINT SOCIETIES OF
WASHINGTON, 1889-1897*

	1889	1890	1891	1892	1893	1894	1895	1896	1897
Anthropological	152	181	210	214	222	205	191	149	138
Biological	145	155	154	149	195	190	134	166	156
Chemical	58	69	82	90	97	102	101	117	89
Entomological	...	...	...	53	109	38	34	34	41
Geographic	204	236	397	473	682	619	778	1040	1040
Geological	...	...	...	...	...	137	156	151	144
Philosophical	246	248	260	250	250	233	215	201	120
Total membership	805	889	1103	1229	1555	1524	1609	1858	1728
Total of persons	579	662	851	962	1259	1138	1234	1536	1450

* *Directory of the Scientific Societies of Washington; Comprising the Anthropological, Biological, Chemical, Entomological, Geological, National Geographic, and Philosophical Societies, 1897* (Washington, 1897), 51. By comparing the figures in the total membership column with those in the column showing the total of persons, one readily sees that throughout the period about one-fifth or more of the persons had membership in more than one society.

THE NUMBER OF PERSONS IN EACH OF THE JOINT SOCIETIES OF WASHINGTON
WHO ARE MEMBERS OF ONE, TWO, THREE, ETC., SOCIETIES*

	1	2	3	4	5	6	7	Sum
Anthropological	81	31	14	7	3	1	1	138
Biological	87	45	10	9	3	1	1	156
Chemical	68	12	5	2	0	1	1	89
Entomological	15	20	1	4	0	0	1	41
Geographic	881	113	29	12	3	1	1	1040
Geological	68	49	13	9	3	1	1	144
Philosophical	47	38	21	9	3	1	1	120
Sum	1247	308	93	52	15	6	7	1728
No. of persons	1247	154	31	13	3	1	1	1450

* *Directory of the Scientific Societies of Washington; Comprising the Anthropological, Biological, Chemical, Entomological, Geological, National Geographic, and Philosophical Societies, 1897* (Washington, 1897), 51.

societies, reflecting, for example, the rise of psychology and membership in psychological societies. The distribution of four thousand names in Cattell's *American Men of Science* (first edition, 1906), was as follows: mathematics, 340: physics, 672; chemistry, 677; astronomy, 160; geology, 444; botany, 401; zoology, 441; physiology, 105; anatomy, 118; pathology, 357; anthropology, 91; psychology, 104.[21] In Cattell's se-

21 J. McK. Cattell, "A Statistical Study of American Men of Science: the Selection of a Group of One Thousand Scientific Men," *Science*, n. s., XXIV (1906), 660.

lected thousand scientific workers, as of data referring approximately
to 1903, the number in each science being taken as roughly proportional
to the total number of investigators in that science, the distribution
was as follows: chemistry, 175; physics, 150; zoology, 150; botany, 100;
geology, 100; mathematics, 80; pathology, 60; astronomy, 50; psychology, 50; physiology, 40; anatomy, 25; anthropology, 20.[22]

Cattell, who attempted the first complete survey of the scientific
activity of the United States at a given period, gives the following
table showing the number and distribution of American men of science
as of January 1, 1903.

THE NUMBER OF AMERICAN MEN OF SCIENCE AND THEIR
DISTRIBUTION AMONG THE SCIENCES*

	Special Societies	Fellows of Association	Members of Academy	University Professors	Who's Who
Mathematics	375	81	1	136	46
Physics	149	167	23	105	73
Chemistry	1933	174	12	143	166
Astronomy	125	40	12	41	51
Geology	256	121	13	55	174
Botany	169	120	7	57	70
Zoology	237	146	17	83	131
Physiology	96	10	2	53	25
Anatomy	136	10	0	56	18
Pathology	138	14	5	68	56
Anthropology	60	60	3	4	37
Psychology	127	40	1	37	21
Total	3801	983	96	838.	868

* J. McK. Cattell, "A Statistical Study of American Men of Science: the Selection of a Group of
One Thousand Scientific Men," *Science*, n. s., XXIV (1906), 660. The author has somewhat abridged
Cattell's table.

It is difficult to pick a list of the outstanding names associated with
the rise of scientific societies in the United States, for in each there
was a whole host of men, mostly university professors, engaged in the
founding of societies, who were about equally prominent in their own
day, but whose names are passing into oblivion with the years. Only
a few names can here be mentioned which, for various reasons, seem
to stand a little higher than the others. The great name in the eighteenth century was Benjamin Franklin, father of the American Philosophical Society. In the first half of the nineteenth century the foremost

22 *Ibid.*

figure was Benjamin Silliman, founder of the *American Journal of Science* in 1818, a remarkable publicist whose labors were especially effective in securing the assistance of other writers in the work of building up scientific publications, and who was a lifelong leader in the movement to found scientific societies.

Two names deserve especial mention for the latter half of the nineteenth century. The first is Joseph Henry, secretary of the Smithsonian Institution from 1846 to 1878, and president of the National Academy of Sciences from 1868 until his death in 1878. The other name is Simon Newcomb, author of over three hundred scientific papers, president of the American Association for the Advancement of Science 1877-1878, and member of many American and foreign scientific bodies. Mention ought also to be made of G. B. Goode, author of the first extended historical monograph on American scientific societies. James McKeen Cattell, editor of numerous scientific publications and a writer for scientific organizations, George Ellery Hale, foreign secretary of the National Research Council and prominent in international scientific organizations, and Vernon L. Kellogg, former permanent secretary of the National Research Council, deserve mention as outstanding contemporary authorities on American scientific societies and on international scientific organizations. The most prominent of the living historians of American scientific societies is probably Herman Leroy Fairchild, who has written histories of the American Association for the Advancement of Science, the Geological Society of America, the New York Academy of Sciences, and other societies.

Men of the caliber mentioned above have produced a long list of notable papers and have made a number of outstanding discoveries. The average member of an American scientific society might be expected to attend meetings with none too exemplary regularity, to take a small part in the discussions, and even to contribute an occasional paper; but, if his importance is obscured by the more brilliant members, it is not to be forgotten that the interest and support of the rank and file, in the last analysis, have made the societies possible.

In conclusion, the scientific societies are among the foremost agencies for advancing scientific research. The great national scientific societies and councils bring together scientific workers from all parts of the nation, and, made up as they are of men from the universities, the professions, research institutions, government bureaus, industrial laboratories, and from the scientifically inclined among the public at large, they afford the best means of coordinating every facility for research throughout the nation. Similarly, state and municipal academies of science and other scientific bodies bring about the coordination of research facilities within smaller areas. Not only have the scientific societies—national, state, and local—been the means of stimulating the esoteric devotees of science, but through their museums, libraries, and open meetings they have exerted a wide educational influence upon the general public. The study of the history of American science involves not only discoveries and the men who made them but also an appreciation of the role played by associations of scientists in stimulating research and in affording the opportunities and media for publicizing and conveying the achievements of science.

With the coming of total war, the scientific societies, as has been mentioned in the preceding chapter, have undergone profound changes. On the one hand, "pure research" must to a considerable extent be put aside until the return of peace. On the other hand, "applied science," such as industrial chemistry, applied physics, and surgery—to take but a few examples—receives a tremendous impetus. The principal scientific societies of the nation have all been called upon in one way or another, often through the National Roster of Scientific Personnel, to play their part in the "all-out" war effort.

When peace returns, the scientific societies of America, and of every other country, will be faced with a challenging call to exert leadership in a new international intellectual reconstruction of the world.

Chapter VI

THE ATOMIC AGE, 1945-1955

Science in America since the close of World War II has been conditioned by such things as the impact of the dawning atomic era, swiftly evolving technological changes, the Cold War, the Korean War, Antarctic exploration, supersonic flight, projected space flight, and mass attack on disease. This period has seen the rise of many new scientific societies in the physical, the earth, the biological, and the social sciences. The societies have had increasing occasion to integrate their activities with each other and with those of governmental agencies. This period has also seen specialization carried to new heights. Significantly, and despite much international tension, it has seen international organization and cooperation in science carried on to an unprecedented degree. This has come about especially through the activities of the United Nations.

In the post-war decade the general national scientific organizations were faced with a variety of problems. Among these were: resumption of normal peacetime activities, integrating a host of new specialized organizations, cooperating with projects of a vastly expanded federal government, meeting the demands of a rapidly increasing population, and taking part in world-wide cooperative scientific projects.

By 1955 the American Association for the Advancement of Science, with 265 affiliated and associated societies including forty-two academies of science,[1] had an aggregate membership of more than 2,000,000[2] and was by far the largest and most important scientific organization in the world. Yet the A.A.A.S. faced grave problems. One of the most acute was the growing national shortage of scientific personnel, which was made the theme of the annual Sigma Xi address at the 1955 A.A.A.S. meeting by Dr. James R. Killian, Jr., president of M.I.T. Another problem concerned the attitude of men of science toward racial segregation. There have been nine meetings of the A.A.A.S. in the South: Charleston, 1850, Nashville, 1877, New Orleans, 1905, Atlanta, 1913, Nashville, 1927, New Orleans, 1931, Richmond, 1938,

[1] *New York Times,* December 27, 1955.
[2] *Ibid.,* December 31, 1955.

Dallas, 1941, and the Atlanta meeting of 1955. The section on anthropology boycotted the latter meeting "in protest because the A.A.A.S. was meeting at a place where there is segregation as to hotels, restaurants and public transportation."[3] Other organizations, including the Eastern Psychological Association, adopted resolutions protesting the Atlanta choice. At the closing meeting of the A.A.A.S. Council members present at Atlanta, it was voted to submit to the full council of 328 members the question of deciding whether to hold future A.A.A.S. annual meetings in any city in which racial segregation is practiced.[4]

The Committee on the Social Aspects of Science, appointed at the December meeting in Atlanta in 1955, and reporting at the New York meeting in 1956, found itself faced with a situation in which the world now lives in a "new social revolution, the Scientific Revolution, even greater in its effect than the Industrial Revolution," and one in which "scientific organizations may be obliged to accept a social responsibility commensurate with the importance of the social effect of science."[5] The Committee noted that "in marked contrast to other associations, scientific societies seldom consider the social and economic position of their group."[6] It closed its report by stressing "the pressing need that scientists concern themselves with social action,"[7] and that "in this situation the A.A.A.S. carries special responsibility."[8]

In 1951 the American Association for the Advancement of Science

[3] *Ibid.*, December 27, 1955.

[4] *Ibid.*, December 31, 1955. The text read as follows:

"The A.A.A.S. is a democratic association of all its members; no one is barred from election because of race or creed. All members are privileged to cooperate freely in the fulfillment of the Association's high objectives, which are the furtherance of science and human welfare. No member is limited in his service because of race or creed.

"In order that the Association may attain its objectives it is necessary and desirable that all members may freely meet for scientific discussions, the exchange of ideas and the diffusion of established knowledge. This they must be able to do in formal meeting and informal social gatherings. These objectives cannot be fulfilled if free association of the members is hindered by unnatural barriers.

"Therefore be it resolved that the annual meeting of the American Association for the Advancement of Science be held under conditions which will make possible the satisfaction of those ideals and requirements."

[5] "Society in the Scientific Revolution," *Science*, CXXIV (1956), 123. Discusses A.A.A.S., Interim Committee on the Social Aspects of Science, *Provisional Report.*

[6] Interim Committee on the Social Aspects of Science, Report to the Council of the American Association for the Advancement of Science, printed in *New York Times*, December 31, 1956, C-6.

[7] *Ibid.*

[8] *Ibid.*

joined with seven major national scientific groups[9] in organizing the Scientific Manpower Commission, to "study the nation's needs for scientists in education, industry, government service and the armed forces."[10] The Commission is a "good example of cooperation among the sciences and between scientific associations and industry."[11] The Scientific Manpower Commission has worked closely with the Engineering Manpower Commission. They have published jointly the *Engineering and Scientific Manpower Newsletter*.

The Scientific Manpower Commission has worked especially as a consultant to federal agencies that deal with manpower problems. Its advice has been sought in "framing legislation, in formulating policy, and in designating occupations in which manpower shortages are of such a severity as to recommend that military service be required only when there are specific military requirements for the technical skills of men in those occupations."[12]

The experience during World War II of the four national research councils (American Council of Learned Societies, American Council on Education, National Research Council, and Social Science Research Council) in seeking out specialists for government and industry revealed both serious shortages in personnel and the importance in the national defense effort of highly trained individuals. The Conference Board of the Associated Research Councils requested and received from the Rockefeller Foundation in 1947 a grant to make a preliminary survey. This work was carried out by the Office of Scientific Personnel of the National Research Council. Additional grants in 1949 and 1952 made it possible for a Commission on Human Resources and Advanced Training, set up by the councils, and functioning with a professional staff directed by Dael Wolfle, to prepare and publish its report: *America's Resources of Specialized Talent*.[13] Certain conclusions in this report are of particular interest to the student of recent science and scientific societies in America.

[9] These were: American Chemical Society, American Geological Institute, American Institute of Biological Sciences, American Institute of Physics, American Psychological Association, Federation of American Societies for Experimental Biology, and the Liaison Committee of the Mathematical Societies. The Scientific Manpower Commission, *Science*, CXVII (1951), 495-496.

[10] *Ibid.*, 495.

[11] Dael Wolfle, "Scientific Manpower Commission," *Science*, CXXII (1955) , 1213.

[12] *Ibid.*

[13] *America's Resources of Specialized Talent; a Current Appraisal and a Look Ahead.* The Report of the Commission on Human Resources and Advanced Training. Prepared by Dael Wolfle, Director (New York, 1954).

The number of people holding doctors' degrees in the sciences has approximately doubled since 1940. The number of research scientists and engineers and the total membership of scientific societies increased in approximately the same relative amount. All these estimates support the generalization that the nation's total supply of scientists doubled between 1940 and 1950.[14]

President Eisenhower, reacting to reports that the Soviet Union was training more scientists and engineers than the United States, set up on April 3, 1956, a nineteen-member National Committee to foster the training of such specialists. Among those named to the committee was Dr. Howard M. Jones of Harvard, and chairman of the American Council of Learned Societies.

The American Academy of Arts and Sciences in 1955 began to issue a new journal called *Daedalus,* designed for the publication of articles of broad interest. Following a session at the sixty-ninth annual meeting of the American Historical Association, December 30, 1954, which recommended microfilming the entire collection of more than three hundred thousand pages of the Adams Papers, the Academy voted a thousand dollars from its Permanent Science Fund to further the project; this money, with two thousand dollars voted by the American Philosophical Society, enabled the Massachusetts Historical Society to buy equipment and get the work started, and thus "the three oldest learned societies in the United States joined in an effort for the common good."[15]

The International Conference on Scientific Information met November 11, 1956, at the National Academy of Sciences to consider problems facing the custodians of scientific reference services.[16]

Quite a number of new national organizations more or less generally devoted to the advancement of science came into existence in the forties and fifties. The Science Clubs of America (1941) were especially active in pursuit of the discovery and development of scientific talent among youth. The Southern Association of Science and Industry, Incorporated, was commenced in 1941 and incorporated in 1945. The Polish Institute of Arts and Sciences in America, Incorporated, was in the process of being set up in 1942 and thereafter; the New York Chapter of the Science Society of China was organized in 1943,[17] the National Institute

[14] *Ibid.,* 83.

[15] "Publishing the Papers of Great Men," *Daedalus; Proceedings of the American Academy of Arts and Sciences, LXXXVI,* no. 1 (1955), 47-79.

[16] "Conference on Scientific Information," *Science, CXXIV* (1956), 1259-1260.

[17] *Science, XCV* (1943), 597.

of Science in 1943, the Shell Development Research Club (affiliated with Sigma Xi) in 1945,[18] the American Swiss Foundation for Scientific Exchange, Incorporated, in 1945, the American Soviet Science Society in 1945, and the Association of Research Directors in 1945. The important Federation of American Scientists, organized as the Federation of Atomic Scientists in 1945, took its present name in 1946; it reorganized in 1950 with ten scientific centers throughout the country.[19] The National Microfilm Association dates from 1946, and the Natural Resources Council of America from the same year. The Scientific Research Society of America (Resa), formed at Yale in 1947, and sponsored by Sigma Xi,[20] was organized "primarily to meet the needs of research workers in industrial and governmental laboratories, but it may function also in educational institutions."[21] Dr. George A. Baitsell, Sigma Xi's national executive secretary, said of the new society, "It is believed that the combination of the Scientific Research Society of America in industry with Sigma Xi in educational institutions will provide encouragement and assistance to research scientists all over the United States."[22] The Society for Social Responsibility in Science was formed in 1949.[23] The Southwestern Association of Naturalists (SWAN) was formed in 1953 to further the study of living and fossil plants and animals in the southwestern United States and Mexico. The Society of Technical Writers was formed late in 1953 in Boston "by a group of technical writers and editors engaged in engineering and educational projects who recognized that a professional organization could serve many interests of technical writers."[24]

Although state academies of science had been formed in most of the states prior to 1940, a few were founded then or thereafter. The Montana Academy of Sciences was organized in 1940. The Philadelphia Council of Amateur Scientists, formed in the same year, represents an interesting attempt at integrating amateur societies.[25] The Oregon Academy of Science dates from 1943. In 1944 the Wisconsin Junior Academy of Science was formed under the joint sponsorship of the

[18] *Ibid.*, CI (1945), 373-374.

[19] *Ibid.*, CXI (1950), 78.

[20] D. B. Prentice, "The Scientific Research Society of America," *Supplement to American Scientist*, XXXVII, no. 1 (1949).

[21] D. B. Prentice, "Scientific Research Society of America," *Science*, CXXI (1955), 7A.

[22] *Science*, CVII (1947), 245.

[23] "Society for Social Responsibility in Science," *Science*, CXVIII (1953), 3.

[24] *Science*, CXXI (1955), 328; also see *ibid.*, CXX (1954), 481.

[25] *The Sky*, XII (1940), 16.

University of Wisconsin and the Wisconsin Academy of Sciences, Arts and Letters.[26] The New Jersey Academy of Science was organized in 1954. Regional organizations included: the Southern Association of Science and Industry in 1941, and the Pacific Science Board in 1946.

In 1955 H. R. Skallerup found that there were some thirty-five state academies of science sponsoring publications. After a detailed study of them he reached the conclusion that "many academy journals still serve to a large extent as vehicles for the publication of papers in zoology and botany at the expense of the other sciences."[27] At any rate the earlier complaint of J. McK. Cattell that although "proceedings and transactions were an important function of an Academy of the eighteenth century . . . there is no longer any excuse for presenting researches on utterly diverse subjects in one volume because the authors happen to be members of the same society,"[28] seemed not altogether justified. It is true, however, that what Cattell had earlier favored, namely that "the National society for each science should directly or indirectly have charge of the publications of that science,"[29] had come about.

Among the regional and local societies founded in the period were: Sequoia Natural History Association (1942), the Olympic Natural History Association (Washington, State, 1948), the Shenardoah Natural History Association (1950), the Black Hills Area Association (South Dakota, 1952), Lake Mead Natural History Association (Nevada, 1953), and the Southwestern Association of Naturalists (1953).

Science teachers' associations came into existence in the forties and fifties as follows: California Science Teachers Association (1941), Tennessee Science Teachers Association (1946), Puerto Rico Science Teachers Association (1946), West Virginia Science Teachers Association (1950), Connecticut Science Teachers Association (1952), Science Teachers of New England (1952), and Ohio Science Education Association (1953).

The founding of specialized societies went on at an accelerated rate in the post-war years. The new societies in the physical sciences will first be considered.

The Society for Research on Meteorites became the Meteoritical Society in 1946. The Amateur Astronomers League, which came into

[26] *Science,* CI (1945), 555.

[27] H. R. Skallerup, "Some Aspects of State Academy of Science Publications," *Science,* CXXI (1955), 905.

[28] J. McK. Cattell, "The Academy of Sciences," *Science,* XVI (1902), 973.

[29] *Ibid.*

being in 1946,[30] had come by 1954 to be a federation of some seventy-seven astronomical clubs with more than 4,000 members.[31] Meanwhile numerous new local astronomical societies were being formed.[32] In 1955 the Chicago Astronomical Society and the Burnham Astronomical Society merged under the name of the first.[33] In 1947 the Association of Lunar and Planetary Observers came into being, an organization which collected data not only from American observers but from astronomers in many other countries. New telescope makers' organizations, too, were being formed, including especially the Eastern Telescope Makers Association (New York), 1949.[34] By the 1950's the possibilities of space travel in rocket ships were being explored by a number of societies in various countries including the American Astronautical Society, formed in 1954 "to promote the achievement of space travel, and to disseminate information for scientists and the public."[35] By 1955 this body, now known as the American Astronautical Federation, had come to include as member societies: the Chicago Rocket Society, M.I.T. Rocket Research Society, Pacific Rocket Society, Philadelphia Astronautical Society (1953), Reaction Research Society, and the Detroit Rocket Society. The Intermountain Rocket Society had also been formed in 1952. In July, 1955, when the United States government announced a space flight project, delegates, Americans among them, from some seventeen astronautical societies federated into the International Astronautical Federation, and held at Copenhagen the sixth International Astronautical Congress. November, 1955, saw the American Rocket Society (originally organized in 1930 as the American Interplanetary Society) scheduled to celebrate its twenty-fifth anniversary in connection with the diamond jubilee of the American Society of Mechanical Engineers with which it was affiliated.

The increasing application of mathematics to science, business, and industry led to such organizations as: Inter American Statistical Institute (1940), Duodecimal Society of America (incorporated 1944),

[30] *Popular Astronomy*, LIV (1946), 492-493.

[31] *Science*, CXIX (June 4, 1954), 3A. The sections of the Astronomical League in 1955 were Great Lakes, Middle East, Middle States, North Central, Northeast, Northwest, Southeast, and Southwest.

[32] For example: Astronomical Society (Rocky Mount, N. C.), 1947; Greensboro (N. C.) Astronomy Club, 1948; Memphis Astronomical Society, 1952; Astronomical Society (Roselle Park, N. J.), 1949; Astronomy Club of Wheeling, West Virginia, 1949; Corning (N. Y.) Astronomy Club, 1952; and Amateur Astronomers of Roanoke, 1954.

[33] *Sky and Telescope*, XIV, no. 10 (1955), 419.

[34] *Popular Astronomy*, LVII, no. 3 (1949), 146.

[35] *Sky and Telescope*, XIII, no. 7 (1954), 217.

Association for Computing Machinery (1947), and the Society of Actuaries (1949). Among other mathematical organizations of the period, one might note the Union Internationale de Mathématiques, founded in New York in 1950. Also, by 1947, there was in existence "The Cooperative Committee on the Teaching of Science and Mathematics" which, according to *Science* (July 11, 1947), was made up of representatives of some thirteen societies. The Society for Industrial and Applied Mathematics (SIAM), incorporated in Delaware in 1952, is a new and rapidly growing society. Its aims are given as "to further the application of mathematics to industry and science, to promote basic research in mathematics leading to new methods and techniques useful to industry and science, and to provide media for the exchange of information and ideas between mathematicians and other technical and scientific personnel."[36] Operational sections include: Delaware Valley, Boston-Cambridge, Southern California, Central Pennsylvania, Northern California,[37] and Baltimore.[38] Also, by the summer of 1955, academic sections either existed or were in the process of being set up at Drexel Institute of Technology, St. Johns College of St. Johns University, Brown University, and Clark University.[39] The society has received financial backing from a number of industrial firms.

The 1940's and 1950's have witnessed a vast expansion of the activities of the nation's physicists. This has brought about a large number of new and highly specialized societies of which a few prominent examples may be given. The American Society for X-Ray and Electron Diffraction dates from 1941.[40] The Electron Microscope Society of America was organized in 1942. The American Physical Society formed a Division of Electron and Ion Optics in 1943.[41] The Society for Experimental Stress Analysis was formed in 1943.[42] As the measurement of time became ever more important, it was understandable that new societies dealing with horology should come into being: the Academy of Time (1946),[43]

[36] SIAM Form 100 (Rev. December, 1954).

[37] *Ibid.*

[38] SIAM *Newsletter*, III, no. 6 (June 1955), 17.

[39] *Ibid.*, Recent Growth of the Society, 3.

[40] *Science*, XCIV (1941), 16; *Journal of Applied Physics*, XIII (1943), 141.

[41] "The New Division of Electron and Ion Optics in the American Physical Society," *Scientific Monthly*, LVII, no. 6 (1943), 570-572; "Formation of a Division of Electron and Ion Optics in the American Physical Society," *Journal of Applied Physics*, XIV (1943), 406.

[42] Marshall Holt, "Society for Experimental Stress Analysis," *Science*, CXIX (1954), 3.A.

[43] *Science*, CVI (1947), 340.

and, even earlier, the National Association of Watch and Clock Collectors (1943). The Ohio Valley Spectrographic Society was formed in 1944, as was the Milwaukee Society for Applied Spectroscopy; the Society for Applied Spectroscopy was formed in New York in 1945 by a group of practicing spectroscopists,[44] and the Spectroscopy Society of Pittsburgh in 1946. The Physics Club of Lehigh Valley dates from 1944. The Cleveland Physics Society was formed in 1945.[45] In 1945, at the Manhattan Laboratory at Los Alamos, New Mexico, an association of investigators working on the atomic bomb formed the Association of Los Alamos Scientists.[46] The Armed Forces Communications Association was formed in 1946. In 1946 the Crystallographic Society (organized originally at Harvard and M.I.T. in 1939 and concerned with "the science of crystallography and its applications to such fields as physics, chemistry, metallurgy, ceramics and biology") resumed its activities after a period of suspension.[47] In January, 1950, the American Crystallographic Association came into being "to . . . carry on the activities of the American Society for X-Ray and Electron Diffraction and the Crystallographic Society of America whose activities were officially ended on that date."[48] The New York Society of Electron Microscopists was formed in 1951. The year 1952 saw the formation of the Radiation Research Society, which soon came to publish *Radiation Research*.[49] In 1954 there was organized the American Nuclear Society, which was described as "the first professional society of scientists and engineers representative of all scientific disciplines engaged in research, development and application of nuclear technology."[50] Its purposes were set forth as "to foster the integration and advancement of nuclear science and technology through the interchange of information and ideas in every field of research involving nuclear techniques."[51] The announcement of its formation went on to state: "Although nuclear power will be of primary interest, other subjects of interest to the membership will include uses of radioisotopes, effects of radiation on materials, radiation-sterilization of foods, and so forth."[52] In 1954, also, came the creation

[44] *Ibid.*, CII (1945), 528.
[45] *Ibid.*, CI (1945), 349.
[46] "The Association of Los Alamos Scientists," *Science*, CII (1945), 608-609.
[47] *Science*, CIII (1946), 18.
[48] "American Crystallographic Association," *Science*, CXI (1950), 214.
[49] G. Failla, "The Radiation Research Society," *Science*, CXIV (1952), 27.
[50] *Science*, CXX (1954), 697.
[51] *Ibid.*
[52] *Ibid.*

of the General Committee on Nuclear Engineering and Science, organized under the sponsorship of the Engineers Joint Council to "meet the pressing problems of nuclear engineering and the related sciences"[53] emphasizing the "industrial usefulness of atomic power."[54] The Engineers Joint Council was at the time "constituted of eight leading engineering societies with a total membership of 170,000," and the American Chemical Society which joined the engineers in the nuclear project had "70,000 members."[55] Indicative of the renewed interest in utilizing the power of the sun was the Association for Applied Solar Research formed in 1954. The formation of the Health Physics Society was announced June 14, 1955.[56]

The Association of Vitamin Chemists was organized in 1943, the Society of Cosmetic Chemists in 1945, the Armed Forces Chemical Association in 1946; the Chemical Warfare Association, organized in 1947, changed to its present name of Chemical Corps Association, Incorporated, in 1947. Among the more active local chemical societies formed in this period was the Trenton Chemical Society (1945). The Catalysis Club of Philadelphia was formed in 1950. The American Chemical Society celebrated its diamond jubilee in 1951 and published in 1952 an elaborate history of its affairs.[57] The Commercial Chemical Development Association dates from 1947.

The Engineers Joint Council, formed in 1941, by 1955 was composed of eight constituent societies having a total membership of 170,000. The constituent societies were: American Institute of Mining and Metallurgical Engineers, American Society of Mechanical Engineers, American Water Works Association, American Institute of Electrical Engineers, Society of Naval Architects and Marine Engineers, American Society for Education, and American Institute for Chemical Engineers.[58] Among the specialized societies formed in 1941 were the Society of Plastic Engineers and the Society for Non-Destructive Testing. Other new societies in the next few years included the Coordinating Research Council (1942) to study fuels, lubricants, and engines; the National

[53] Ibid.   [54] Ibid.   ·   [55] Ibid.

[56] Ibid., CXXII (1955), 112.

[57] C. A. Browne and M. E. Weeks, A History of the American Chemical Society: Seventy-five Eventful Years (Washington, 1952).

[58] Scientific and Technical Societies of the United States and Canada, 6th ed., National Academy of Sciences—National Research Council, Publication no. 369 (Washington, 1955), 171.

Association of Corrosion Engineers (1943); the Reaction Research Society (1943) ; and the American Helicopter Society (1943). In 1946 were formed the Association of Senior Engineers of the Bureau of Ships and the American Society for Quality Control. Other engineering societies were formed as follows: Society of Photographic Engineers (1948)', Audio Engineering Society (1948), Association of Federal Communications Consulting Engineers (1948), American Institute of Industrial Engineers (1948), the Society of Women Engineers (1949), the American Material Handling Society (1949), the New Hampshire Society of Engineers (1949), the Society of Fire Protection Engineers (1950), the Gulf Institute of Consulting Engineers (1950), the Structural Engineers Association of Oregon (1950), the American Lithuanian Engineers and Architects Association (1951). JETS (Junior Engineering Training for Schools), an organization that originated at Michigan State University in 1950, had come by 1956 to include 77 clubs in 19 states with an active membership of more than 2,000 boys and girls of high school age. It is doing in engineering what 4-H and F.F.A. clubs long have done in agriculture. Especially interesting were the circumstances leading to the formation of the Operations Research Society of America (1952). For several years "the activity variously known as operations research, operational research or operations analysis, originally devised for military needs," had been a "subject of increasing interest in non-military circles," and a "number of workers in the field . . . felt that the time was ripe for some organization . . . to provide a means for the advancement of diffusion of knowledge concerning operations research. The National Research Council . . . was unofficially in agreement with this view. Consequently ten interested persons met in Cambridge, Massachusetts, in January 1952 to work out preliminaries . . . a constitution . . . discussed. The group met again in New York in March."[59] The Society held its first meeting in New York on May 26, 1952,[60] when a Constitution and By Laws were adopted. "The object of the Society," according to its Constitution, "shall be the advancement of the science of operational research, through exchange of information, the establishment and maintenance of professional standards of competence for work known as operations research, and the encouragement and development of students of operations re-

[59] P. M. Morse, "The Operations Research Society of America," *Journal of the Operations Research Society of America*, I, no. 1 (1952), 1.
[60] "The Founding Meeting of the Society," *ibid.*, 18.

search."[61] In 1952 the American Society of Civil Engineers, the oldest national organization of engineers in the United States, celebrated its centennial.[62] The year 1953 saw the formation of the Reaction Missile Research Society and the Atomic Industrial Forum, Incorporated. In November, 1955, the American Society of Mechanical Engineers celebrated its diamond jubilee in Chicago. The American Society of Lubrication Engineers was formed in 1955.

Many of the older technical societies had come to find a new usefulness in the national life as a result of their activities in World War II and could look forward to an active future at its close.[63] New technological societies, too, came into existence during and after the war period: the Fiber Society, Incorporated (1941), the United Inventors and Scientists of America (1942), the Northeastern Wood Utilization Council (1942), the Society of Industrial Packaging and Materials Handling Engineers (1945), the Air Pollution Control Association (1950), and the Building Research Institute (1951), organized as a unit of the Division of Engineering and Industrial Research of the National Research Council. Among the new societies formed after the war, one of the most interesting was the Society for the Social Study of Invention, organized at the American Association for the Advancement of Science at its Chicago meeting in 1947.[64] Its aims were given as "to study, promote, rationalize, and economize invention and its utilization, and incidentally to build the structure of culture generally."[65] Another of the more important organizations is the Institute of Management Sciences, a "national professional society that was established in December 1953 by a nationwide group of management analysts, social scientists, mathematicians and engineers with a common interest in the scientific analysis of management problems."[66] In 1954 the Technical Publishing Society, "a professional group to encourage the interchange of information among individuals engaged in preparing, editing, or publishing technical and scientific documents," was organized in Los Angeles.[67] Many of the trade associations sponsor scientific research in

[61] *Ibid.*, 28.

[62] E. L. Chandler, "Centennial of Engineering Convocation," *Science*, CXVI (1952), 3.

[63] See, e.g., C. F. Kettering, "The Future of Science," *Science*, CIV (1946), 609-614, especially "World War II in Retrospect: The Role of Our Technical Societies."

[64] *Science*, CVII (1948), 243-244.

[65] *Ibid.*, 244.

[66] *Ibid.*, CXX (1954), 530-531.

[67] *Ibid.*, 530.

one way or another. Trade associations have grown by leaps and bounds according to a Department of Commerce document. "By 1900 there were approximately 100 national and interstate trade associations. In 1920 the number had increased to more than 1,000 and in 1949 to 2,000."[68]

The scientists in the earth sciences were active, too, in the setting up of new specialized agencies.

Among the new societies devoted to the study of geography were: the American Congress of Surveying and Mapping (1941), the Arctic Institute of North America (1944), and the American Antarctic Association (1944). The National Council on Asian Affairs, organized at Philadelphia in 1955, had for its purpose to stimulate interest in the Asian people, "to promote an East-West exchange of teachers and make available more graphic material for schools, and to promote understanding between Asia and the United States."[69]

The National Geographic Society, founded in 1888, is cooperating with Mount Palomar in bringing to completion a sky atlas begun in 1949 which will include maps on about 70 percent of the sky.

The geologists, both professional and amateur, have been especially active in forming new societies and in reorganizing or federating existing ones to such an extent that any strictly chronological treatment becomes inadequate. Some of the more active organizations are mentioned here. The Midwest Federation of Mineralogical and Geological Societies was formed in 1940, the Geological Society of Kentucky in the same year, the Rocky Mountain Federation of Mineralogical Societies in 1941, the New Orleans Geological Society in the same year, the Corpus Christi Geological Society in 1942, the Northern California Geological Society in 1944, the Southeastern Geological Society in the same year, the Southwest Federation of Mineral Societies about the same time, the Wyoming Geological Society in 1945, the Pittsburgh Geological Society in the same year, the Utah Geological Society and the Abilene Geological Society in 1946. The National Speleological Society, begun about 1939 as the Speleological Society of the District of Columbia, reorganized and took its present title in 1946.[70] Its members have, among other things, ferreted out caves that might be useful refuges in case of atom warfare. The year 1947 saw the formation of the New

---

[68] Jay Judkins and H. E. Wells, *National Associations of the United States* (Washington, 1949), viii.

[69] *New York Times,* October 23, 1955.

[70] For its work see G. Nicholas, "National Speleological Society," *Science,* CXX (1954), 5A.

Mexico Geological Society and the Arizona Geological Society, and also the creation of a rather important organization of amateur mineralogists, the American Federation of Mineralogical Societies. The Coast Geological Society was formed in 1948. The American Geological Institute, comprising originally eleven national societies with a combined membership of more than 10,000 professional geologists, was organized at Washington in 1948,[71] and sponsored by the National Research Council. It included the following member societies: Geological Society of America, American Association of Petroleum Geologists, American Institute of Mining and Metallurgical Engineers, American Geophysical Union, Mineralogical Society of America, Society of Economic Geologists, Society of Exploration Geophysicists, Society of Economic Paleontologists and Mineralogists, Seismological Society of America, Paleontological Society, and the Society of Vertebrate Paleontology. Other geological organizations came along as follows: the Intermountain Association of Petroleum Geologists (1949), the Eastern Nevada Geological Society (1950), the Eastern Federation of Mineralogical and Lapidary Societies (1950), the Roswell (New Mexico) Geological Society (1950), the Billings (Montana) Geological Society (1950), the North Dakota Geological Society (1951), the Lafayette (Louisiana) Geological Society (1952), and the Lubbock (Texas) Geological Society (1952).

In May, 1938, a group of midwestern professors of geology organized the Association of College Geology Teachers, for the exchange of ideas and teaching techniques. A few years later the word "College" was dropped from its name. In November, 1951, the Association was organized on a national basis. Its purpose, according to its Constitution, is "to foster improvement in the teaching of the earth sciences at all levels of formal and informal instruction, to emphasize the cultural significance of the earth sciences, and to disseminate knowledge in this field to the general public." In 1955 membership in the organization, according to its secretary, was about 400.[72]

The American Meteorological Society celebrated its twenty-fifth anniversary at Kansas City in 1945.[73] The Amateur Weathermen of America, formed in 1947 at the Franklin Institute in Philadelphia, had as its main purpose to encourage the study of meteorology among the youth

[71] *Science,* CVIII (1948), 640.

[72] R. L. Bates, "Association of Geology Teachers, a New Affiliated Society of the A.A.A.S.," *Science,* CXXI (June 17, 1955), 7A.

[73] "The Twenty-fifth Anniversary of the American Meteorological Society." *Science,* CI (1945), 217-218.

of America. It started publication of *Weatherwise* in 1948. On January 1, 1952, the activities of the Amateur Weathermen of America became merged with those of the American Meteorological Society, the latter organization creating the associate membership class. Local groups, like the Amateur Weathermen of Pittsburgh, had also come into existence.

The World Meteorological Organization (WMO), a specialized agency of the United Nations, formally came into being on April 4, 1951, following ratification by many nations of a convention adopted in Washington, October 11, 1947, by the twelfth Conference of Directors of the International Meteorological Organization (IMO). The World Meteorological Organization has as its aims the facilitating of world-wide cooperation in observations, standardization of observation techniques, and the application of meteorology to human activities.

The American Society of Limnology and Oceanography was formed in 1948.[74] The Caribbean Research Council was set up in 1943 by the Anglo-American Caribbean Commission.[75]

Workers in the biological sciences were no less active than those in the physical and earth sciences; numerous new societies both of a general and of a specialized character came into existence.

The Society of General Physiologists was founded at the Marine Biological Laboratory, Woods Hole, in 1946.[76] In the same year there came into being the Society for the Study of Evolution, which was an outgrowth of an informal organization called Society for the Study of Speciation (1940) and the National Research Council Committee on Common Problems of Genetics, Paleontology and Systematics (1943) The year 1946 also saw the formation of the Ecologists Union. In 1947 a meeting of the Organizing Board of the American Institute of Biological Sciences was held at the National Research Council in Washington, and the Institute was formally established in 1948. This is described as a "voluntary association of organizations which have in common an interest in the life sciences."[77] Other organizations coming into being about the same time were: American Society of Professional Biologists (1947); the Biometric Society, an international society formed at the

[74] *Science*, CVII (1948), 318.
[75] "Organization of the Caribbean Research Council," *Science*, CIII (1946), 452-453.
[76] John Buck, "Society of General Physiologists," *Science*, CXVIII (1953), 3.
[77] S. L. Meyer, "The American Institute of Biological Sciences," *Science*, CXVII (1953), 3.

Marine Biological Laboratory, Woods Hole, in 1947;[78] the American Society of Human Genetics (1948), which began in that year to publish a periodical, *Human Genetics;*[79] the Atlantic Estuarine Research Society (1949);[80] and the Society for Industrial Microbiology (1949). The Genetics Society of America (1900) celebrated its Golden Jubilee in 1950.[81] The Biology Council of the National Academy of Sciences— National Research Council prepared for distribution in 1956 to every junior high school, high school, and college in the country a booklet entitled *Career Opportunities in Biology.*[82] The Biophysical Society was brought into existence early in March, 1957, by the National Biophysical Conference meeting at Columbus, Ohio, "to encourage biophysical research by increasing communication among all those working in this field, by furthering the training of biophysicists, and by upholding standards in the physical analysis of biological problems."[83]

Among the botanical societies formed in this period were: the American Rhododendron Society (1944); the National Chrysanthemum Society (1944); the North American Gladiolus Council (1945); the Holly Society of America, Incorporated (1945); the American Camelia Society (1945); the American Penstemon Society (1946); the Phycological Society of America (1946); the Hemerocollis Society (1946); the National Tulip Society (1947); the California Arboretum Foundation, Incorporated (1948); the National Snapdragon Society (1949); the American Hibiscus Society (1950); the Bromeliad Society (Incorporated) (1950); the Plant Propagators Society (1951); and the American Gloxinia Society (1951). The well-known Sullivant Moss Society (1898) changed its name to become the American Bryological Society in 1948.[84]

The Society of Vertebrate Paleontology came into existence in 1940. In the same year the National Association of Audubon Societies (incorporated 1905) took the name National Audubon Society. The Hawaiian Malocological Society was formed in 1941. The American Association of Zoological Parks and Aquariums dates from 1942. The

[78] *Science,* CVI (1947), 413.

[79] *Ibid.,* CX (1949), 289.

[80] J. D. Andrews, "The Atlantic Estuarine Research Society," *Science,* CXVI (1952), 153-154.

[81] *Science,* CXII (1950), 795-796; see also L. C. Dunn, edit., *Genetics in the 20th Century* (New York, 1951).

[82] M. B. Visscher, "Role of Council for International Organizations in Medical Sciences," *Science,* CXXIII (1956), 337.

[83] S. A. Talbot, "Biophysical Conference and Society," *Science,* CXXV (1957), 754.

[84] *Science,* CVIII (1948), 700.

Society of Professional Zoologists was founded at Atlanta in 1945.[85] The American Society of Protozoologists was organized in Chicago in 1947.[86] In the next few years other societies came into being: the Lepidopterists' Society (1947) ;[87] the Association of American Pesticide Control Officials, Incorporated (1947) ; the Society of Systematic Zoology (1948) ; and the New Haven Entomological Society (1949). In 1949 the Cambridge Entomological Club held its seventy-fifth anniversary meeting. The Entomological Society of America was formed in 1953[88] as a result of the merger of the former Entomological Society of America (1906) and the American Association of Economic Entomologists (1889). The Southeastern Society of Ichthyologists and Herpetologists came into existence about 1951. The Honolulu Aquarium Society was formed in 1951; so also was the Sweander Entomological Society of Pittsburgh.

The Soil Conservation Society of America was founded in 1941; the American Farm Research Association in 1944; and the American Horticultural Council, Incorporated, in 1945; the Agricultural Research Institute was organized in 1951-52 under the auspices of the National Academy of Sciences and the National Research Council.[89] The American Society for Horticultural Science celebrated its fiftieth anniversary in 1953.[90]

The Forest Products Research Society was founded in 1947; the Association of Consulting Foresters in 1948; the International Society of Tropical Foresters in 1950 at Washington, D. C.;[91] and the Forest Conservation Society of America in 1954.

The conservation movement, which received a new impetus in the New Deal period, went steadily forward in the forties and fifties with many new or reorganized societies taking up the work. Included among these were: the National Council for Stream Improvement, Incorporated (1943) ; the American Waterfowl Association (1945) ; the American Wildlife Foundation (1946), founded in 1935 as the American Wildlife Institute; Wildlife Management Institute, incorporated in New York in 1946; Natural Resources Council of America (1946) ;

[85] "A Society for All Biologists," *Science*, CIV (1946) , 325-326.
[86] *American Scientist*, XXXVI, no. 2 (1948) , 200; N. D. Levine, "Society of Protozoologists," *Science*, CXXI (1955) , 9A.
[87] *Science*, CIX (1949), 243.
[88] C. F. W. Muesebeck, "National Entomological Societies Merge," *Science*, CXVII (1953), 546-547.
[89] Le Roy Voris, "Agricultural Research Institute," *Science*, CXVI (1952) , 3.
[90] F. S. Howlett, "The American Society for Horticultural Science," *Science*, CXVIII (1953), 617.
[91] *Science*, CXIII (1951) , 306.

American Society of Range Management (1947) ; American Watershed Council, Incorporated (1950) ; Conservation Council for Hawaii (1950) ; Wildlife Disease Association, formally organized in Miami in 1952;[92] Citizens Club for Conservation, Incorporated (1953) ; and Conservation Education Association (1953) .

The medical profession has been especially active in the formation of new, highly specialized societies in the past ten or fifteen years. For example, early in the forties among the new national societies that came into being were: the American Diabetes Association, Incorporated (1940) ; the American Federation for Clinical Research (1941) ; the American Fracture Association (1941) ; the American Gastroscopic Society (1941) ; the Public Health Cancer Association of America (1941) ; the Association of Schools of Public Health (1941) ; the Conference of Professors of Preventive Medicine (1942) ; the Association of State and Territorial Health Officers (1942) ; the American College of Allergists (1942) ; the American Geriatrics Society (1942) ; the American Oto-rhinologic Society for Plastic Surgery (1942) ; the American Academy of Allergy (1943) ; the American Society for the Study of Sterility (1944) ; the United Epilepsy Association (1944) ; Hay Fever Prevention Society, Incorporated (1945) ; Gerontological Society, Incorporated (1945) ; National Society for Medical Research (1945) ; and the Society for the Study of Blood (1945) . Among the more active national medical societies formed in 1946 were: Tissue Culture Association; American Electroencephalographic Society; Conference of Public Health Veterinarians; American Academy of Compensation Medicine, Incorporated; Society of Medical Consultants to the Armed Forces; Association of Food Industry Sanitarians; American Society for Surgery of the Hand; Society of General Physiologists; National Council to Combat Blindness; National Multiple Sclerosis Society (organized as Association for Advancement of Research on Multiple Sclerosis, Incorporated, present name since 1947) ; and the National Vitamin Foundation, Incorporated. The American Medical Association, celebrating its centenary in 1947, put out an admirable centennial volume: Morris Fishbein, *A History of the American Medical Association, 1847-1947* (Philadelphia, 1947) , 1226 pp. Active national societies formed in 1947 included: American Association of Blood Banks; American Blood Irradiation Society; American Board of Medical Technologists; American Society for the Study of Arteriosclerosis; National Association of Clinical Laboratories, Incorporated; Women's Veterinary Medical Association; American Col-

[92] *Ibid.*, CXVI (1952) , 269-270.

lege of Anesthesiologists; Society for Vascular Surgery; Coordinating Council for Cerebral Palsy; American Academy of General Practice; and College of American Pathologists. The American Physical Therapy Association (1921) took its present name in 1947. In 1948 the following national societies came into being: American Academy for Cerebral Palsy; American Academy of Neurology; American College of Veterinary Pathologists; American Association of Clinical Chemists; Neurosurgical Society of America; and United Cerebral Palsy Associations, Incorporated. The year 1949 saw the formation of the American College of Cardiology and the Commission on Chronic Illness. In 1950 the Muscular Dystrophy Associations of America, Incorporated, the National Association for Retarded Children, and the Student American Medical Association came into being. The following national societies came into existence in 1951: American Society of Tropical Medicine and Hygiene (formed by the amalgamation of the American Society of Tropical Medicine [1903] and the National Malaria Society [1916]) ; American Academy of Obstetrics and Gynecology; Association for the Advancement of Instruction About Alcohol and Narcotics;[93] American Naturopathic Physicians and Surgeons Association; National Medical Veterans Society; and Walter Reed Society. The year 1952 saw the creation of these societies: Association of State and Territorial Public Health Directors; National League for Nursing, formed by joining together three national nursing organizations and four national committees and representing some 20,000 professional nurses; the Society of Pelvic Surgeons; and the National Association of Recreational Therapists. The Psoriasis Research Association came into existence in 1953. The year 1954 saw the formation of the Society of Nuclear Medicine and the American Society for Artificial Internal Organs.[94] The local medical groups formed in the forties and fifties proved to be far too numerous to be mentioned here. A few general trends may, however, be noted. One is the rapid spreading of radiological societies at the state and local level. Another is a similar spreading of societies of anesthesiologists. Still another is the growth of societies for the study of circulatory diseases. A fourth is the rapid increase of public health associations.

New national dental societies of recent years include: American Association of Industrial Dentists (1943) ; American Academy of Dental Medicine (1946) ; American Academy of Oral Pathology (1946) ; American Association of Endodontists (1947) ; American Academy of Pedo-

[93] *Ibid.*, CXIV (1951) , 225.
[94] *Ibid.*, CXXI (1955) , 456.

dontics (1947) ; American Academy of the History of Dentistry (1950) ; American Academy of Implant Dentures (1951) ; and the National Association of Dental Laboratories (1952). By 1954 the American Dental Association, nearly a century old (1859), had 54 constituent state and local societies carrying on·active programs, and as of October, 1953, with 70,000 dentist members and 10,000 student members, represented 80 percent of the dentists of the nation and claimed to be the largest dental organization in the world.[95]

The American College of Apothecaries was founded in 1940. The American Institute of the History of Pharmacy, with headquarters at the University of Wisconsin, was founded in 1941.[96] The American Pharmaceutical Association celebrated its centennial in 1952.[97] The International Academy of the History of Pharmacy, with headquarters at The Hague (Holland), was founded June 13, 1952, on the seventieth birthday of George Urdang, professor of the history of pharmacy at the University of Wisconsin, who became the society's first president. Its purpose was given as "to stimulate international cooperation in the history of pharmacy."[98] The National Pharmaceutical Council, Incorporated, was formed in 1953.[99] The Nevada State Pharmaceutical Association dates from 1950.

New societies for dealing with problems of mental hygiene that came into existence in the forties included: the American Psychosomatic Society, Incorporated (1943) ; the Bronx Society of Neurology and Psychiatry (1946) ; the Association for Physical and Mental Rehabilitation (1946) ; the American Association of Psychiatric Clinics for Children (1946) ; the Society of Biological Psychiatry (1946) ; the Oklahoma Society of Neurologists and Psychiatrists (1947) ; the Southeastern Society of Neurology and Psychiatry (1947; the Society for Clinical and Experimental Hypnosis (1949) ; the Maryland Psychiatric Society (1949) ; the Brooklyn Psychiatric Society (1949) ; the Washington Psychiatric Society (1949) ; and the Finger Lakes Neuropsychiatric Society of New York (1949). In the fifties, among the similar organizations

[95] H. T. Dean, "The American Dental Association: Status Changed from an Associated to an Affiliated, AAAS," *Science*, CXIX (1954), 394.

[96] "The American Institute of the History of Pharmacy," Tenth Anniversary, *Science*, CXIII (1951), 570.

[97] "Centennial Convention, A. Ph. A.," *Science*, CXVI (1952), 582-584.

[98] "Founding of an International Academy of the History of Pharmacy" (Académie Internationale de l'Histoire de la Pharmacie), *Bulletin of the History of Medicine*, XXVI (1952), 385.

[99] *Science*, CXIX (1954), 34.

coming into existence were: the National Association for Mental Health (1950); the American Association of Psychiatric Workers (1950); the Arkansas Psychiatric Society (1950); the Intermountain Psychiatric Association (1951); the East Bay Psychiatric Association (California) (1951); the Academy of Psychosomatic Medicine (1952);[100] the American Academy of Child Psychiatry (1953); and the Northern California Psychiatric Society (1954).

The scholars working in the social sciences and humanities found themselves organizing and reorganizing as never before to meet the demands of a swiftly moving society. One of their most common complaints was that relatively little in the way of research funds came their way. This imposed on them an additional need for wisely correlating their activities through their learned societies and councils.

In this period the Social Science Research Council, in addition to furnishing many grants-in-aid for research, undertook a number of publications of its own, such as: *A Directory of Social Science Research Organizations in Universities and Colleges* (1950) and *Fellows of the Social Science Research Council* (1951). The Society for Social Responsibility in Science was formed in 1949. It aims to "foster throughout the world a functioning cooperative tradition of personal moral responsibility for humanity of professional activity, with emphasis on constructive alternatives to militarism."[101] The American Studies Association was coming into being in 1951-1952.[102] According to its constitution this society hoped to accomplish its aims by "the improvement of communication across those disciplines which deal with phases of American civilization" and by the "fostering of interdisciplinary research and of courses and programs in American civilization." Since 1952 the *American Quarterly* has been endorsed by the American Studies Association as its official journal.[103] By 1952 the American Academy of Fine Arts was rounding out a century and a half of activity and was looking forward to publishing a lengthy history of its accomplishments.[104]

The Iowa Archaeological Society was formed in 1951. Three years

[100] *Ibid.*, 282.

[101] *Scientific and Technical Societies of the United States and Canada*, 6th ed., National Academy of Sciences—National Research Council, *Publication* no. 369 (Washington, 1955), 316.

[102] *American Historical Review*, LVI, no. 4 (1951), 1030.

[103] *American Quarterly*, IV, no. 1 (1952), 93.

[104] M. B. Coudrey, *American Academy of Fine Arts and American Art-Union* (2 vols., New York, 1953).

earlier a local archaeological society, the Trowel and Brush Society, Angel Mounds, Newburgh, Indiana, was formed.

The Society for the Advancement of Criminology was organized in 1940.

The Society for Applied Anthropology, Incorporated, was formed in 1941. The American Institute of Human Paleontology was founded in 1949, and plans were set up in 1954 for expansion.[105] The American Association of Physical Anthropologists (1930) became affiliated with the American Association for the Advancement of Science in 1953.[106] Among the more active new local anthropological societies that came into existence in the period were: the Florida Anthropological Society (1948) ; the Seattle Anthropological Society (1948) ; the Kroeber Anthropological Society (1949) at the University of California; and the Oklahoma Anthropological Society (1952) .

Among the economic associations formed in this period that were somewhat concerned with science, mention might be made of the American International Association for Economic and Social Development (1947) to encourage practical programs of international cooperation in such fields as agriculture, public health, and industry.

The American Catholic Psychological Association was organized in 1948. The Inter-American Society of Psychology was formed in 1951.[107] The Society of Correctional Psychologists dates from 1953. A number of new state, local, and regional psychological associations came into existence all over the United States in this period. Among the more active were: Connecticut State Psychological Society (1944) ; Ohio Psychological Association (1945) ; District of Columbia Psychological Association, Incorporated (1946) ; Georgia Psychological Association (1946) ; Colorado Psychological Association (1946) ; Texas Psychological Association (1947) ; Minnesota Psychological Association (1947) ; Missouri Psychological Association (1947) ; Oklahoma State Psychological Association (1947) ; Delaware Psychological Association (1947) ; Washington State Psychological Association (1947) ; Louisiana Psychological Association (1948) ; California State Psychological Association (1948) ; Tennessee Psychological Association (1948) ; North Carolina Psychological Association (1948) ; Florida Psychological Association (1949) ; Hawaii Psychological Association (1949) ; Vermont Psychologi-

[105] T. D. Steward, "American Institute of Human Paleontology," *Science*, CXX (1954), 7A.

[106] J. L. Angel, "American Association of Physical Anthropologists," *Science*, CXVIII (1953), 3.

[107] *Science*, CXVI (1952) , 55.

cal Association (1949) ; New Jersey Psychological Association (1950) ; West Virginia Psychological Association (1950) ; Arizona Psychological Association (1950) ; Alabama Psychological Association (1950) ; Maine Psychological Association (1950) ; Wisconsin Psychological Association (1950), previously organized in 1938 as the Wisconsin Association for Applied Psychology; Utah Psychological Association (1951) ; Nebraska Psychological Association (1952) ; Brooklyn Psychological Association (1953) ; New Hampshire Psychological Association (1953) ; Southwestern Psychological Association (1954) ; and the South Carolina Psychological Association (1954).

With the ending of World War II, the way was again opened for a resumption of the trend toward scientific organization on an international, even a global, scale. The United Nations was soon to give focus to such efforts, particularly through one of its agencies, UNESCO.

The bases of UNESCO (the United Nations Educational, Scientific and Cultural Organization) were laid by the conference of Allied Ministers of Education in London between 1942 and 1945. Following the United Nations Conference in San Francisco, a conference convened in London attended by forty-three countries and adopted a constitution on November 16, 1945. "The purpose of the Organization is to contribute to peace and security by promoting collaboration among the nations through education, science and culture in order to further universal respect for justice, for the rule of law and for the human rights and fundamental freedoms which are affirmed for the peoples of the world, without distinction of race, sex, language or religion by the Charter of the United Nations."[108] UNESCO came into being on November 4, 1946. By July, 1954, UNESCO had seventy-two member states. UNESCO headquarters are in Paris.

UNESCO is concerned primarily with education. In furthering this UNESCO has three objectives: increasing education, improving education, and training for international good citizenship. In its struggle against illiteracy, disease, and poverty it stresses "fundamental education," by attempting to give those without benefit of any scholastic training the basic information necessary for improving their standard of living.[109] UNESCO established a regional training center in 1951 in Mexico for the Latin American countries, and another in 1953 in Egypt for the Arab countries.

In the natural sciences, UNESCO promotes international scientific

[108] UNESCO, Constitution, Article I.
[109] *Index Generalis, 1954-1955* (Paris, 1955), vi.

cooperation by "initiating meetings between scientists and aiding the work of international scientific organizations." It further "encourages scientific research designed to improve the living conditions of mankind. It seeks to help improve techniques for the teaching and popularization of science." Regarding the arid zones UNESCO seeks "to secure international co-operation and the full exchange of experience in order to make more than a quarter of the land surface of the earth productive to man."[110] An international council for the study of nuclear research has been set up in Geneva. National Research Councils for Pure and Applied Science in the member states of UNESCO include for the United States: the National Academy of Sciences–National Research Council (NAS-NRC) and the National Science Foundation (NSF).[111] Biologists from nine countries (including the United States) who attended a meeting convened in Paris in December, 1955, by the Council for International Organizations of Medical Sciences at the request of UNESCO submitted to UNESCO's Director General a plan of action to encourage international and regional research on normal and abnormal cell growth. The meeting recommended that an international committee be formed to collect information on research in the field, pointing out that UNESCO could perform valuable services through such clearing-house facilities.[112] UNESCO maintains science cooperative area offices in Montevideo, Cairo, New Delhi, and Djakarta and branch offices in Istanbul and Manila. UNESCO activities are described in the UNESCO *Courier* and in *Science in UNESCO,* issued by the Office of International Relations of the National Academy of Sciences–National Research Council.

In the social studies, UNESCO seeks to apply scientific knowledge to human relationships within and between states. It seeks, too, to study social tensions and the attainment of human rights for men and women throughout the world.

UNESCO's program of cultural activities seeks to promote cultural exchanges among peoples and to foster cooperation among writers throughout the world. It assists in the development of museums and libraries and in the dissemination of literary masterpieces throughout the world. An international copyright convention was signed at Geneva in 1952 by thirty-five countries. A convention providing for the protection of cultural treasures in case of war was adopted at The Hague by thirty-seven countries in 1954.

[110] *Ibid.,* vi.
[111] *Ibid.,* xv.
[112] *Science,* CXXIII (1956), 112; *UNESCO Chronicle,* II, no. 3 (1956) , 96.

UNESCO participates in the United Nations Assistance program for economic development in such ways as sending out teams of teachers and experts, granting scholarships, and furnishing materials. At its headquarters it maintains a clearing house of educational, scientific, and cultural information.

Delegates from UNESCO's member states meet in general conference every two years to determine its policies. UNESCO cooperates with the United Nations, with the other U.N. specialized agencies, with governmental bodies, and with about a hundred international non-governmental organizations. National commissions or cooperating bodies set up in the member states serve as connecting links between the U.N. and the government departments in the various countries. By July 1, 1952, sixty-three such bodies had been set up.[113] In 1950 UNESCO brought out the valuable *Directory of International Scientific Organizations*.[114]

In December, 1946, an agreement was signed between UNESCO and the International Council of Scientific Unions (ICSU) recognizing the Council as the coordinating body between UNESCO and the Unions. UNESCO has aided ICSU with funds.

UNESCO has assumed leadership in the creation of new international professional organizations. Included among these are: the International Association of Universities (1950) ; the Union of International Engineering Associations, set up in March, 1951, which comprises twelve international member associations; the International Council for Philosophy and Humanistic Studies, with a membership of twelve international associations; the International Theatre Institute; the International Music Council; the International Sociological Association; the International Political Science Association; the International Economic Association; and the International Committee of Comparative Law.[115] Of growing importance is the Council for International Organizations of Medical Sciences (CIOMS), founded in Brussels in 1949 under the auspices of UNESCO and the World Health Organization, its object being "to facilitate the dissemination of knowledge in such ways as to promote advances in medical science and to encourage the utilization of such knowledge throughout the world."[116] UNESCO gives subventions to many of these organizations.

UNESCO also aids the work of the International Social Science Council (1952), as well as that of the World Braille Council (1952). The

[113] *Index Generalis, 1954-1955* (Paris, 1955), vii.
[114] UNESCO, *Directory of International Scientific Organizations* (Paris, 1950).
[115] *Science*, CXXII (1955), 1196.
[116] *The World of Learning, 1954* (6th ed., London, 1955), 7.

International PEN Club, founded in 1921, and now with about forty international centers, has come to associate more and more closely with the work of UNESCO. UNESCO cooperates with the International Bureau of Education, and with a World Confederation of the Teaching Profession which it has set up.

Early in 1956, UNESCO issued a comprehensive handbook of educational organization and statistics under the title of *World Survey of Education*. This contained data from 200 countries and territories and was the first of a series to be issued at three-year intervals.

UNESCO in 1956 planned a two-year "broad program for international cooperation in arid zone research," involving among other things: reviews of the status of research, an international symposium on climatology, and studies of the arid zone from Egypt to India.[117]

The International Advisory Committee for Documentation and Terminology in Pure and Applied Science, meeting in London in November, 1955, dealt with UNESCO's projects relating to "methods of retrieval and processing of recorded scientific data and information."[118] The First International Congress on Documentation of Applied Chemistry, meeting also in London soon thereafter, "thought that UNESCO should be asked for information on the possibilities for collaboration between national associations in improving the uniformity and equivalence of chemical terminology."[119]

Early in 1956, UNESCO reported the formation of a committee at New Delhi to form an Asian Federation of Library Associations to begin functioning in 1957.

A meeting of Directors of National Cultural Relations Services was held at UNESCO House, Paris, in December, 1955. Its purpose as defined in resolution IV.1.7.31, adopted by the General Conference at its eighth session, was to advise the Director General as to the best way "of co-ordinating Unesco's activities more closely with those conducted by Member States in pursuance of bilateral and multilateral agreements."[120] In addition to representatives present from nations and other political organizations, there were representatives from the following international societies: Council for International Organizations of Medical Sciences, International Council for Philosophy and Human-

[117] G. F. White, "International Cooperation in Arid Zone Research," *Science*, CXXIII (1956), 537-538.

[118] *UNESCO Chronicle*, II, no. 2 (1956), 61.

[119] *Ibid.*, 62.

[120] René Maheu, "UNESCO and the National Cultural Relations Services," *UNESCO Chronicle*, II, no. 3 (1956), 75-79.

istic Studies, International Association of Universities, International Association of Plastic Arts, and the International Social Science Council.

The fifth Conference of International Non-Governmental Organizations Having Consultative Arrangements with UNESCO met in June, 1956, to consider UNESCO's draft program for 1957-1958, and also, among other things, to discuss "Education for International Understanding and Cooperation."[121]

The statutes for a new scientific organization, the International Society of Bioclimatology, were approved at a symposium held at UNESCO House, Paris, at the end of August, 1956.[122]

The World Congress of Sociology, arranged by the International Sociological Association with the assistance of UNESCO, took place at Amsterdam in August, 1956.[123]

Concerning plans for 1957-58, UNESCO stated:

> It is indispensable that non-governmental organizations directly concerned with education, science or cultural problems should take part in Unesco's programme. . . . [UNESCO] will maintain . . . the relations it has established with some 400 international institutions . . . ranging from restricted groups of experts to associations of more than a million members. . . . among the 36 organizations likely to benefit next year from financial aid may be . . . the International Council for Philosophy and Humanistic Studies, grouping 12 international specialist associations—and the International Council of Scientific Unions which coordinates at world level work done in 13 different branches of learning.[124]

The Council for International Organizations of Medical Sciences (CIOMS) was set up at Brussels in April, 1949, by an assembly acting under the aegis of UNESCO and the World Health Organization. The members as listed in UNESCO, *Directory of International Scientific Organizations* (2d ed., 1953) included over forty international medical societies. CIOMS was described as "the first world-wide interdisciplinary organization in the medical sciences ever to be established."[125]

The International Council of Scientific Unions (ICSU) came into being in 1931 in Brussels as successor to the International Research

---

[121] *UNESCO Chronicle*, II, no. 3 (1956), 97.

[122] *Ibid.*, II, no. 11 (1956), 346.

[123] "World Congress of Sociology," *UNESCO Chronicle*, II, no. 11 (1956), 327-334.

[124] "The Proposed Program of UNESCO for 1957-58," *UNESCO Chronicle*, II, no. 11 (1956), 327-334.

[125] M. B. Visscher, "The Role of Council for International Organizations in Medical Sciences," *Science*, CXXIII (1956), 337.

Council (1919) .[126] The International Council of Scientific Unions is an intergovernmental body, whose objectives are set forth in articles 1.1 and 1.2 of its statutes:

1. The International Council of Scientific Unions has two principal objectives:

    a) To coordinate and facilitate the action of international scientific unions in the sphere of the exact and natural sciences.

    b) To serve as a center of coordination of the national organizations belonging to the Council.

2. The Council has the additional aims:

    a) To encourage international scientific activity in spheres where appropriate international organizations do not exist.

    b) To enter into relations, by the intermediary of the national member organizations, with the governments of countries adhering to the Council, with a view to developing scientific research in these countries.

    c) To keep in contact with the United Nations and with its specialized institutions.

With more especial reference to the international scientific activity of the various scientific unions, ICSU aims at coordinating "the national adhering organizations, and also the various international unions, to direct international scientific activity in subjects which do not fall within the purview of any existing international associations; to enter, through the national adhering organizations, into relation with the governments of the countries adhering to the Council in order to promote scientific investigation in these countries."[127] A formal agreement was concluded in December, 1946, between the Council and UNESCO, by which UNESCO recognizes the various scientific unions "as providing a natural and appropriate form for the international organization of science and ICSU as their co-ordinating and representative body."[128] The office of the secretary of ICSU is in London at the Royal Society. On December 31, 1954, ICSU was composed of the following members:[129]

[126] W. M. Rudolph, "International Council of Scientific Unions," *Science*, CXXII (1955), 652, 654, describes important recent undertakings of the ICSU and its relation to national governments.

[127] UNESCO, *Directory of International Scientific Organizations* (Paris, 1950), 22-23.

[128] *Ibid.*, 25.

[129] *Index Generalis, 1954-1955* (Paris, 1955), xii.

National members: the academies or research councils of 40 countries
Scientific members: the eleven international scientific unions which
  compose the Council, namely:

> Astronomy
> Geodesy and Geophysics
> Physics, pure and applied
> Chemistry, pure and applied
> Mathematics
> Biological sciences
> Geography
> Mechanics
> Radio science
> Crystallography
> History of Science

The Soviet Union was admitted to the International Council of Scientific Unions in 1955.[130] The United States adheres to ICSU through the National Academy of Sciences–National Research Council, the principal liaison being provided by the Office of International Relations of these two organizations. American scientists participate in the activities of the unions either as members-at-large or as members of specific commissions of the unions.

The activities of ICSU are not confined to the major international unions alone. It also has seven permanent joint commissions. In addition there are two special committees: one for the International Geophysical Year, and the other the ICSU Abstracting Board.[131]

The Abstracting Board of the International Council of Scientific Unions,[132] controlled by ICSU and financed by funds from UNESCO, began operations on June 1, 1952. It has sought to have all original articles dealing with physics in journals in the United States, the United Kingdom, Canada, Belgium, France, Italy, Germany, Japan, etc., summarized in French or English or both according to rules approved by the Royal Society and UNESCO. In this it has worked with the International Union of Pure and Applied Physics (IUPAP). The board is helping set up a similar system in conjunction with the International Union of Pure and Applied Chemistry (IUPAC). The International

[130] *Science*, CLII (1955), 652.

[131] See B. J. Bok, "Science in International Cooperation," *Science*, CXXI (1955), 843-847.

[132] See G. A. Bounty, "Abstracting Board of International Council of Scientific Unions," *Science*, CXXIII (1956), 423-424.

Union of Mechanics and the International Union of Biology as of 1956 had also expressed interest in the board's work. Exchange of proofs with Russian journals of physics began early in 1955.

The International Council for Philosophy and Humanistic Studies, with headquarters in Paris, was founded January 19, 1950. Among its aims are the following: "To encourage the formation of international organizations in spheres where they do not exist"; and "To stimulate international meetings, congresses, discussions or committees of experts; to facilitate mutual understanding among nations and the knowledge of man by fostering the broadest international cooperation in the sphere of philosophy and the humanities."[133] On July 1, 1954, the twelve member organizations of the Council were: International Academic Union, International Federation of Societies of Philosophy, International Committee on Historical Sciences, Permanent International Committee of Linguistics, International Federation of the Societies of Classical Studies, International Congress of Anthropological and Ethnological Studies, International Commission on Folk Arts and Folklore, International Committee on the History of Art, International Association for the Study of the History of Religion, International Federation of Modern Languages and Literature, International Union of Orientalists, and International Musicological Society. Nearly sixty international meetings have been held since 1950 under the auspices of the Council. The Council works in close collaboration with UNESCO.

By 1955 intensive plans were being made in many countries to participate in the so-called International Geophysical Year (1957-1958) "when scientists will conduct the most comprehensive study of the earth ever undertaken."[134] Investigations will be conducted in meteorology, latitude and longitude determinations, geomagnetism, gravity measurements, the ionosphere, aurorae, solar activity, cosmic rays, glaciology, oceanography, seismology, and upper atmosphere exploration by means of rockets.[135] Nations participating in the IGY program include: Argentina, Australia, Austria, Belgium, Brazil, Burma, Canada, Chile, Czechoslovakia, Denmark, Finland, France, East Germany, West Germany, Great Britain, Greece, Hungary, Iceland, India, Ireland, Israel, Italy, Japan, Mexico, Morocco, Netherlands, New Zealand, Norway, Pakistan, Peru, Philippines, Spain, Sweden, Switzerland, Thailand, Tunisia, Union of South Africa, USSR, United States, and Yugo-

[133] *Index Generalis, 1954-1955* (Paris, 1955), ii.
[134] *Science*, CXXII (1955), 322.
[135] *Ibid.*

slavia.[136] Each country will carry out its own plan under a general plan coordinated by the Special Committee for the International Geophysical Year. The U. S. National Committee for the IGY was established by the National Academy of Sciences. It is in charge of the planning and executing of this country's IGY program, and it functions in cooperation with panels of scientists representing universities and research institutions. Additional federal sponsorship and support has been obtained through the National Science Foundation. The most spectacular part of the IGY program concerns the earth satellite program. This was initiated in compliance with a resolution passed by the Special Committee for the International Geophysical Year (CSAGI) at its meeting in Rome in 1954. The National Academy of Sciences, working through the U. S. National Committee for the International Geophysical Year, is the sponsor of the American satellite effort.[137] By December, 1955, according to a news report of the National Academy of Sciences–National Research Council, a team of scientists made up of "specialists from the Department of Defense, the Academy-Research Council, and many universities and Government and private research institutions"[138] were planning to construct and launch twelve earth satellites to circle the earth at heights of from 200 to 800 miles, accompanied by instruments for recording observations of such things as ultraviolet radiation, cosmic rays, and outer air densities.

By October 4, 1957, when the Soviet Union announced that it had successfully launched the first earth satellite, IGY satellite tracking stations had been organized into a world-wide network.[139]

[136] *Ibid.*

[137] Joseph Kaplan and Hugh Odishaw. "Satellite Program," *Science*, CXXII (1955), 1003-1005.

[138] *New York Times*, December 25, 1955.

[139] Map published in *ibid.*, October 6, 1957, E5.

Chapter VII

SCIENTIFIC SOCIETIES IN THE SPACE AGE, 1955–1965

Historians in the future will doubtlessly take particular note of a striking series of scientific events that took place around the middle of the twentieth century. The uranium atom was first split in 1938 by two German refugees, Otto Hahn and Lise Meitner, working in Denmark. The first controlled nuclear chain reaction (fission of uranium isotope U-235) took place on December 2, 1942, at the University of Chicago. The first test atomic bomb was detonated at Alamogordo, New Mexico on July 16, 1945. The first nuclear created electricity was produced at the National Reactor Testing station, Idaho, in December 1951. The first satellite, Sputnik I, was shot into space on October 4, 1957. The world's first use of nuclear power in space came on June 29, 1961, when an atomic battery was used to provide power for transmitters in the Transit IV-A satellite. Thus, within a few years the so-called "Atomic Age" and the so-called "Space Age" had both not only come into being but also had merged. Moreover, in the short space of a few years, mankind found itself confronted with the possibilities of undreamed of advancement or utter destruction. It was perhaps the scientists themselves that were the group most cognizant of the awesome new problems that were raised.[1] A host of new scientific societies were called into being by the technical demands of atomic science and space science. Workers in the earth, biological, and the social sciences also felt the impact and the need to set up new organizations of their own.

General Societies

National crises have a way of calling forth new scientific agencies. Thus, the American Revolution helped spark the American Academy of Arts and Sciences, the Civil War called the National Academy of Sciences into being, World War I brought about the National Re-

---

[1] See, for example, J. E. Burchard, edit., *Mid-Century: The Social Implications of Scientific Progress* (New York, 1950).

search Council, and World War II called forth the Office of Scientific Research and Development and paved the way for the Atomic Energy Commission (1946) and the National Science Foundation (1950). The international tensions of the "Cold War" and the Space Age called into being, among other things, the President's Science Advisory Committee (1957) and a vast new civilian body, the National Aeronautics and Space Agency (1958), as well as the Federal Council on Science and Technology (1959).

A number of new, broadly integrative organizations, too, were coming into being or extending their scope in the atomic age and the space age. Among them were the Federation of American Scientists (1946), the Scientific Research Society of America (1947), Society for Social Responsibility in Science (1949), Operations Research Society of America (1952), Scientific Manpower Commission (1953), Society of Technical Writers and Publishers (1953), International Science Foundation (1954), Scientists of Tomorrow (1955), National Federation of Science Abstracting and Indexing Services (1958), Aviation/Space Writers Association, an outgrowth of Aviation Writers Association (1960), International Federation of Information Processing Societies (1960), and the Future Scientists of America (1960).

Physical Science

It was the workers in the physical sciences who appeared to bear the brunt of the attack on the microcosmic world of the atom and the macrocosmic world of the universe. Quite naturally, as groups of them pursued their special advances, they found the older organizations insufficient and came to demand new ones.

Mathematicians, burdened with ever-lengthening computations, turned increasingly to mechanical computers, sometimes to the super-devices popularly called "electronic brains." Smithsonian Astrophysical Observatory scientists, for example, found that they could carry out computations in a few hours that had taken Kepler several years and with far less errors than he had made. Or again, to take another example, a few seconds of computer time could help unravel the mysteries of ancient ruins at Stonehenge. Among the new societies devoted to mathematical computations were Association for Computing Machinery (1947), Industrial Mathematics Society (1949), National Joint Computer Committee (1950), and the Society for Industrial and Applied Mathematics (1952).

Amateur astronomers and telescope makers in the "Space Age" continued to form many local astronomical and telescope makers socie-ties.[2] These often provided nuclei for the formation of "Moonwatch" teams, organized by the Smithsonian Astrophysical Observatory for the optical observation of the numerous artificial satellites put into space by the American and Russian governments. These teams were especially useful in (1) observing the satellites when they were first put into orbit, (2) tracking them when they were coming down, and (3) in locating lost satellites.

Mention should also be made of some rather specialized societies that were organized to cope with the technicalities of the space age. In this category were Rocket City Astronomical Association, Hunts-ville, Alabama (1954), which stressed the observational, computational, and applied phases of astronomy and related sciences; Astronautical Society of America (1953); Association for Applied Solar Energy (1954); and the American Institute of Aeronautics and Astronautics formed in 1963 by a merger of the Institute of Aerospace Science and the American Rocket Society. The Air Force Association (1946) continued strong advocacy of American air and space powers in its monthly review called *Air Force and Space Digest: The Magazine of Aerospace Power.*

The emphasis in physics was on nuclear studies and radiation. Indic-ative of this were the several new organizations dealing with the structure of matter: the Educational Foundation for Nuclear Science (1945), the Oak Ridge Institute of Nuclear Studies (1946), Associated Universities, Inc. (1946) to operate Brookhaven National Laboratory under contract from the Atomic Energy Commission, American Crystal-lographic Association (1949), Radiation Research Society (1952), Coblenz Society (1954) for infrared spectroscopy, and the American Nuclear Society (1955). Other physics societies included Association of Professional Photogammetrists (1951) and Experimental Research So-ciety (1955). In 1956 the American Institute of Physics began a project under which eight Soviet physics journals were regularly translated into English. "Before the project, about one article in a hundred, appearing in the American Physical Society publication the *Physical Review,* cited data from Russian sources. Now one of every five articles carries such data."[3]

Expanding civilian construction and government projects placed a greater premium than ever before on engineers and architects. The

---

2 For a list, see *Sky and Telescope,* XXVIII, no. 4 (October, 1964), 220-225.
3 *The New York Times,* Sunday June 20, 1964, E-7.

"space industry" alone, accounting for some five billion dollars a year of spending by NASA, plus much other spending by private concerns, was heralded as a "new industry" comparable in magnitude to the auto industry. Cities and states were soon vying for such things as a "moon center" or an "electronics center," as Presidents Kennedy and Johnson talked hopefully of placing an American first on the moon by 1970. In another trend, automation was making rapid inroads into American industry.

Among the more active of the newer engineering and related societies in the space age and the years immediately preceding it were Engineers Joint Council (1945), Institute of Navigation (1945), Society of Packaging and Handling Engineers (1945), Water Conditioning Association International (1945), American Society of Body Engineers (1945), Pacific Rocket Society (1946), Armed Forces Communications and Electronics Association (1946), Standards Engineering Society (1947), Foundation for Instrumentation and Research (1947), Society of Photographic Scientists and Engineers (1948), U.S. National Committee on Theoretical & Applied Mechanics (1948), Association of Federal Communications Consulting Engineers (1948), Audio Engineering Society (1948), Water Conditioning Foundation (1948), Pan-American Federation of Engineering Societies (1949), Society of Women Engineers (1949), Society of Fire Protection Engineers (1950), Water Conditioning Research Council (1950), Methods Time Measurement Association for Standards and Research (1951), Experimental Aircraft Association (1953), Combustion Institute (1954), Society of Photographic and Instrumentation Engineers (1954), Supersonic Tunnel Association (1954), Gas Appliance Engineers Society (1954), Society of Die Casting Engineers (1955), Society of Experimental Test Pilots (1955), Society of Reproduction Engineers (1956), American Association of Cost Engineers (1956), National Association of County Engineers (1956), Consulting Engineers Council (1956), Space Education Foundation (1956), Society of American Registered Architects (1956), Metallurgical Society (1957), Metals Engineering Institute (1957), American Automatic Control Council (1957), Etched Circuit Society (1957), Association of Professional Draughtsmen (1959), Manufacturing Engineering Council (1959), and the Fluid Power Society (1960).

A most important development in engineering took place toward the end of 1964 with the creation of the National Academy of Engineering. Established under authority granted to the National Academy of Sciences in its act of incorporation in 1863, the NAE shares with the NAS the responsibility for advising the government on matters of science and engineering, working closely with the NAS in conducting the affairs of the National Research Council.[4]

Among the more important of the newer technological societies appearing in listings of such organizations in the space age were: Fuels Research Council (1945), Reinforced Concrete Research Council (1948), American Material Handling Society (1949), Record Industry Association of America (1952), Building Research Institute (1952), Atomic Industrial Forum (1953), American Vacuum Society (1953), Ultrasonic Manufacturers Association (1956), Printing Paper Manufacturers Association (1957), and the American Society for Abrasives (1958). Such older organizations as the American Society for Testing Materials now had new tasks in such fields as the ceramics of nose cones for satellites reentering the earth's atmosphere.

Earth Science

Entry into the space age posed many new problems for the earth scientists — geographers, geologists, meteorologists, and oceanographers — and it also furnished them with new research facilities. In particular, the earth satellites were soon affording novel means of obtaining information. Thus, for example, the Vanguards and subsequent satellites gave new information concerning the shape of the world, and scientists were soon talking about a "pear-shaped earth." Geodesy, too, was being revolutionized by these new triangulation points in the skies. The Tiros satellites enabled meteorologists to photograph weather conditions over oceans, deserts, and arctic wastes hitherto not readily accessible to meteorologists and to secure simultaneous data from large areas of the earth's surface. Information on radiation from the sun that came to be called the "solar wind" and on the hitherto unknown Van Allen belts, and data on upper air density were other early products of the satellite era.

Reminiscent of the operations of the earth scientists in the IGY, operations which incidentally were continued on a small scale after the IGY officially ceased, were new and far-flung programs in which American scientists took part in the so-called International Years of the Quiet Sun, 1964–1965. Arctic and antarctic studies not only threw new light on the powerhouses of atmospheric physics but also reached

---

[4] John Walsh, "NAE: Search for a Form Produces a National Academy of Engineering in a 'Partnership' with NAS," *Science*, CXLVI (1964), 1661-1662, gives the background of NAE and the "objects and purposes" as set forth in its articles of organization.

back into the glacial epochs of the earth's history. A newcomer, fighting for respectability among the earth sciences, was "oceanology."

The older National Geographic Society (1888), in this period, continued to map the earth and the heavens, finance mountain climbing expeditions, and with almost daily press releases keep the public informed on geographical advances. The Association of American Geographers, too, continued to be very active. Among the newer agencies entering the field were: the American Congress on Surveying and Mapping (1944), the American Antarctic Association (1944), and the Arctic Institute of North America (1945). The Antarctic Treaty of 1960, signed by twelve countries including the United States and the U.S.S.R., provided agreement on the continued use of Antarctica for scientific research on "the last geographic frontier."[5]

In an era of ever-increasing geological specialization, among the newer organizations were: the Geochemical Society (1955), The Cave Research Associates (1959), and the Society of Professional Well Log Analysts (1959). The American Geological Institute (1948) sought a broadly popular approach to geological problems through its publication *Geotimes*. Fighting for the better teaching of geology was the National Association of Geology Teachers (1951).

In the realm of meteorology, while the American Meteorological Society and the World Meteorological Organization gave cohesion to the workers in the conventional branches of the science, a host of new matters were coming to call for attention. Among them were such things as unprecedented air pollution, radioactive contamination of the atmosphere, climatological cycles, and the reduction of upper air data obtained from aircraft, rockets, and satellites. Then too, there were persons, often looked upon with polite scepticism in scientific circles, like the founders of the Aerial Phenomena Research Organization (1952), who sought to challenge the traditional explanations of astronomers and meterologists for what they now came to call UFO (unidentified flying objects). As earthling scientists were busy hurtling rockets toward the moon, sun, Venus, and Mars and receiving radio waves from Jupiter, they came to take a new look at the solar system and the planetary atmospheres. The new investigations involved problems that could be solved, if at all, only by "teams" of astronomers, meteorologists, physicists, and engineers.

Oceanography, which had its infancy in America in the days of

[5] *Scientific American*, CCVII, no. 3 (September, 1962), was devoted to the antarctic scientific programs with special reference to American participation therein.

Franklin's studies of the Gulf Stream, and the early marine societies of New York (1770), Newburyport (1772), and Salem (1799), and its youth in the time of the great Matthew Maury, seemed now to come of age in the work of such organizations as the American Society of Limnology and Oceanography (1935) and the Institute of Navigation (1946). Oceanographers, by the 1950's and 1960's, were plotting the course of newly discovered networks of mountain ranges submerged under the seven seas, were finding deep ocean currents and off-shore canyons, and were working with geologists to take core samples from the ocean floors to unravel the mysteries of long-gone ages. Daring scientists, aided by federal funds and oil company derricks, were planning various "Moholes" in the Caribbean and in the Pacific, borings under the seas and through the earth's crust and into its mantle. The earth scientists frequently were known to express the belief that too large a proportion of the federal funds for science were going into space exploration, which was very costly, and that too little was spent to learn what is hidden in the earth under men's feet.

Biological Science

Whereas in years past, biologists would perhaps not be expected to have much more than a speaking acquaintance with astronomers, in the space age the two groups of scientists were now often found working together on such problems as that of life elsewhere in the universe or Man's possibilities of survival in outer space or on Mars, Mercury, Venus, or Jupiter. Especially, Project Apollo, the project designed to put men on the moon by about 1970, called forth the combined labors of a whole gamut of American scientists. Radiation effects, particularly radiation hazards, became of great concern to both astronomers and biologists. Biologists, already engrossed in revolutionary advances in genetics and cell structure, now were to be found disagreeing over the effects of nuclear and solar radiation on living organisms and on generations as yet unborn. The many new problems confronting the broad spectrum of workers known collectively as "biologists" naturally called forth new scientific societies or redirected the activities of some of the older ones.

General societies dealing with biological science included the American Society of Professional Biologists (1947), Biometric Society, Eastern North American Region (1947), Biometric Society, Western North American Region (1947), American Institute of Biological Sciences (1948), Institute of Environmental Sciences (1957), Biophysical Society (1958), and the American Board of Bio-Analysts (1959). In 1964, the American Institute of Biological Sciences, a federation of biologists, appealed to the National Aeronautics and Space Administration to start planning for landing life-detecting instruments on Mars in 1969. This appeal was contained in an editorial in the June issue of its publication called *Bioscience* and expressed concern lest NASA let the opportunity slip for what the society considered one of the grandest, most momentous experiments of all time — finding out whether life exists on other planets.

Among the societies dealing one way or another with things zoological that were founded either in the space age or shortly before it might be mentioned the following: Tissue Culture Association (1946), Society of General Physiologists (1946), Society of Protozoologists (1947), American Society of Human Geriatrics (1948), Society of Systematic Zoology (1948), Wild Life Disease Association (1951), Society for the Investigation of Human Ecology (1955), and the Institute of Fisheries Research (1956). In an era of ultra-high-powered microscopes, there sprang up the Society for Industrial Microbiology (1948), the Foundation for Microbiology (1951), and the American Academy of Microbiology (1955).

American agriculture in the space age was carried on by fewer farmers than previously but with such efficiency that more was raised than ever before, and huge surpluses became available to meet the ever-expanding needs of the "population explosion" of Latin America, Europe, and Asia. Even Russia made large purchases of American wheat.

Among the more active of the newer agricultural, conservational, and forestry associations concerned with these trends over the last two decades or so, the following might be particularly mentioned: American Farm Research Association (1944), the American Grassland Council (1944), National Flying Farmers Association (1944), National Peach Council (1945), Agricultural Aircraft Association (1945), National Association of Soil Conservation Districts (1946), American Society of Range Management (1947), Association of American Pesticide Control Officials (1947), Cotton History Group (1947), Conservation Education Association (1947), Defenders of Wildlife (1947), Association of Consulting Foresters (1948), International Union for Conservation of Nature and Natural Resources (1948), Conservation Foundation (1948), National Potato Council (1948), Council of Conservation (1949), American Watershed Council (1950), National Dried Bean Council (1950), Weed Society of America (1950), Agricultural Research Institute (affiliated with NAS-NRC) (1951), American Poultry Historical Society (1952), U.S. National Committee, International Commission on Irrigation and Drainage (1952), Agricultural Relations Council (1953), American Association of Feed Microscopists (1953), Rare Breeds Poultry Club of America (1953), Crop Science Society of America (1954), Citizens Committee on Natural Resources (1954), American Mushroom Institute (1955), American Sheep Producers Council (1955), Food Facilities Engineering Society (1955), Grain Sorghum Producers Association (1955), International Association of Agricultural Librarians and Documentalists (1955), National Beef Council (1955), Hog Growers of America (1956), American Conservation Association (1958), Trout Unlimited (1959), and the American Federation of Poultry Producers Associations (1960).

In the space age, the profession of medicine, probably the most highly organized of all the professions, had not only its local and state medical societies and the always very active American Medical Association now over a century old, and a long legacy of specialized societies from former eras, but also now came to support a bewildering array of new, highly specialized medical societies. A somewhat arbitrary list of some of the newer societies might include the following: American Association of Clinical Chemists (1949), the American Academy for the History of Dentistry (1950), American College of Foot Orthopedists (1950), American Registry of Medical Assistants (1950), Chiropody Bibliographical Research Society (1950), Muscular Dystrophy Associations of America (1950), Student American Medical Association (1950), Academy of Dentistry for the Handicapped (1951), American Academy of Implant Dentures (1951), American Association of Cleft Palate Rehabilitation (1951), American Congress of Neurological Surgeons (1951), American College of Obstetricians and Gynecologists (1951), American Medical Education Foundation (1951), Inter-Society Cytological Council (1951), National Association of Dental Laboratories (1951), Mended Hearts (1951), Walter Reed Society (1951), American Nutrition Society (1952), American Osteopathic College of Anesthesiologists (1952), National Committee Against Fluoridation (1952), International Fertility Association (1952), Pan American Cancer Cytology Society (1952), Society of Pelvic Surgeons (1952), Society of Public Health Educators (1952), Affiliated Chiropodists-Podists of America (1953), American Association of Correctional Psychologists (1953), Association of State and Territorial Public Health Nutrition Directors (1953), National Citizens Committee for the World Health Organization (1953), National Student Nurses Association (1953), Psoriasis Research Association (1953), American College of Preventive Medicine (1954), Association of Hospital Directors of Medical Education (1954), Association of Professors of Medicine (1954), Dysantonomia Association (1954), National Committee for Careers in Medical Technology (1954), Society for Nuclear Medicine (1954), American Academy of Psychotherapists (1955), American Sanitary Engineering Intersociety Board (1955), American Society for Artificial Internal Organs (1955), American Association of Hospital Purchasing Agents (1955), Joint Blood Council (1955), Cardiac Society (1955), Gastroenterology Research Group (1955), American Society of Internal Medicine (1956), American Association of Medical Assistants (1956), American Association of Bioanalysts (1956), American Association of Veterinary Nutritionists (1956), Health Physics Society (1956), Medico-Alert Foundation International (1956), National Eye Research Foundation (1956), National Medical Foundation for Eye Care (1956), Educational Council for Foreign Medical Graduates (1957), National Tay-Sachs Association (1957), Physicians Council, Inc. (1957), American Association for Social Psychiatry (1958), Association of Medical Record Consultants (1958), Contact Lens Association for Optometry (1958), Institute for Advancement of Medical Communication (1958), Medical International Cooperation Organization (1958), National Association of Physically Handicapped (1958), Society for Adolescent Psychiatry (1958), American Patients Association (1959), Group Health Association of America (1959), Joint Council to Improve Health Care of the Aged (1959), American Society for Chemical Nutrition (1960), National Association of Optometrists and Opticians (1960), and the International Society of Professional Ambulances Services (1960).

Social Science

In the space age, workers in the social sciences often felt themselves somewhat on the defensive as they sought to cope with the ever-mounting problems of the United States and the world at large. For one thing, nineteenth-century faith in constructing a "science of society" now seemed a bit naive. Rightists and leftists were pitted against each other in the Congress of the United States and in the United Nations and its subsidiary, UNESCO. Social scientists often differed sharply among themselves as to the remedies for society's ills. Successive congresses that voted billions for research in space science and hundreds of millions for medical and biological research generally remained unimpressed by the appeals of workers in the social sciences and humanities for even much smaller amounts of the federal largesses. Moreover, historians, sociologists, and archaeologists often found their time-honored preserves invaded by chemists and geologists, who, using C-14 and other techniques, were revolutionizing Man's conception of the antiquity of Man. Perhaps even more disturbing to traditional ideas of the "evolution of the race" or the "perfectibility of Man" were the creations of the atomic scientists and the forebodings of the geneticists concerning the impact of nuclear warfare or even testing in the atmosphere. Many of the social scientists were in the forefront of those who campaigned successfully for the ratification of the Nuclear Test Ban Treaty ratified by the United States, the Soviet Union, and Great Britain in 1963 and which prohibited nuclear testing in the atmosphere, in outer space, and under water, and even more significantly pointed to the easing of Cold War tensions. Social scientists were now a bit freer to reflect that the peaceful uses of atomic energy might augur a great new society for all mankind. In any event, the social scientists, finding much work cut out for them, lost no time in forming new scientific and learned societies to help them attack their problems.

The sociologists seemed to be especially active. Among their newer organizations were the Society for the Study of Social Problems (1949), the Council on Social Work Education (1952), the National Association of Social Workers (1955), and the Population Association of America (1958).

While historical societies lie mainly outside the purview of the present study, it might be noted in passing that many new historical societies came into being in the 1950's and 1960's and that historical societies often, incidentally, collect and record for posterity much data on "natural history." The History of Science Society and the Philosophy of Science Association came to find it expedient to have joint meetings. The Renaissance Society of America (1954) was organized to deal broadly with an epoch of great advance in human history. The National Council on Asian Affairs (1955), adopting the regional approach, announced that it was formed to stimulate "knowledge of Asia."

With an increasing sense of horror, the nation's archaeologists came to realize that many of their most cherished sites were being menaced with total and irrevocable destruction by the advance of large engineering projects in America and abroad. Great dams like the Aswan in Egypt and many in America threatened to flood whole valleys and sent archaeologists hurrying and scurrying to preserve as best they could such artifacts of the past as still remained accessible, while impatient bulldozer crews were forced to wait. Numerous local chapters of state archaeological societies were formed. The work of the archaeologists, many of them devoted week-end amateurs, made it necessary to rewrite the history of early America from Plymouth to the Southwest and from Bering Strait to Yucatan and Peru. In particular, C-14 dating revolutionized the chronology of Indian archaeology. Colonial archaeology came to buttress, or sometimes refute, colonial tradition and history. Among the leading enterprises involving colonial archaeology were the Colonial Williamsburg Restoration and Plimoth Plantation. Americans were not only interested in the buried past of their own land and the New World but helped finance diggings in Italy, Iran, Iraq, Egypt, and many other countries. The Council for Old World Archaeology was set up in 1953. World War II had opened the eyes of Americans to the possibilities of ethnological and archaeological research in the Southwest Pacific, and now, a number of American expeditions went there too.

Even in an age when men talked glibly about the "expanding universe," Man remained very much the proper study of Mankind. Psychologists, who had been three-quarters of a century earlier a breed like social scientists, who had been compelled to fight for public recognition to be considered among the ranks of "scientists" and not as "philosophers," now found themselves called upon, as never before, to give advice on many problems of personal or social adjustment or maladjustment. In particular, the stresses of modern life seemed to, unhappily, bring about a so-called "plague of decomposing personalities." It was, in part, against this kind of background that new organizations were called for, among them being: the American Group Psychotherapy Association (1942), the Individual Psychology Association of New York (1943), the National Psychological Association for Psychoanalysis (1948), the American Catholic Psychological Association (1948), the Interamerican Society for Psychology (1951), the National Council on Psychological Aspects of Disability (1952), and the Human Factors Society (1957).

Epilogue

More than two centuries have elapsed since colonial Americans began their experience with the first scientific society to serve them, the Royal Society of London. In every period of American history — colonial, early national, later national, and recent — the scientific societies have been a potent force in the intellectual life of the American people. In times of national crisis — Civil War, World War I, World War II, and the Cold War — they have been called upon to help mobilize the nation's scientific manpower resources. At all times, they have served as citadels on the "advancing front of science."

A CHRONOLOGY OF SCIENCE AND TECHNOLOGY IN THE UNITED STATES

1000? Voyage to Vinland by Lief Ericson
1492 Columbus discovered America
1513 Balboa reached the Pacific Ocean
1541 de Soto discovered the Mississippi River
1588 Thomas Hariot's *A Briefe and True Report of the New Found Land of Virginia*
1607 The Virginia of Sagadahock, Maine — first vessel launched by the English in America
Jamestown — first permanent settlement in the South
Glassmaking at Jamestown
1612 John Smith's *A Map of Virginia*
1614 Tobacco planted in Virginia
1616 John Smith's *A Description of New England*
1620 Plymouth — first permanent settlement in the North
First physician arrived in America — Samuel Fuller, at Plymouth
1621 First windmill in America — Virginia
Cotton introduced into Virginia
1622 First gristmill in Virginia
1623 Leather tanning introduced, Plymouth, Mass.
1624 Captain John Smith's *General History of Virginia and New England*
1626 First flour mill in colonies built at New Amsterdam
1631 First sawmill in America, Portsmouth, N.H.
1636 Harvard College founded
1639 First printing press in English Colonies, at Cambridge, Mass.
1643 Tide mill, Hingham, Mass.
1644 First complete ironworks in America, Saugus, Mass.
1645 First American patent — granted by Massachusetts to Joseph Jenks for scythe-grinding machinery
1658 First hospital in America — New York
1662 Royal Society of London — served as scientific society for colonists
1677 First medical work printed in America — *Guide to the Common People of New England*
1683 Boston Philosophical Society
1687 Sir Isaac Newton's *Principia*
1691 Yellow-fever epidemic in Boston
1693 College of William and Mary founded

1699 Yellow-fever epidemic in Philadelphia
1701 Yale founded
1721 Cotton Mather and Dr. Nathaniel Boylston introduced inoculation for smallpox in America, during Boston epidemic
1724 *Angel of Bethesda*, by Cotton Mather — first general treatise of medicine in the colonies
1727 Franklin's Junto
Colden's *History of the Five Indian Nations*
Harvard established professorship of natural philosophy — Isaac Greenwood, first encumbent
1728 First Botanic Garden established by John Bartram in Philadelphia
1729 Isaac Greenwood's *Arithmetick Vulgar and Decimal* — first arithmetic textbook by an American
1730 Thomas Godfrey invented a marine quadrant
Thomas Cadwalader gave the earliest-known American anatomical lectures in Philadelphia
Philosophical Society — Newport, R.I.
1735 Roland Houghton's "improved theodolite"
Scarlet fever epidemic in New England
Medical Society formed in Boston
1736 Dr. William Douglass published *The Practical History of a New Epidemical Eruptive Military Fever in . . . Boston . . . 1735 and 1736.* Earliest clinical description of Scarlet fever
1741 Indigo culture initiated in South Carolina, Eliza Lucas
1742 Franklin stove or Pennsylvania fireplace
1743 American Philosophical Society
Cadwallader Colden published first American public health articles
1746 College of New Jersey (Princeton) founded
1748 *Essays on Field-Husbandry in New England* — Rev. Jared Eliot
1751 Franklin's *New Experiments and Observations on Electricity*
The Pennsylvania Hospital, the first large-scale hospital in colonial America, opens its doors in Philadelphia
First sugarcane grown in America brought to Louisiana
John Bartram's *Observations on American Plants*
1752 First dispensary in America opened in Pennsylvania Hospital in Philadelphia
Franklin's kite experiment proving the electrical nature of lightning. Lightning rod invented by Franklin
Gregorian calendar adopted in England and the colonies
1754 King's College (Columbia) founded
Franklin's Academy chartered (became University of Pennsylvania, 1791)
1755 Great New England Earthquake
1759 Halley's Comet reappeared. John Winthrop of Harvard lectured on comets

1761 Transit of Venus observations — Winthrop expedition to Newfoundland
1762 William Shippen's anatomical lectures
1763–1767 Survey of Mason-Dixon line
1764 Rhode Island College (Brown) founded
1765 First medical school in the United States opened at the University of Pennsylvania
   Survey of the Mississippi River by Lt. Ross
   John Bartram appointed royal botanist in America
1766 The New Jersey Medical Society established — first state medical society in America
   Queen's College (Rutgers) founded
1767 David Rittenhouse manufactured an orrery — first planetarium in America
1769 Stiegel glass works opened at Manheim, Pa.
   Dartmouth founded
   Anthracite coal used at Wilkes-Barre, Pa.
   Transit of Venus observations, David Rittenhouse, etc.
1770 Boston Medical Society
   Benjamin Rush, *Syllabus of a Course of Lectures on Chemistry* — first American textbook on chemistry
1773 Oliver Evans — experiments with steam propulsion
   First insane asylum in America opened at Williamsburg, Va.
1774 First Continental Congress, Philadelphia, encouraged domestic manufactures
1775 *American Turtle,* first submarine built by David Bushnell. He invented marine torpedo
   First American book on surgery published by John Jones
1776 Phi Beta Kappa founded at College of William and Mary
   Thomas Bond presented first systematic course of clinical lectures in America
1777 *Directions for Preserving the Health of Soldiers,* by Benjamin Rush
1778 Captain James Cook explored the Pacific Coast
   Pharmacopoeia (first in United States) written by William Brown
1780 Harvard Eclipse Expedition — state aided
   Constitution of Massachusetts. Provided for encouragement of science
   American Academy of Arts and Sciences
1781 Thomas Hutchins designated Geographer of the United States
   Benjamin Rush, *Inquiry into the Effects of Spiritous Liquors on the Human Body and Mind*
   Massachusetts Medical Society
1782 Harvard Medical School founded
1784 Franklin invented bifocals
   Notes on Virginia, by Thomas Jefferson
   Geography Made Easy, by Jedediah Morse

1785 Philadelphia Society for Promoting Agriculture
South Carolina Agricultural Society
Land Ordinance — provided for rectangular surveys
University of Georgia (first state university to be chartered, but not established until 1801)
Oliver Evans developed automatic flour mill

1786 Académie des Etats-Unis de l'Amérique
First machine-made nails in United States by E. Reed
Franklin's study of the Gulf Stream

1787 Northwest Ordinance provided for public education
Constitutional Convention
The College of Physicians of Philadelphia founded
John Fitch invented steamboat
Pennsylvania Society for the Encouragement of Manufactures and Useful Arts organized
High-pressure steam engine — Oliver Evans

1789 Federal Government organized
A Survey of the Roads of the United States of America, compiled by Christopher Colles, published in New York — contained first known road maps published in the United States
University of North Carolina founded (first state university to begin instruction, 1795)

1790 System of weights and measures
First U.S. patent law. Samuel Hopkins received first patent for method of making "pot-ash and pearl-ash"
First U.S. copyright act
U.S. Patent Office opened
First U.S. decennial census
Samuel Slater built first successful cotton mill in United States at Pawtucket

1791 Hamilton's *Report on Manufactures*
New York Society for the Promotion of Agriculture, Arts and Manufactures
Travels through North and South Carolina, Georgia, East and West Florida, by William Bartram

1792 U.S. mint
Cotton gin invented by Eli Whitney
Captain Gray discovered the Columbia
Screw propeller invented by John Stevens
Chemical Society of Philadelphia

1793 Yellow-fever epidemic in Philadelphia, described in *A Short Account of the Malignant Fever Lately Prevalent in Philadelphia and a List of the Dead from August 1 to the middle of December 1793,* by Mathew Carey, published in 1794
Blanchard made first successful balloon ascent in United States
Plow mouldboard invented by Thomas Jefferson

1794 Peale's museum started in Philadelphia
First U.S. turnpike completed between Philadelphia and Lancaster
1796 First medical patent in United States granted to Elisha Perkins
1797 *The Medical Repository* founded, first U.S. medical journal
Cast-iron plow patented by Charles Newbold
1798 Marine Hospital Service
Eli Whitney developed the concept of the interchangeable part in arms production
1799 Connecticut Academy of Arts and Sciences
Nathaniel Bowditch's *Practical Navigator*
1800 Washington became the U.S. capital
Library of Congress
Benjamin Waterhouse introduced cowpox vaccination into United States
1801 Robert Hare invented oxyhydrogen blowpipe
1802 Army Engineers
U.S. Military Academy
1803 Lewis and Clark Expedition
1807 U.S. Coast Survey
Fulton's steamboat
1812 American Antiquarian Society
Academy of Natural Sciences, Philadelphia
1816 Columbian Institute
1818 Army Medical Department
1819 Iron plow, Jethro Wood
1824 Survey Act
1825 Babbitt metal
Erie Canal opened
1828 Electromagnetic telegraph, Joseph Henry
1830 Navy's Depot of Charts and Instruments established
1831 McCormick reaper
1842 Frémont's expedition to the Rockies
U.S. Botanic Garden
Naval Observatory
1843 Captain Sumner's method of finding a ship's position at sea
1844 First successful telegraph line, Baltimore to Washington — Morse
1845 U.S. Naval Academy
1846 Smithsonian Institution
Use of ether as an anaesthetic by William Morton
Rotary printing press invented by Hoe
Sewing machine invented by Elias Howe
1847 American Medical Association
Owen's survey of federal lands
1848 American Association for the Advancement of Science
1849 Nautical Almanac Office
1850 W. C. Bond at Harvard took first stellar photograph

1856   Matthew F. Maury's *Physical Geography of the Sea*
1861   Government Printing Office
      Massachusetts Institute of Technology
1862   Department of Agriculture
      Morrill Act, established Land-Grant Colleges
      Battle of the "ironclads" — the Merrimac and the Monitor
1863   National Academy of Sciences
1866   Navy Hydrographic Office
      Metric system of weights and measures authorized by Congress
      Atlantic cable to Britain, Cyrus W. Field
1867   King's Geological Survey of the Fortieth Parallel
      Army Medical Museum
1868   Army Medical Library
      Airbrake perfected by Westinghouse
1869   Wheeler's Geographical Surveys
      First state board of health — Massachusetts
1870   Army Signal Corps commenced meteorological work
      Powell's Survey of the Colorado River
1871   U.S. Fish Commission
1872   *Popular Science Monthly*
      American Public Health Association
1873   Hayden's Territorial Surveys
1874   Barbed wire patent, Glidden
1876   Telephone invented by Bell
      American Chemical Society
      Rule of Phase — J. W. Gibbs
      American Library Association
      Centennial Exposition
1877   Moons of Mars discovered by Asaph Hall
1878   Interferometer invented by A. A. Michelson
1879   Incandescent lamp invented by Edison
      U.S. Geological Survey
      National Board of Health
      Archaeological Institute of America
      Michelson measured the speed of light
1880   Bureau of Ethnology
1882   American Forestry Association
1883   Civil Service Commission
      Standard time established for whole country
1884   American Institute of Electrical Engineers
      International Prime Meridian Conference at Washington decided on
          Greenwich as common prime meridian for world
      Bureau of Animal Industry
1886   Division of Economic Ornithology and Mammalogy
1887   Hatch Act — Agricultural Experiment Stations
      Marine Laboratory established at Woods Hole, Mass.

1888   Geological Society of America
       National Geographic Society
1889   American Academy of Political and Social Science
1890   Weather Bureau
1891   Smithsonian Astrophysical Observatory
1893   U.S. Army Medical School
1896   National Academy Committee on Forestry
1897   Yerkes Observatory — world's largest refracting telescope
1899   Astronomical and Astrophysical Society of America (since 1914, American Astronomical Society)
       American Physical Society
1901   Bureau of Chemistry
       Bureau of Plant Industry
       Bureau of Soils
       National Bureau of Standards
1902   Newlands Reclamation Act
1903   First successful airplane flights — Wright Brothers
       Philippine Bureau of Science
       Committee on Organization of Scientific Work
1905   Forest Service
       De Forest invented three-electrode vacuum tube
1906   Pure Food and Drug Act
1908   Governors Conference on Conservation
1909   Peary reached North Pole
1910   Bureau of Mines
1912   Public Health Service
1914   Panama Canal opened
       Smith-Lever Act — Agricultural extension program
1915   National Advisory Committee on Aeronautics
1916   National Park Service
       National Research Council
1917   Smith-Hughes Vocational Education Act
1918   Chemical Warfare Service
1926   National Research Fund
1927   Lindbergh flew across the Atlantic
1930   National Health Institute
       American Rocket Society
1932   Cyclotron built by E. O. Lawrence
1933   Science Advisory Board
1934   Agricultural Research Center
1935   National Resources Committee
1937   National Cancer Institute
1941   Office of Scientific Research and Development
1942   Agricultural Research Administration
1945   First atomic bomb
1946   Atomic Energy Commission

1948 Completion of 200″ Hale reflecting telescope at Mt. Palomar
1950 National Science Foundation
1953 Salk polio vaccine
1957 Director of Research and Engineering, Department of Defense
Science Advisor to the President and Science Advisory Committee
1958 National Aeronautics and Space Administration
National Defense Education Act
Explorer I — first successful American satellite in space
Vanguard I — detected Van Allen radiation belts
1959 Federal Council on Science and Technology
1960 Tiros I — first weather satellite
Transit I B — first navigational satellite
1962 Telstar -— first active communications satellite
Orbital flight in space by Lt. Col. John H. Glenn, Jr.
1963 Syncom II successfully orbited — stationary satellite
Mariner II sent to Venus
1964 National Academy of Engineering
Ranger VII photographed the moon
Mariner IV launched to Mars
1965 Gemini IV Flight — Major Edward White, II, "walk in space"

BIBLIOGRAPHY

Principal Guides to American Scientific Societies

1. The most up-to-date guide to American scientific societies is *Scientific and Technical Societies of the United States and Canada*, 6th ed., comp. by the Library of the National Academy of Sciences–National Research Council. National Research Council, *Publication* no. 369 (Washington, 1955). See also earlier editions.

2. An older work, also of the greatest value, is J. D. Thompson, comp., *Handbook of Learned Societies and Institutions: America* (Carnegie Institution, *Publications*, no. 39, Washington, 1908). In this work, the national societies are listed alphabetically and the local societies are arranged alphabetically by states, making the study of the geographical distribution of the smaller societies easier. This work also gives the brief history, object, membership, time and place of meeting, research funds, medals and awards, library and laboratory facilities, and the publications of each society.

3. Still older, and inferior to both the above as a list, is S. B. Weeks, comp., "A Preliminary List of American Learned and Educational Societies," U. S. Commissioner of Education, *Report for 1893-1894*, II, 1493-1661. In this work, the societies are arranged according to the fields of knowledge to which they are devoted; thus, for example, all the medical societies are listed together. It has a good historical introduction.

4. A brief list of American scientific societies which also contains an excellent historical introduction is J. McK. Cattell, "Scientific Societies and Associations," in N. M. Butler, edit., *Education in the United States* (Albany, New York, 1900), II, 865-892.

5. Of some service, also, is G. B. Goode, "The Origin of the National Scientific and Educational Institutions of the United States," American Historical Association, *Papers*, IV (1890), pt. 2, 93-202; also issued in American Historical Association, *Annual Report for 1889*, 53-161, and in Smithsonian Institution, *Annual Report for 1897*, pt. 2, 263-354.

6. A work of great value for the period which it covers is Max Meisel, *Bibliography of American Natural History* (3 vols., Brooklyn, New York, 1924-1929). This contains a chronological list of the principal scientific societies founded in the United States down to 1865. Each society is discussed in detail, many valuable historical data are given, and biographical and bibliographical data as well.

7. A work which gives useful information concerning many of the contempo-

rary scientific, technical, and professional societies of the United States is *Public Administration Organizations; a Directory of Unofficial Organizations in the Field of Public Administration in the United States and Canada* (Chicago, 1954). Earlier editions.

8. The various editions of the *World Almanac,* 1897–, contain valuable directory lists of the principal learned and scientific societies of America.

9. Judkins, Jay. *National Associations of the United States* (U. S. Department of Commerce, Washington, 1949). Includes lists of scientific, technical, and educational associations.

10. A recent work of great usefulness is Encyclopedia of Associations (3rd ed., Detroit, Mich., 1961) which contains a number of sections dealing with scientific and technical societies.

The most complete assortment of the proceedings and transactions of American scientific and other learned societies is the unrivaled collection in the Library of Congress.

Other Lists of Societies

American Almanac (New York and Washington, 1880-1889). See "Societies and Institutions."

American Council of Learned Societies, *Bulletin,* 11 (June, 1929), "Directory of Constituent Societies," 86-90.

——*Bulletin,* 17 (May, 1932), Directory Number. This contains information about the origin, membership, meetings, publications, administration, finances, and activities of a number of the leading societies. For its treatment of the administrative details of the societies, this bulletin has a unique value.

American Dental Directory, 1956. Published by American Dental Association (Chicago, 1956). Lists national, state, and international dental organizations.

American Medical Directory (19th ed., Chicago, 1956). "National Medical Organizations," 79-81. "American Medical Association," 19-28. "World Medical Association," 29-30.

American Nature Association. The Nature Almanac (Washington, 1927–). See "Associations and Clubs Interested in the Promotion of Nature Education."

American Tree Association. Forestry Almanac (Washington, 1926–). Lists national and state associations interested in forestry and conservation.

Associations and Societies: A List of Sources. Library of Congress (Washington, 1948).

Black, A. D., comp. Index of Periodical Dental Literature (Buffalo and London, 1929). See "Dental Periodicals," 21, and "Dental Societies," 22-24.

Bolton, H. C., edit. Catalogue of Scientific and Technical Periodicals, 1665-1895 (Smithsonian Miscellaneous Collections, XL, 2d ed., Washington, 1898).

Boston Public Library. Transactions and Other Serial Publications of Societies and Institutions, Currently Taken in the Library (Boston, 1893).

Bowker, R. R., edit. Publications of Societies: A Provisional List of American Scientific, Literary and Other Societies (New York, 1899).

Cassino, S. E., edit. The Naturalists' Universal Directory, 1877-1930. The title varies: 1877-1880, 1884, 1886, 1898, 1914, The Naturalists' Directory; 1881-1883, 1885, 1888, The International Scientists' Directory; 1904, The Naturalists' Universal Directory (Salem, Mass., 1877, and Boston, 1879–).

Chicago Daily News. Almanac and Yearbook, 1885—. See "National Associations and Societies."

Crane, E. J., and A. M. Patterson. Guide to the Literature of Chemistry (New York, 1927). See Appendix 5, 273-278, "Scientific and Technical Organizations," which is a world list of chemical and related societies. Appendix 6, 279-342, is a world list of periodicals devoted to chemistry.

Delawnay, H., edit. Annuaire international des sociétés savantes (Paris, 1903). This contains considerable information about American scientific societies, but American sources are generally more reliable and should be used whenever possible.

"Directory of Science and Mathematics Societies," *School Science and Mathematics* (March, 1915), 272-276; (June, 1915), 458-564; and (October, 1915), 649-650.

Dunbar, G. S. A Preliminary Checklist of Virginia's Scientific and Technical Societies and Their Publications, 1772-1959. A mimeographed list published by the Department of Geography, University of Virginia.

Good, C. V. Teaching in College and University (Baltimore, 1929). See Appendix, 443-498, for references to American learned and professional societies.

Handbook of American Museums. Published by the American Association of Museums (Washington, 1932).

Handbook of Scientific Societies in Japan (1954) and Handbook of Scientific Societies in Japan—Humanities & Social Sciences (1955). Published by Japanese National Commission for UNESCO.

Handbook of the Learned and Scientific Societies and Institutions of Latin America. Comp. by H. O. Severance (Washington, 1940).

Historical Societies in the United States and Canada: A Handbook (Washington, 1944).

Index Generalis, General Year Book of the Universities, High Schools, Academies, Archives, Libraries, Scientific Institutions, Botanical and Zoological Gardens, Museums, Observatories, and Learned Societies (Paris, 1919–).

Industrial Arts Index (New York, 1913–). Annual volumes contain valuable lists of technical societies and many references to the publications of scientific and purely technical societies.

International Directory of Anthropological Institutions. Ed. by W. L. Thomas, Jr. and A. M. Pikelis (New York, 1953). "United States of America," by E. W. Voegelin, 331-427.

Joint Commission. Directory of Scientific Societies of Washington (Washington, 1889).

League of Nations, International Bureau. Handbook of International Organizations (Geneva, 1926, 1931).

Mental Hygiene, I (1917) 663-665, "Directory of Societies and Committees for Mental Hygiene."

Minerva, Jahrbuch der Gelehrten Welt (Strasbourg, 1891–). This work lists thousands of learned and scientific societies and institutions throughout the world, including many American.

Patterson's American Education (Chicago, 1956–). Published as Patterson's American Educational Directory, 1904-1953. See "Educational Associations and Societies."

Reverdin, Henri. The Principal Academies and Learned Societies. League of Nations Committee on Intellectual·Cooperation (Imp. Réunies, Chambéry). Contains fairly detailed accounts of the American Philosophical Society, the American Academy of Arts and Sciences, the American Association for the Advancement of Science, the National Academy of Sciences, and the National Research Council.

Rhees, W. J., edit. Manual of Publications, Libraries, Institutions and Societies in the United States and British Province of North America (Baltimore, 1859).

Rich, W. S. American Foundations and Their Fields (7th ed., New York, 1955).

Scudder, S. H., edit. Catalogue of Scientific Serials of All Countries Including the Transactions of Learned Societies, in the Natural, Physical, and Mathematical Sciences, 1663-1876 (Cambridge, Mass., 1879). See American scientific serials.

Skallerup, H. R. "Bibliography of the Histories of American Academies of Science," *Transactions of the Kansas Academy of Science,* LXVI, no. 2 (1963), 274-281.

Smith, R. C. A Bibliography of Museums and Museum Work (American Association of Museums, Washington, 1928).

Smithsonian Institution. Catalogue of the Publications of Societies. Smithsonian Miscellaneous Collections, no. 179 (Washington, 1896).

——List of the Principal Scientific and Literary Institutions in the United States (Washington, 1879).

Spahr, E. S., and R. J. Swenson. Methods and Status of Scientific Research (New York, 1930). See ch. XV, "International Research," and ch. XVII, "Councils, Learned Societies, Universities and Foundations in Research."

UNESCO. Directory of International Scientific Organizations (Paris, 1950).

——International Organizations in the Social Sciences (Paris, 1956).

——World Handbook of Educational Organization and Statistics (Paris, 1952). "United States," 428-437.

Union List of Serials in Libraries of the United States and Canada (New York, 1927); Supplements. Since this publication lists the periodicals subscribed to by about 200 cooperating libraries and gives the bound volumes of each periodical in each library, it is of the utmost importance in locating the copies of the proceedings of the various scientific societies.

United States Bureau of Education. Education Directory (Washington, 1928). Lists educational, learned and scientific associations.

——Handbook of Educational Associations and Foundations in the United States (Washington, 1926). At times of value.

United States Bureau of Standards. Standards Yearbook (Washington, 1928–). Lists a very large number of trade, scientific, and technical associations interested in standardization activities.

United States Catalogue; Books in Print, 1899 (2 vols. in 1, Minneapolis, 1900). Brought to date by editions of Cumulative Book Index. Lists publications of many learned and scientific societies.

World List of Scientific Periodicals (Oxford, 1925).

World of Learning, The, 1956 (7th ed., London, 1957). Excellent recent lists.

Book References

Bates, R. S. "Chemical Societies," in Encyclopaedia Britannica (24 vols., Chicago, 1956), V, 353-354.

——"The Rise of Scientific Societies in the United States," Harvard University, Summaries of Ph.D. Theses (Cambridge, Mass., 1937).

——Scientific Societies in the United States (New York, 1945).

——Scientific Societies in the United States (New York, 1958).

Brasch, F. E. "Early American Learned Societies," Report of the Librarian of Congress for the Fiscal Year Ending June 30, 1932 (Washington, 1932), 263-265.

Caullery, Maurice. Universities and Scientific Life in the United States (Cambridge, Mass., and London, 1922), ch. XVII, "Academies and Scientific Societies."

Dexter, E. G. History of Education in the United States (New York, 1904), ch. XXVIII, "Learned Societies and Associations."

Dunlap, L. W. American Historial Societies, 1790-1860 (Madison, Wisconsin, 1944).

Kandel, I. L. "Scientific Societies," in Paul Monroe, edit., A Cyclopedia of Education (5 vols., New York, 1911-1913), V, 293-300.

Ogg, F. A. Research in the Humanistic and Social Sciences (New York and London, 1928).

Steeves, H. R. Learned Societies and English Literary Scholarship in Great Britain and the United States (Columbia University, Studies in English and Comparative Literature, New York, 1913), ch. VII, "American Societies and Clubs."

Periodical References

Aikens, H. A. "The Government of Learned Societies," *Science,* n. s., XXXIX (1914), 711-716.

Allen, F. H. "Bird Clubs in America," *Bird Lore,* V (1902), 12-17.

Angell, J. R. "The Organization of Research," Association of American Universities, *Proceedings,* XXI (1919), 27-41.

Armsby, H. P. "The Organization of Research," *Science,* n. s., II (1920), 33-38.

Astell, L. A. "How State Academies of Science May Encourage Scientific Endeavor among High School Students," *Science,* n. s., LXXI (1930), 445-449.

——"Inspiration Which the Junior Academy of Science Has Brought to the High School Science Clubs in the State of Illinois," *School Science and Mathematics,* XXXIII (1932), 747-757.

Bache, A. D. "On the Conditions and Achievements of American Scientific Societies," American Association for the Advancement of Science, *Proceedings,* VI (1851), preface, xli-lx.

Barker, J. H. "Scientific Societies, a Retrospect and a Prospect," Institute of Mechanical Engineers, *Proceedings,* no. 2 (1928), 499-504.

Bayley, W. S. "The Place of State Academies of Science among Scientific Organizations," *Science,* n. s., LVII (1923), 623-629.

Bentham, George. "Address . . . at the Anniversary Meeting of the Linnaean Society on Friday, May 24, 1867," *American Journal of Science and Arts,* 2d ser., XLIV (1867), 297-316.

Bernard, L. L., and J. S. Bernard. "A Century of Progress in the Social Sciences," *Social Forces,* XI (1933), 488-505.

Bogert, W. T. "American Chemical Societies," American Chemical Society, *Journal,* XXX (1908), 163-182.

Bolton, H. C. "Chemical Societies of the Nineteenth Century," American Chemical Society, Supplement to *Journal,* Twenty-fifth Anniversary Number (1902).

——"Early American Chemical Societies," *Popular Science Monthly,* LI (1897), 819-826.

Boyd, J. P. "State and Local Historical Societies in the United States," *American Historical Review,* XL (1934), 10-37.

Boyd, P. P. "The Future of the State Academy of Science," *Science,* n. s., LI (1920), 575-580.

Brewer, W. H. "The Debt of the Century to Learned Societies," Connecticut Academy of Arts and Sciences, *Transactions,* XI (1901), pt. I, preface, xlvi-lii.

Cattell, J. McK. "The Organization of Scientific Men," *Scientific Monthly,* XIV (1922), 568-578.

Clark, A. H., and L. G. Forbes. "Science in Chicago," *Scientific Monthly,* XXXVI (1933), 556-567.

Clinton, De Witt. A Discourse Delivered before the Literary and Philosophical Society of New York, July 4th, 1814 . . . (New York, 1815). Reviewed in *North American Review,* I (1815), 390-402. Treats of the achievements of the scientific societies up to that time.

Cogswell, William. "Literary and Scientific Associations in the United States," *American Quarterly Register,* XV (1842), 175-181.

Cooke, M. L. "On the Organization of an Engineering Society," *Mechanical Engineering,* XLIII (1921), 323-325, 356.

Dwight, S. E. "Notice of Scientific Societies in the United States," *Silliman's Journal*, X (1826), 369-376.

Farrar, John. "Learned Societies," *North American Review*, VIII (1818), 157-168.

Fäy, Bernard. "Learned Societies in Europe and America in the Eighteenth Century," *American Historial Review*, XXXVII (1932), 255-266.

Fink, C. G. "Scientific Meetings in War Times," *Science*, n. s., XLV (1917), 661.

German, F. E. E. "Cooperation in Research," *Science*, n. s., LXIII (1926), 324-327.

Greene, E. B. "Our Pioneer Historical Societies," Indiana Historical Society, *Publication*, X (1931), 83.

Griffin, A. P. C., edit. "Bibliography of American Historical Societies," American Historical Association, *Report, 1905*, II.

Hale, G. E. "Cooperation in Research," *Science*, n. s., LI (1920), 149-154.

——"National Academies of Science and the Progress of Research," *Science*, n. s., XXXVIII (1913), 681-698; XXXIX (1914), 189-200; XL (1914), 907-919.

Hall, Asaph. "American Scientific Societies," American Philosophical Society of Washington, *Bulletin*, VIII (1885), preface, 33-47.

Heatwole, Thelma. "Junior Academies of Science," *Science*, CXXIII (1956), 976.

Henderson, L. J. "Universities and Learned Societies," *Science*, n. s., LIX (1914), 477-478.

Henry, Joseph. "On the Organization of Local Scientific Societies," Smithsonian Institution, *Annual Report of the Board of Regents of the Smithsonian Institution, 1875*, 217-219.

Hirschfelder, J. O. "Organizing Scientists to Meet a National Emergency," *Science*, n. s., CXXI (1955), 809-810.

Johnson, G. E. "Methods to Finance the Work of the Academies," *Science*, n. s., LXXVI (1932), 373-375.

Kaplan, Joseph, and Hugh Odishaw. "Satellite Program," *Science*, CXXII (1955), 1003-1005.

Kettering, C. F. "The Future of Science," *Science*, CIV (1946), 609-614, especially "World War II in Restrospect: The Role of Our Technical Societies."

Large, Thomas. "A Function of Regional Scientific Societies," *Science*, n. s., LXVII (1928), 272-273.

Lehman, H. C. "Ages at Time of First Election of Presidents of Professional Organizations," *Scientific Monthly*, LXXX, no. 5 (1955), 293-298.

Lieb, J. W. "The Organization and Administration of National Engineering Societies," *Science*, n. s., XXII (1905), 65-73.

Livingston, B. E., J. B. Overton, and Walter Thomas. "Incorporation of Scientific Societies," *Science*, n. s., LXXV (1932), 438-439.

"Loyalty and Research," *Science*, CXXIII (1956), 660-662. Report of the

Committee on Loyalty in Relation to Government Support of Unclassified Research. Recommendations of the National Academy of Sciences.

MacDonald, William. "The Intellectual Worker," *Science*, n. s., LIII (1926), 317-321.

Mathematics Clubs. "Organizations Affiliated with the National Council of Teachers of Mathematics," *Mathematics Teacher*, XXV (1932), 419.

Miller, A. G. "Election of Officers by Scientific Societies," *Science*, n. s., XLVII (1918), 191-192.

Muesebeck, C. F. W. "National Entomological Societies Merge," *Science*, CXVII (1953), 546-547.

Munroe, C. E. "Organization of Chemists in the United States," *Science*, n. s., LXII (1925), 313-317.

Newcomb, Simon. "The Organization of Scientific Research," *North American Review*, CLXXXII (1902), 32-43.

Oliver, J. W. "A Significant Decade in Science," *Scientific Monthly*, LXVII, no. 2 (1948), 83-86.

"Organization of an Engineering Society Discussed at A. S. M. E. Spring Meeting," *Mechanical Engineering*, XLIII (1921), 538-540.

Pickering, E. C. "Associate Members of American Societies," *Popular Science Monthly*, LXXVII (1910), 286-291.

——"Foreign Associates of National Societies," *Harvard Graduate Magazine*, XVII (1908), 254-257.

——"Foreign Associates of National Societies," *Popular Science Monthly*, LXXIII (1908), 372-379; LXXIV (1909), 80-83; LXXXVI (1915), 187-192.

"Publishing the Papers of Great Men," *Daedalus; Proceedings of the American Academy of Arts and Sciences*, LXXXVI, no. 1 (1955), 47-79.

Quincy, Josiah. "Massachusetts Institution," *North American Review*, II (1816), 309-310.

Rice, C. W. "War Activities of Our Technical Societies," *Scientific American*, CXVIII (1918), 586.

Ritter, W. E. "Organization of Scientific Research," *Popular Science Monthly*, LXVII (1905), 49-53.

Roberts, E. L. "Study of Existing Science Clubs as Portrayed by Current Science Magazine Articles," *School Science and Mathematics*, XXXII (1932), 948-953.

"Scientific Institutions of Boston and Vicinity," American Association for the Advancement of Science, *Proceedings*, XXIX (1880), 767-791.

"Scientific Societies and the Government," *Science*, n. s., XLIV (1916), 312-314.

Segerblom, Wilhelm. "The State Academies of Science Affiliated with the American Association," *Science*, n. s., LXVI (1927), 517-579.

Skallerup, H. R. "Some Aspects of State Academy of Science Publications," *Science*, n. s., CXXI (1955), 904-905.

"Societies for the Diffusion of Science," *Popular Science Monthly*, X (1876-1877), 497-500.

"Society for All Biologists, A," *Science*, CIV (1946), 325-326.

"State Academies of Science," *Science*, n. s., LXVII (1928), 195-196.

Thwaites, R. G. "Bibliographical Activities of Historical Societies of the United States," Bibliographical Society of America, *Papers*, I (1906-1907).

——"State-supported Historical Societies and Their Functions," American Historical Association, *Report, 1897*, 61.

True, R. H. "The Early Development of Agricultural Societies in the United States," American Historical Association, *Report, 1920*, 295-306.

Trytten, M. H. "Advisory Committee on Scientific Personnel," *Science*, CIII (1946), 437.

"Twenty-fifth Anniversary of the Founding of the Mental Hygiene Movement, The," *Mental Hygiene*, XVII (1933), 529-568.

Webb, H. A. "Some First-hand Information Concerning Science Clubs," *School Science and Mathematics*, XXIX (1929), 273-276.

Wheeler, W. M. "The Organization of Research," *Science*, n. s., LII (1920), 53-67.

Whetzel, H. H. "Democratic Coordination of Scientific Efforts," *Science*, n. s., L (1919), 51-55.

Whitney, D. D. "State Academies of Science," *Science*, n. s., LI (1919), 517-518.

Winchell, N. H. "Review of the Formation of Geological Societies in the United States," Geological Society of America, *Bulletin*, XXV (1914), 27-30.

References to International Scientific Organizations

Abraham, H. J. "The Improvement of History Textbooks in the Interests of International Understanding," *UNESCO Chronicle*, II, no. 1 (1956), 9-14.

"Advancement of Science and Society, The: Proposed World Association," *Nature*, CXLI (1938), 169.

American Council of Learned Societies. A Program for the Improvement of American Understanding of Asian Civilizations (Washington, 1951).

Auger, Pierre. "UNESCO and the Development of Research in the Field of the Natural Sciences," *UNESCO Chronicle*, no. 2 (1955), 3-8.

Avias, Jacques. "International Organization of Scientific Documentation Based on Legislation," *Science*, CXV (1952), 250-251.

Baskerville, Charles. "International Congresses," *Science*, n. s., XXXII (1910), 652-659.

Berkner, L. V. "International Scientific Action: The International Geophysical Year 1957-58," *Science*, CXIX (1954), 569-575.

Besterman, Theodore, UNESCO: Peace in the Minds of Men (New York, 1951).

Bok, B. J. "Science in International Cooperation," *Science*, CXXI (1955), 843-847.

——"Science in UNESCO," *Scientific Monthly*, LXIII (1946), 327-332.

Boutry, G. A. "Abstracting Board of International Council of Scientific Unions," *Science*, CXXIII (1956), 423-424.

Campbell, N. R. "National Representation upon International Councils," *Nature*, CIV (1919), 72-73.

Compton, A. H. "The Place of Science in the Program of UNESCO," American Philosophical Society, *Proceedings*, XCII, no. 4 (1947), 303-306.

Conference for the Establishment of the United Nations Educational, Scientific and Cultural Organization Preparatory Commission UNESCO (London, 1946).

"Contribution of the Social Sciences to Peaceful Co-operation, The," *UNESCO Chronicle*, II, no. 10 (1956), 303-304.

Coolidge, H. J. "International Union for the Protection of Nature," *Science*, CXIX (1954), 3A.

"Co-operation with International Non-governmental Organizations," *UNESCO Chronicle*, II, no. 10 (1956), 310-313.

Cunningham, E. R. "UNESCO Initiates Cooperation in the Abstracting of Biological and Medical Sciences," *Science*, CVI (1947), 609-610.

Darboux, Gaston. "The International Association of Academies," *Nature*, LXII (1900), 249-250. Translation of "Communication de M. Darboux relative à l'Association internationale des Académies," l'Académie des Sciences, *Comptes rendus des séances de l'Académie des sciences*, CXXXI (1900), 6-9.

Documents Establishing the United Nations Educational, Scientific and Cultural Organization (London, 1945).

Evans, Luther. "UNESCO and Books," *UNESCO Chronicle*, II, no. 7 (1956), 211-218.

——"The Universal Copyright Convention," *UNESCO Chronicle*, I, no. 3 (1955), 3-5.

"First Six Months of the International Institute of Intellectual Cooperation," *Science*, n. s., LXIV (1926), 175.

"Founding of an International Academy of the History of Pharmacy" (Académie Internationale de l'Histoire de la Pharmacie), *Bulletin of the History of Medicine*, XXVI (1952), 385.

"Future of International Congresses," *Geographical Journal*, LXI (1923), 440-443.

Gibboney, C. N. "The United Nations Scientific Conference for the Conservation and Utilization of Resources," *Science*, CX (1949), 675-678.

Gray, G. W. Education on an International Scale; a History of the International Education Board, 1923-1938 (New York, 1941).

Greaves, H. R. G. The League Committees and World Order (London, 1931), ch. V, "The Committee on Intellectual Co-operation."

Hale, G. E. "The International Organization of Scientific Research," *International Conciliation*, CLIV (1920), 431-441.

Hughes, E. C. "Social Sciences and International Peaceful Cooperation," *Science*, CXXIV (1956), 1157-1158.

"Intellectual Cooperation," *Nature,* CXXVI (1930), 981-982.

"International Atomic Energy Agency," *Science,* CXXIV (1956), 973-974.

"International Campaign for Museums, The," *UNESCO Chronicle,* II, no. 10 (1956), 287-288.

International Congresses. Checklist of United States Public Documents, 1789-1909 (Washington, 1911), 944-953, "International Congresses, Conferences, and Commissions."

"International Congresses," *Science,* n. s., LXVI (1927), 626-627.

International Congress of the History of Science. Science at the Cross Roads (New York, 1931).

"International Council of Scientific Unions," *Nature,* CXXXIX (1938), 869-870.

International Institute of Intellectual Cooperation. International Institute of Intellectual Cooperation (Paris, 1933), 81-82.

"Internationalism in Science: A Movement for Centralization," *Scientific American Supplement,* LXXXIII (1912), 235.

"International Organization of Science," *Nature,* CII (1919), 341-342.

"International Scientific Centers in Paris," *Science,* n. s., LXXVI (1932), 186-187.

"International Scientific Congresses Held since 1930, or Announced for 1941 or Later." Comp. in the Library of the National Research Council (Washington, 1941).

"International Scientific Service," *Popular Science Monthly,* XII (1877-1878), 249-250.

Kellogg, Vernon. "The League of Nations Committee and Institute of International Intellectual Cooperation," *Science,* n. s., LXIV (1926), 291-292.

League of Nations. Annuaire de la société des nations, 1920-1927 (Geneva, 1927), Section K, "Coopération intellectuelle."

——Handbook of International Organizations (Geneva, 1926, 1929) and Supplement (1931).

——International Institute of Intellectual Cooperation (Paris, 1926).

——Publications, XIIA, Intellectual Cooperation, 1922-1933 (Geneva, 1922-1933).

League of Nations, Secretariat. Ten Years of World Cooperation (Geneva, 1930), ch. IX, "Intellectual Co-operation."

League of Nations Starts (London, 1920), ch. XIII, "International Associations of Various Types."

Leake, C. D. "Proposed International Association of Scientists," *Science,* CIII (1946), 148.

Leland, W. G. "The International Union of Academies and the American Council of Learned Societies," *International Conciliation,* CLIV (1920), 442-457.

——"The International Union of Academies and the American Council of Learned Societies," Institute of Historical Research, *Bulletin,* IV (1926-1927), 65-72.

Lewis, J. O. D. "Notes on the Geneva Convention," *UNESCO Chronicle,* I, no. 3 (1955), 5-8.

MacDonald, William. The Intellectual Worker and His Work (New York, 1924).

Maheu, René. "UNESCO and the National Cultural Relations Services," *UNESCO Chronicle,* II, no. 3 (1956), 75-79.

Mather, K. F. "United Nations Research Laboratories," *Science,* CXI (1950), 397-399.

"Memorandum on the Distribution of American Learned Periodicals and Publications Abroad," American Council of Learned Societies, *Bulletin,* IV (1925), 52-53.

Merriam, J. C. "International Coöperation in Research," Pan American Union, *Bulletin,* LX (1926), 219-222.

Moulton, F. R. "Scientists and International Relations," American Association for the Advancement of Science, *Bulletin,* V, no. 3 (1946), 21-23.

Murra, K. O. International Scientific Organizations; a Guide to Their Library (Documentation and Information Services, Library of Congress, Washington, D.C., 1962).

Myers, D. P. Handbook of the League of Nations since 1920 (Boston, 1930), ch. III, 4, "Intellectual Cooperation."

Newell, H. E. "International Geophysical Year Earth Satellite Program," *Scientific Monthly,* LXXXIII, no. 1 (1956), 13-21.

Research Councils in the Social Sciences, UNESCO, Series and Papers in the Social Sciences, no. 3 (1955).

Rudolph, W. M. "International Council of Scientific Unions," *Science,* CXXII (1955), 652, 654.

Ruttan, R. F. "International Cooperation in Science," *Canadian Chemistry & Metallurgy,* V (1921), 17-19.

Schuster, Arthur. "The International Association of Academies," *Nature,* LXXXIII (1910), 370-372.

——"International Cooperation in Research," *Science,* n. s., XXXVII (1913), 691-701.

"Science and International Relations," *Science,* CXXIII (1956), 1067.

"Science's Magna Charta," *New York Times* (Dec. 28, 1937).

"Scientific Organizations of the Allied Nations," *Science,* n. s., XLVIII (1918), 291.

Shimkin, M. B. "The World Health Organization," *Science,* CIV (1946), 281-283.

Spahr, W. E., and R. J. Swenson. Methods and Status of Scientific Research (New York, 1930), ch. XV, "International Research."

UNESCO. Bilateral Consultations for the Improvement of History Textbooks, Educational Studies and Documents, no. 4 (1953).

——Directory of International Scientific Organizations (Paris, 1950).

"UNESCO and the Development of Education," *UNESCO Chronicle,* II, nos. 8-9 (1956), 247-250.

Union des Organisations Internationales. Annuaire des Organisations Internationales, 1956-1957 (6th ed., Brussels, 1957).

"United Nations Committee on Effects of Atomic Radiation," *Science,* CXXIV (1956), 1019.

Visscher, M. B. "Role of Council for International Organizations in Medical Sciences," *Science,* CXXIII (1956), 337.

Wendt, Gerald. "Informal Meeting of Representatives of Associations for the Advancement of Science," *Science,* CXVII (1953), 94-96.

White, G. F. "International Cooperation in Arid Zone Research," *Science,* CXXIII (1956), 537-538.

———"Work of the UNESCO Advisory Committee on Arid Zone Research," *Science,* CXX (1954), 15.

Winslow, C. E. A. "The Movement for Scientific Internationalism at The Hague," *Science,* n. s., XXXV (1912), 293-296.

World of Learning, The, 1956 (7th ed., London, 1957).

Year Book of the International Council of Scientific Unions, The, 1956 (London, 1956).

Yerkes, R. M. The New World of Science (New York, 1920), ch. XXIII, "International Organization of Research."

Yoshuda, Masao. "The International Advisory Committee on Marine Sciences," *UNESCO Chronicle,* II, no. 7 (1956), 219-221.

Zimmern, A. E. Learning and Leadership: A Study of the Needs and Possibilities of International Intellectual Co-operation (London, 1930).

Histories of Scientific Societies

ACADÉMIE DES ÉTATS-UNIS D'AMÉRIQUE:

Adams, H. B. "Sketch of l'Académie des Sciences et Beaux-Arts des États-Unis d'Amérique, établie à Richmond," *The Academy,* II, no. 9 (1887), 403, 412.

Gaines, R. H. "Richmond's First Academy, Projected by M. Quesnay de Beaurepaire in 1786," *Collection of the Virginia Historical Society,* n. s., XI (1891), 167-175.

Hinsdale, B. A. "Notes on the History of Foreign Influences upon Education in the United States," Report of the Commissioner of Education for the Year 1897-1898 (Washington, 1899), I, ch. XIII.

Jones, H. M. America and French Culture, 1750-1848 (Chapel Hill, 1927), 478-479.

Quesnay de Beaurepaire. Memoir concerning the Academy of Arts and Sciences for the United States of America, translated by Rosewell Page (Richmond, 1922). Published in the Report of the Virginia State Library for 1920-1921.

Rosengarten, J. G. French Colonists and Exiles in the United States (Philadelphia and London, 1907), 75-76.

ACADEMY OF NATURAL SCIENCES OF PHILADELPHIA:

Dixon, S. G. "Report of the President," Academy of Natural Sciences of Philadelphia, *Proceedings,* LIII (1901), 741-748.

Morton, S. G. "The Academy of Natural Sciences of Philadelphia," *American Quarterly Register,* XIII (1841), 433-438.

——"Brief History of the Institution," Academy of Natural Sciences of Philadelphia, *Proceedings,* III (1847), 207-208.

Nolan, E. J. "Celebration of the One Hundredth Anniversary of the Founding of the Academy," *Proceedings,* LXIV (1912), 129-130.

——Short History of the Academy of Natural Sciences of Philadelphia (Philadelphia, 1909).

Phillips, M. E. "A Brief History of Academy Publications," *Proceedings,* L (1948), i-xl.

Ruschenberger, W. S. W. A Notice of the Origin, Progress and Present Condition of the Academy of Natural Sciences of Philadelphia (Philadelphia, 1852).

——Report on the Condition of the Academy (Philadelphia, 1876).

Ruschenberger, W. S. W., and G. W. Tryon. Guide to the Museum of the Academy of Natural Sciences of Philadelphia (Philadelphia, 1876), 99-199, "Summary History of the Academy."

Scharf, J. T., and Thompson Wescott. History of Philadelphia (2 vols., Philadelphia, 1884), II, 1173-1229, "Libraries and Historical and Scientific Societies of Philadelphia."

ACADEMY OF SCIENCE AND LETTERS OF SIOUX CITY, IOWA:
Iowa Journal of History and Politics, III (1905), 342-343.

ACADEMY OF SCIENCE OF ST. LOUIS:
Hendrickson, Walter B. "Natural Science and Urban Culture in the Nineteenth Century Middle West," *Transactions,* XXX, no. 9 (1958), 232-248.

Klem, M. J. "The History of Science in St. Louis," Academy of Science of St. Louis, *Transactions,* XXIII (1914), no. 10.

Meiners, E. P. "A Centennial History of the Academy of Science at St. Louis," *Transactions,* XXXI, no. 8 (1956), 1-20.

Starr, Frederick. "The Academy of Natural Science of St. Louis," *Popular Science Monthly,* LII (1898), 629-647.

Trelease, William. "The Academy of Science of St. Louis," *Popular Science Monthly,* LIV (1903), 117-131.

Whelpley, H. M. "A Sketch of the History of the Academy," Academy of Science of St. Louis, *Transactions,* XVI (1906), xx-xxx.

ACTUARIAL SOCIETY OF AMERICA:
Papers and Transactions, VI (Decennial Number), 117-135.

AGASSIZ ASSOCIATION:
Guyot, A. H. Memoir of Louis Agassiz (Princeton, 1883).

ALABAMA ACADEMY OF SCIENCE:
Barker, S. B. "Thirty-five years of the Alabama Academy of Science . . . ," *Journal,* XXXI, no. 1 (1959), 19-28.

Gardner, W. A. "The Organization of the Alabama Academy of Science," *Journal,* III (1932), 7-9.

ALBANY INSTITUTE AND HISTORICAL AND ART SOCIETY:

"History of the Institute, with an Abstract of Its Proceedings," Albany Institute, *Transactions,* I, pt. 2 (1830), appendix, 1-65.

Meads, Orlando. "Annual Address . . . on Some of the Leading Facts in the History of the Society," Albany Institute, *Transactions,* VII (1872), 1-34.

Pratt, D. J. "Manual of the Albany Institute," Albany Institute, *Transactions,* VI (1863-1865), 299-344.

AMERICAN ACADEMY OF ARTS AND SCIENCES:

American Academy of Arts and Sciences, The, 1780-1940 (Boston, 1941).

Bates, R. S. "The American Academy of Arts and Sciences," *Scientific Monthly,* LIV (1942), 265-268.

——"Baily's Beads or Williams' Beads," *Telescope,* VIII (1941), 36-38. Harvard-American Academy Eclipse Expedition of 1780.

"Centennial Celebration," *Memoirs of the American Academy of Arts and Sciences,* XI (Centennial Volume), pt. I (1882), 31.

Cross, C. C. "The Rumford Fund," *Proceedings of the American Academy of Arts and Sciences,* LVI (1920-1921), 355-371.

Jones, H. M. The Future of the Academy. Presidential Address, American Academy of Arts and Sciences (Boston, 1944).

Memoirs of the American Academy of Arts and Sciences, I (1785), preface, iii-xi.

"Records of Meetings," *Proceedings of the American Academy of Arts and Sciences,* XLVI, no. 25 (1912), 683-735.

Rumford Fund of the American Academy of Arts and Sciences, The (Boston, 1905).

"Scientific Institutions of Boston and Vicinity," American Association for the Advancement of Science, *Proceedings,* XXIX (1880), 767-791.

AMERICAN ACADEMY OF FINE ARTS:

Coudrey, M. B. American Academy of Fine Arts and American Art-Union (2 vols., New York, 1953).

AMERICAN ANTHROPOLOGICAL ASSOCIATION:

American Anthropologist, n. s., V (1903), 178-192.

AMERICAN ASSOCIATION FOR THE ADVANCEMENT OF SCIENCE:

"A. A. A. S. Bulletin, The," American Association for the Advancement of Science, *Bulletin,* I, no. 1 (1942), 1-2.

"American Association for the Advancement of Science and Scientific Organization," *Science,* n. s., LI (1919), 112-114.

"American Association for the Advancement of Science Buys Washington Building Site," *Science,* CIII (1946), 591.

"American Association for the Advancement of Science, The: the Secretaries'

Conference and the Academy Conference," *Science,* n. s., LXXVIII (1933), 581-582.

"Association from 1848 to 1860, The," American Association for the Advancement of Science, *Bulletin,* I, no. 2 (1942), 13-14.

"Association from 1861 to 1870, The," American Association for the Advancement of Science, *Bulletin,* I, no. 4 (1942), 29-30.

Benjamin, Marcus. "The Early Presidents of the American Association," *Science,* n. s., X (1899), 625-637, 705-713, 759-766.

"Centennial Celebration—Washington, D. C., September 13-17, 1948," *Science,* CVIII (1948), 99.

"Civil Liberties of Scientists," Report by Special Committee of the A. A. A. S., *Science,* CX (1949), 177-179.

Fairchild, H. L. "History of the American Association for the Advancement of Science," *Science,* n. s., LIX (1924), 365-369, 385-390, 410-415.

"First President of the American Association for the Advancement of Science, The," *Scientific Monthly,* LXVII (1948), iii, Portrait.

Forman, Sidney. "West Point and the American Association for the Advancement of Science," *Science,* CIV (1946), 47-48.

Germann, F. E. E. "The Southwestern Division of the A. A. A. S.," *Science,* CVIII (1948), 224-226.

Hale, W. H. "Early Years of the American Association," *Popular Science Monthly,* LI (1896), 501-507.

"Historical Sketch of the Association," *Science,* CVI (1947), 463-464.

Lark-Horovitz, Karl. "The Cooperative Committee for the Teaching of Science," *Science,* CXI (1950), 197-200.

Lark-Horovitz, Karl, and Eleanor Carmichael. A Chronology of Scientific Development, 1848-1948 (New York, 1948).

Livingston, B. E. "Members of the American Association for the Advancement of Science per Million of Population in the United States," *Science,* n. s., LX (1924), 467-469.

Martin, D. S. "The First Half Century of the American Association," *Popular Science Monthly,* LII (1898), 422-435.

Meyerhoff, H. A. "American Association for the Advancement of Science Headquarters," *Science,* CXVI (1952), 3.

——"The Association's Journals," *Science,* CXV (1952), 3.

——"Revision of the American Association for the Advancement of Science Constitution and Bylaws," *Science,* CXVI (1952), 403-407.

Miller, R. C. "The American Association for the Advancement of Science on the Pacific Slope," *Science,* CVIII (1948), 220-223.

Moulton, F. R. "Affiliated and Associated Societies," American Association for the Advancement of Science, *Bulletin,* I, no. 4 (1942), 30-31.

——"The American Association for the Advancement of Science: A Brief Historical Sketch 1848-1948," *Science,* CVIII (1948), 217-218.

——"The American Association for the Advancement of Science and Organized American Science," *Science,* CVIII (1948), 573-577.

——"The Association in Its New Home," American Association for the Advancement of Science, *Bulletin,* V, no. 9 (1946), 65.

——"Constitution of the Association," American Association for the Advancement of Science, *Bulletin,* V, no. 4 (1946), 29-30.

——"Duration of Association Memberships," American Association for the Advancement of Science, *Bulletin,* I, no. 4 (1942), 27-28.

"Origin of the Association," American Association for the Advancement of Science, *Bulletin,* I, no. 1 (1942), 2-3.

Shapley, H., E. W. Sinnott, and E. C. Stakman. "American Association for the Advancement of Science Enters Its Second Century," *Scientific Monthly,* LXVII (1948), 242.

Taylor, R. L. "The Boston Meeting of the Association: A Bit of Background," *Science,* CXVIII (1953), 224-226.

——"New York Meetings of the American Association for the Advancement of Science; 1887-1956," *Science,* CXXIV (1956), 544-546.

Wolfle, Dael. "The Future of the American Association for the Advancement of Science," *Science,* CXIX (1954), 3A.

——"A New Home for the American Association for the Advancement of Science," *Science,* CXX (1954), 358.

——"Paying for the American Association for the Advancement of Science Headquarters Building," *Science,* CXX (1954), 5A.

AMERICAN ASSOCIATION FOR THE ADVANCEMENT OF SCIENCE, COOPERATIVE COMMITTEE ON THE TEACHING OF SCIENCE AND MATHEMATICS:

"Improving Science Teaching," *Science,* CXXII (1955), 145-148.

AMERICAN ASSOCIATION FOR THE ADVANCEMENT OF SCIENCE, INTERIM COMMITTEE ON THE SOCIAL ASPECTS OF SCIENCE:

"Society in the Scientific Revolution," *Science,* CXXIV (1956), 1231.

AMERICAN ASSOCIATION OF MECHANICAL ENGINEERS:

Chandler, E. L. "Centennial of Engineering Convocation," *Science,* CXVI (1952), 3.

AMERICAN ASSOCIATION OF MUSEUMS:

Coleman, L. V. "The Three-Year Experiment of the American Association of Museums," *Museum Work,* VIII (1925), 99-112.

AMERICAN ASSOCIATION OF PHYSICAL ANTHROPOLOGISTS:

Angel, J. L. "American Association of Physical Anthropologists," *Science,* CXVIII (1953), 3.

Hrdlička, Aleš. "American Association of Physical Anthropologists," *Science,* n. s., LXIX (1929), 304-305.

AMERICAN ASSOCIATION OF UNIVERSITY PROFESSORS:

Laprade, W. T. "Ralph E. Hemstead in Context; the Growth and Development of the Association," American Association of University Professors, *Bulletin,* XLI, no. 3 (1941), 407-418.

AMERICAN ASTRONOMICAL SOCIETY:

"Astronomical and Astrophysical Society of America," *Popular Astronomy,* VII (1899), 444-445.

Frost, E. B. "The Astronomical and Astrophysical Society of America," *Science,* n. s., X (1899), 785-795.

AMERICAN CHEMICAL SOCIETY:

Benjamin, Marcus. "Organization and Development of the Chemical Section of the American Association for the Advancement of Science," American Chemical Society, Twenty-fifth Anniversary Volume (1902), 86-98.

Browne, C. A., edit. A Half-Century of Chemistry in America, 1876-1926 (Easton, Pa., 1926). Golden Jubilee Number, Supplement to American Chemical Society, *Journal* (1926). See especially C. A. Browne, "Introduction," v-x; W. H. Nichols, "The Organization of the American Chemical Society," 11-16; F. W. Clarke, "The Evolution of the American Chemical Society," 17-21; C. E. Munroe, "The First General Meeting and the First Local Section of the American Chemical Society," 31-61.

Browne, C. A., and M. E. Weeks. A History of the American Chemical Society; Seventy-five Eventful Years (Washington, 1952).

Hale, A. C. "History of the American Chemical Society," American Chemical Society, Twenty-fifth Anniversary Volume (1902), 86-98.

AMERICAN COUNCIL OF LEARNED SOCIETIES:

"American Council of Learned Societies and Its Activities in 1928, The," American Council of Learned Societies, *Bulletin,* IX (1928), 7-50.

Jameson, J. F. "The American Council of Learned Societies," *American Historical Review,* XXV (1920), 440-446.

AMERICAN CRYSTALLOGRAPHIC ASSOCIATION:

"American Crystallographic Association," *Science,* CXI (1950), 214.

AMERICAN DENTAL ASSOCIATION:

Dean, H. T. "The American Dental Association: Status Changed from an Associated to an Affiliated, American Association for the Advancement of Science," *Science,* CXIX (1954), 394.

AMERICAN ELECTROCHEMICAL SOCIETY:

American Electrochemical Society, *Transactions,* I (1902), 3-39.

AMERICAN ENTOMOLOGICAL SOCIETY:

Cresson, E. T. A History of the American Entomological Society (Philadelphia, 1909).

AMERICAN ETHNOLOGICAL SOCIETY:

American Anthropologist, n. s., II (1900), 785; n. s., XIX (1918), 107.

Anthropological Institute of New York, *Journal,* I (1871-1872), 14-21.

AMERICAN GEOGRAPHICAL SOCIETY OF NEW YORK:

Joerg, W. L. G. "Historical Sketch," *Journal of Geography,* XVI (1918), 384-385.

AMERICAN INSTITUTE OF BIOLOGICAL SCIENCES:

Meyer, S. L. "The American Institute of Biological Sciences," *Science,* CXVII (1953), 3.

AMERICAN INSTITUTE OF ELECTRICAL ENGINEERS:

American Institute of Electrical Engineers. Handbook for the Year 1900, 1-16.

——*Transactions,* VIII (1891), 601-608.

AMERICAN INSTITUTE OF HUMAN PALEONTOLOGY:

Steward, T. D. "American Institute of Human Paleontology," *Science,* CXX (1954), 7A.

AMERICAN INSTITUTE OF THE HISTORY OF PHARMACY:

"American Institute of the History of Pharmacy, The," Tenth Anniversary, *Science,* CXIII (1951), 570.

AMERICAN MATHEMATICAL SOCIETY:

Fiske, T. S. "Mathematical Progress in America," *Science,* n. s., XXI (1905), 209-215.

AMERICAN MEDICAL ASSOCIATION:

"American Medical Association," *Fortune,* Nov., 1938.

Bevan, A. D. "The Organization of the Medical Profession for War," *Science,* n. s., XLVII (1918), 576-603.

Bridgman, D. G. C. "The American Medical Association and the Great Depression" (Thesis, Harvard, 1956).

Davis, N. S. History of the American Medical Association, from Its Organization up to January, 1855 (Philadelphia, 1855).

Fishbein, Morris. A History of the American Medical Association, 1847 to 1947 (Philadelphia, 1947).

Garceau, Oliver. Political Life of American Medical Association (Cambridge, Mass., 1941).

——"Some Aspects of Medical Politics" (Ph.D. Thesis, Harvard, 1940).

Hoffman, E. L. "The Political Syndrome of Organized Medicine; a Survey of the Interaction of American Medicine and the Political Process in Recent Years" (Honors Thesis, Harvard, 1951).

Josephson, E. M. Merchants in Medicine (New York, 1941).

McMurtry, L. S. "The American Medical Association; Its Origin, Progress and Purpose," American Medical Association, *Journal*, XLV (1905), 145-149; reprinted in *Science*, n. s., XXII (1905), 97-105.

Rorty, James. American Medicine Mobilizes (New York, 1939).

AMERICAN METEOROLOGICAL SOCIETY:

"Twenty-fifth Anniversary of the American Meteorological Society, The," *Science*, CI (1945), 217-218.

AMERICAN OTOLOGICAL SOCIETY:

Harris, T. J. "The Early History of the American Otological Society with Special Reference to Its Founders," *Annals of Otology, Rhinology and Laryngology*, XXV (1926), 456-474.

AMERICAN PHARMACEUTICAL ASSOCIATION:

"Centennial Convention, American Pharmaceutical Association," *Science*, CXVI (1952), 582-584.

AMERICAN PHILOSOPHICAL SOCIETY:

American Philosophical Society. Mankind Advancing (Philadelphia, 1929).

American Philosophical Society; an Historical Account of the Origin and Formation of the American Philosophical Society, . . . (Philadelphia, 1914).

"Celebration of the One Hundredth Anniversary, May 25, 1843," American Philosophical Society, *Proceedings*, III (1843), 1-36.

Chinard, Gilbert, "The American Philosophical Society and the World of Science (1768-1800)," American Philosophical Society, *Proceedings*, LXXXVII (1944), 1-11.

——"Jefferson and the American Philosophical Society," American Philosophical Society, *Proceedings*, LXXXVII, no. 3 (1944), 263-276.

"Commemoration of the Centennial Anniversary of the Society's Occupation of Its Present Hall," American Philosophical Society, *Proceedings*, XXVII (1889), 1-52.

Conklin, E. G. "The American Philosophical Society and International Relations," American Philosophical Society, *Proceedings*, XCI, no. 1 (1947), 1-9.

Corner, G. W. "Medical Treasures in the Library of the American Philosophical Society," *Science*, CVI (1947), 120.

Dercum, F. X. "The Origin and Activities of the American Philosophical Society," The Record of the Celebration of the Two Hundredth Anniversary of the Founding of the American Philosophical Society, American Philosophical Society, *Proceedings*, LXVI (1927), 19-30.

Du Ponceau, P. S. An Historical Account of the American Philosophical Society (Philadelphia, 1914).

Dvoichenko-Markoff, Eufrosina. "Benjamin Franklin, the American Philosophical Society, and the Russian Academy of Science," American Philosophical Society, *Proceedings*, XCI, no. 3 (1947), 250-251.

"Early History of Science and Learning in America, The; with Especial References to the Work of the American Philosophical Society during the Eighteenth and Nineteenth Centuries," American Philosophical Society, *Proceedings*, LXXXVI, no. 1 (1942), and *ibid.*, LXXXVII, no. 1 (1943).

Hindle, Brooke. "The American Philosophical Society," in The Pursuit of Science in Revolutionary America, 1735-1789 (Chapel Hill, 1956), 127-145.

——"The Rise of the American Philosophical Society, 1766-1787" (Ph.D. Dissertation, University of Pennsylvania, 1949).

Lingelbach, W. E. "Benjamin Franklin and the Scientific Societies," *Journal of the Franklin Institute*, CCLXI, no. 1 (1956), 9-31.

——"The Story of 'Philosophical Hall' 1769-1949," American Philosophical Society, *Proceedings*, XCIV (1950), 185-213.

Pace, Antonio. "The American Philosophical Society and Italy," American Philosophical Society, *Proceedings*, XC, no. 5 (1946), 387-421.

Phillips, Henry, Jr., comp. "Early Proceedings of the American Philosophical Society, 1744-1838," American Philosophical Society, *Proceedings*, XXII, no. 3 (1884).

Record of the Celebration of the Two Hundredth Anniversary of the Founding of the American Philosophical Society, American Philosophical Society, *Proceedings*, LVI (1927).

Rosengarten, J. G. "The American Philosophical Society, 1743-1903," *Pennsylvania Magazine of History and Biography*, XXVII (1903), 320-336.

——"The Early French Members of the American Philosophical Society," American Philosophical Society, *Proceedings*, XLVI (1907), 87-93.

Scharf, J. T., and Thompson Wescott. History of Philadelphia (2 vols., Philadelphia, 1884), II, 1173-1229, "Libraries and Historical and Scientific Societies of Philadelphia."

Sellers, C. C. "Charles Willson Peale and the American Philosophical Society," American Philosophical Society, Year Book, 1944 (Philadelphia, 1945), 68.

Van Doren, Carl. "The Beginnings of the American Philosophical Society," American Philosophical Society, *Proceedings*, LXXXVII (1944), 277-289.

AMERICAN PHYSICAL SOCIETY:

"Formation of a Division of Electron and Ion Optics in the American Physical Society," *Journal of Applied Physics*, XIV (1943), 406.

"New Division of Electron and Ion Optics in the American Physical Society, The," *Scientific Monthly*, LVII, no. 6 (1943), 570-572.

AMERICAN PSYCHOLOGICAL ASSOCIATION:

"American Psychological Association, The," *Science*, n. s., XX (1892), 104.

Buchner, E. F. "A Quarter Century of Psychology in America," *American Journal of Psychology*, XIV (1903), 402-416.

Cattell, J. McK. "Psychology in America," *Science,* n. s., LXX (1929), 334-347.

Fernberger, S. W. "The American Psychological Association," *Psychological Bulletin,* XXIX (1932), 1-89.

Hall, G. S. "A Reminiscence," *American Journal of Psychology,* XXVIII (1917), 297-300.

Moore, C. B. "Notes on the Presidents of the American Psychological Association," *American Journal of Psychology,* XXIX (1918), 347-349.

Pierce, J. M. "American Psychological Association," *Education,* XVII (1897), 346-350.

AMERICAN PUBLIC HEALTH ASSOCIATION:

Smith, S. "Historical Sketch," *Public Health Papers and Reports,* V (1879), vii-liv.

AMERICAN SOCIETY FOR HORTICULTURAL SCIENCE:

Howlett, F. S. "The American Society for Horticultural Science," *Science,* CXVIII (1953), 617.

AMERICAN SOCIETY OF BIOLOGICAL CHEMISTS:

Gies, W. J. "American Society of Biological Chemists," *Science,* n. s., XXV (1907), 139-142.

AMERICAN SOCIETY OF CIVIL ENGINEERS:

Hunt, C. W. "The Activities of the American Society of Civil Engineers during the Past Twenty-five Years," American Society of Civil Engineers, *Transactions,* LXXXII (1918), 1577-1615.

——"The First Fifty Years of the American Society of Civil Engineers, 1852-1902," American Society of Civil Engineers, *Transactions,* XLVIII (1902), 220-226.

——Historical Sketch of the American Society of Civil Engineers (New York, 1897).

AMERICAN SOCIETY OF MECHANICAL ENGINEERS:

"History of the American Society of Mechanical Engineers, Preliminary Report of the Committee on the History of the Society," *Mechanical Engineering,* XXX (1908).

Hutton, F. R. "History of the American Society of Mechanical Engineers from 1880 to 1915," American Society of Mechanical Engineers, *Extra Publication* (1915).

——"The Mechanical Engineer and the Function of the Engineering Society," American Society of Mechanical Engineers, *Transactions,* XXIX (1907), 727-762.

Thurston, R. H. "President's Inaugural Address," American Society of Mechanical Engineers, *Transactions,* I (1880), 1-15.

AMERICAN SOCIETY OF NATURALISTS:
Minot, C. S., C. B. Davenport, W. J. McGee, William Trelease, S. A. Forbes, and J. McK. Cattell. "The Relation of the American Society of Naturalists to Other Scientific Societies," *Science*, n. s., XV (1902), 241-255.
"Semi-Centennial of the American Society of Naturalists," *Science*, n. s., LXXVIII (1933), 549-550.

AMERICAN SOCIETY OF PARASITOLOGISTS:
"Formation of the American Society of Parasitologists, The," *Journal of Parasitology*, XI (1925), 177-180.

ANTIQUARIAN AND NATURAL HISTORY SOCIETY OF ARKANSAS, LITTLE ROCK:
Hendrickson, Walter B. "Natural Science and Urban Culture in the Nineteenth Century Middle West," *Transactions of the Academy of Science of St. Louis*, XXXI, no. 9 (1958), 232-248.

ARIZONA ACADEMY OF SCIENCE:
Danson, Edward. "From the President," *Journal*, I, no. 11 (1959), 1.

ARKANSAS ACADEMY OF SCIENCE:
Ham, L. B. "Early History [of] the Arkansas Academy of Science," *Proceedings*, I (1941), 3-6.

ASSOCIATION OF AMERICAN GEOGRAPHERS:
Brigham, A. P. "The Association of American Geographers," *Science*, n. s., XXI (1905), 300-302.

ASSOCIATION OF AMERICAN GEOLOGISTS AND NATURALISTS:
Hitchcock, Edward. "Address at the Opening of the Geological Hall at Albany, N.Y., August 27, 1856," New York State Museum, Tenth Annual Report (1856), 20-26.
Tuckerman, Frederick. "Edward Hitchcock and the Origin of the Association of American Geologists," *Science*, n. s., LX (1924), 134-135.

ASSOCIATION OF SCIENTIFIC WORKERS:
"Association of Scientific Workers, The," *Science*, n. s., LXXXVIII (1938), 562-563.

ASSOCIATION TO AID SCIENTIFIC RESEARCH BY WOMEN:
Crawford, H. J. "The Association to Aid Scientific Research by Women," *Science*, n. s., LXXXVI (1932), 492-493.

ATLANTIC ESTUARINE RESEARCH SOCIETY:
Andrews, J. D. "The Atlantic Estuarine Research Society," *Science*, CXVI (1952), 153-154.

BOSTON PHILOSOPHICAL SOCIETY:

Mather, Cotton. Parentator, Memoirs of Remarkables in the Life and Death of the Ever-Memorable Dr. Increase Mather (Boston, 1724), 86.

Murdock, Kenneth. Increase Mather, the Foremost Puritan (Cambridge, 1925), 147-148.

BOSTON SOCIETY OF CIVIL ENGINEERS:

Fitzgerald, Desmond. "Historical Address," Association of Engineering Societies, *Journal,* XXI (1898), 268-280.

BOSTON SOCIETY OF NATURAL HISTORY:

Binney, Amos. Remarks . . . Showing the Origin and History of the Society . . . (Boston, 1845).

Boston Society of Natural History. The Boston Society of Natural History, 1830-1930 (Boston, 1930).

Bouvé, T. T. "Historical Sketch of the Boston Society of Natural History; with a Notice of the Linnaean Society Which Preceded It," Boston Society of Natural History. Anniversary Memoirs: 1830-1880 (1880), 14-250.

——"Some Reminiscences of Earlier Days in the History of the Society," Boston Society of Natural History, *Proceedings,* XVIII (1876), 242-250.

Gould, A. A. "Notice of the Origin, Progress and Present Condition of the Boston Society of Natural History," *American Quarterly Register,* XIV (1842), 236-241.

Warren, J. C. Address to the Society (Boston, 1853).

BOTANICAL SOCIETY OF AMERICA:

"Brief Sketch of the Society," Botanical Society of America, *Publication,* no. 10.

BRIDGEPORT SCIENTIFIC AND HISTORICAL SOCIETY:

Powers, H. N. Annual Address (Bridgeport, 1883).

BROOKLYN INSTITUTE OF ARTS AND SCIENCES:

Brooklyn Botanic Garden Record, I (1912), 76-84; II (1913), 109-114.

Seventeenth Yearbook of the Brooklyn Institute of Arts and Sciences, The (1905), 107-138, "A Brief History of the Brooklyn Institute of Arts and Sciences."

BUFFALO SOCIETY OF NATURAL SCIENCES:

Howland, H. R. "Historical Sketch," Buffalo Society of Natural Sciences, *Bulletin,* VIII, no. 6 (1907).

CALIFORNIA ACADEMY OF SCIENCES:

Century of Progress in the Natural Sciences, A, 1853-1953. Published in Celebration of the Centennial of the California Academy of Sciences (San Francisco, 1955).

Grunsky, C. E. "[Address]," *Proceeding,* 4th ser., VI (1916), 230-233.

CARIBBEAN RESEARCH COUNCIL:

"Organization of the Caribbean Research Council," *Science,* CIII (1946), 452-453.

CHARLESTON BOTANIC SOCIETY AND GARDEN:

"Charleston Botanic Society and Garden, The," *Medical Repository,* IX (1806), 434-436.

"Charleston Botanic Society and Garden, The," *Philadelphia Medical and Physical Journal,* II (1805), pt. 1, 200.

Ramsay, David. History of South Carolina (2 vols., Charleston, 1809), 107-108.

Shecut, J. Medical and Philosophical Essays (Charleston, 1819), 44-47.

CHESTER COUNTY CABINET OF NATURAL SCIENCES, WEST CHESTER, PENNSYLVANIA:

Carpenter, G. W. "The West Chester Cabinet of Natural Sciences," *American Journal of Science,* XIV (1828), 2-3.

Chester County Cabinet of Natural Science. Reports (Annual, 1828-1849, West Chester, Pa.).

CHICAGO ACADEMY OF SCIENCES:

Baker, F. C. The Chicago Academy of Sciences: Its Past History and Present Collections (Chicago Academy of Science, Special Publications, no. 1, Chicago, 1902).

Hendrickson, Walter B. "Natural Science and Urban Culture in the Nineteenth Century Middle West," *Transactions of the Academy of Science of St. Louis,* XXXI, no. 9 (1958), 232-248.

Higley, W. K. The Chicago Academy of Sciences: Historical Sketch of the Academy (Chicago Academy of Sciences, Special Publications, no. 1, Chicago, 1902).

Kennicott, H. L. "Historical Sketch of the Academy," in *Special Publications,* no. 13: "The Chicago Academy of Sciences Centennial Meeting, May 22, 1957" (1958), 7-12.

McCagg, E. B. "The Chicago Academy of Sciences: A Historical Sketch," in Rufus Blanchard, Discovery & Conquests of the North-west with History of Chicago (Chicago, 1880), 583-587.

CHICAGO MEDICAL SOCIETY:

Brief History of the Chicago Medical Society from 1850 to Oct. 1, 1902, A, *Chicago Medical Recorder* (1915).

Chicago Medical Society, Council. History of Medicine and Surgery, and Physicians and Surgeons of Chicago (Chicago, 1922).

CINCINNATI SOCIETY OF NATURAL HISTORY:

Cincinnati Society of Natural History, The: An Account of Its Organization and a Description of Its Collections (Cincinnati, 1902).

Greve, C. T. "The Cincinnati Society of Natural History," in Centennial History of Cincinnati (Cincinnati, 1904), I, 903-904.

"Western Academy of Natural Sciences," in History of Cincinnati & Hamilton County (Cincinnati, 1894), 151-152.

CLEVELAND ACADEMY OF NATURAL SCIENCE:

Hendrickson, Walter B. "Natural Science and Urban Culture in the Nineteenth Century Middle West," *Transactions of the Academy of Science of St. Louis,* XXXI, no. 9 (1958), 232-248.

COLLEGE OF PHYSICIANS OF PHILADELPHIA:

Ruschenberger, W. S. W. An Account of the Institution and Progress of the College of Physicians of Philadelphia (Philadelphia, 1887).

COLORADO STATE MEDICAL SOCIETY:

Colorado State Medical Society. Jubilee Volume (1922).

COLUMBIAN INSTITUTE FOR THE PROMOTION OF ARTS AND SCIENCES:

"Columbian Institute for the Promotion of Arts and Sciences . . . , The," United States National Museum, *Bulletin,* no. 101 (Washington, 1917).

CONNECTICUT ACADEMY OF ARTS AND SCIENCES:

Baldwin, S. E. "The First Century of the Connecticut Academy of Arts and Sciences," Connecticut Academy of Arts and Sciences, *Transactions,* XI (1901-1903), pt. 1, xiii-xxxv.

Herrick, E. C. "Historical Sketch of the Connecticut Academy of Arts and Sciences," *American Quarterly Register,* XIII (1840), 23-28.

Loomis, Elias. "Connecticut Academy of Arts & Sciences," in History of Yale College, W. L. Kingsley, edit. (1879), I, 329-337.

Osterweis, R. G. "The Sesquicentennial History of the Connecticut Academy of Arts and Sciences," *Transactions,* 38 (1949), 103-149.

DAVENPORT ACADEMY OF SCIENCES:

McCowen, Jennie. "Anniversary Address," *Proceedings,* VI (1889-1897), 311-313. 25th Anniversary Address.

Pratt, W. H. "Reminiscences of the Early History of the Academy," *Proceedings,* II (1876-1878), 193-202.

Starr, Frederick. "The Davenport Academy of Natural Sciences," *Popular Science Monthly,* LI (1897), 83-98.

DELAWARE STATE MEDICAL SOCIETY:

Bush, L. P. The Delaware State Medical Society (New York, 1886).

DEUTSCHER GESELLIG-WISSENSCHAFTLICHER VEREIN VON NEW YORK:

Winter, Joseph. "Geschichte der Deutscher Gesellig-Wissenschaftlicher Verein von New York," in Festschrift zum dreissigsten Stiftungsfeste (New York, 1920).

ELISHA MITCHELL SCIENTIFIC SOCIETY:

Henderson, Archibald. "The Elisha Mitchell Scientific Society: Its History and Achievements," *Journal*, Elisha Mitchell Scientific Society, L (1934), 1-13.

Venable, F. P. "Historical Sketch of the Elisha Mitchell Scientific Society," *Journal*, XXXIX (1923-1924), 117-122.

ENGINEERS' CLUB OF PHILADELPHIA:

Engineers' Club of Philadelphia, *Proceedings*, XVIII (1901), 61-67; XX (1903), 7-9.

ENGINEERS' CLUB OF ST. LOUIS:

Bryan, W. H. "The Engineers' Club of St. Louis," Association of Engineering Societies, *Journal*, XXIV (1900), 158-174.

ENGINEERS' SOCIETY OF WESTERN PENNSYLVANIA:

Davison, G. S. "The First Half Century of the Engineers' Society of Western Pennsylvania," Engineers' Society of Western Pennsylvania, *Proceedings*, XLVII (1931), 178-183.

Scott, C. F. Our Society (Engineers' Society of Western Pennsylvania, Pittsburgh, 1902).

ESSEX COUNTY NATURAL HISTORY SOCIETY:

Essex Institute. An Historical Notice of the Essex Institute (Salem, 1865), 8-11, "History of Natural History Society."

Fowler, S. P. "A Historical Sketch: The Essex County Natural History Society, and the Essex Institute," Essex Institute, *Bulletin*, XVI (1884), 141-145.

Wheatland, Henry. "On the History of the Essex County Natural History Society and Some of Its Pioneers in the Pursuit of Natural History," Essex Institute, *Proceedings*, II (1856), 24-28.

ESSEX INSTITUTE:

"First Half-Century of the Essex Institute, The, Commemorated at Salem, March First and Second, 1898," Essex Institute, *Bulletin*, XXX (1898), nos. 1-6.

Phippen, G. D. "Remarks of August 6, 1861," Essex Institute, *Proceedings*, III (1861), 101.

Visitors' Guide to Salem (1922).

FEDERATED AMERICAN ENGINEERING SOCIETIES:

"Engineering Council Merges into Council of F.A.E.S.," *Mechanical Engineering*, XLIII (1921), 145-146.

"Federated American Engineering Societies, The," *Mechanical Engineering*, XLIII (1921), 830-832.

FEDERATION OF AMERICAN SCIENTISTS:

Grobstein, Clifford. "The Federation of American Scientists," *Bulletin of Atomic Scientists*, VI (1950).

FLORIDA ACADEMY OF SCIENCES:

Kusner, J. H. "The Academy During 1936," *Quarterly Journal*, I (1936), 1-2.

GEOGRAPHICAL SOCIETY OF PHILADELPHIA:

Salisbury, R. D. "Historical Sketch," *Journal of Geography*, XVı (1918), 390.

GEOGRAPHIC SOCIETY OF CHICAGO:

Salisbury, R. D. "Historical Sketch," *Journal of Geography*, XVI (1918), 390-391.

GEOGOLOGICAL SOCIETY OF AMERICA:

Fairchild, H. L. The Geological Society of America, 1888-1930 (New York, 1932).

Geology, 1888-1938. Fiftieth Anniversary Volume (Published by the Society, 1941).

Winchell, N. H. "The Foundation of the Geological Society of America," *Science*, n. s., XXXIX (1914), 819-821.

GEOLOGICAL SOCIETY OF PENNSYLVANIA:

Lesley, J. P. "The Geological Society of Pennsylvania and What It Did to Bring About the First Geological Survey of the State," in his "Historical Sketch of the Geological Survey of Pennsylvania, 1874, 1875, 1876," A, 29-52.

GEORGIA ACADEMY OF SCIENCE:

Baker, W. B. "Preliminary Report on History of the Georgia Academy of Science," *Bulletin*, XII (1954), 1-5; XIII (1955), 71-80.

Lagemann, R. T. "The President's Message," *Bulletin*, IX, no. 2 (1951), 1-2.

Redmond, W. B. "The Twenty-fifth Anniversary Meeting," *Bulletin*, V, no. 3 (1947), 1-3.

HAWAIIAN ACADEMY OF SCIENCE:

Cox, D. C. ". . . Hawaiian Academy of Science, A Tricentennial Review," *Proceedings*, XXXIV (1958-1959), 3-11.

"Historical Sketch," *Proceedings*, I (1926), 1-2.

Newcombe, F. C. "The Field of the Hawaiian Academy of Science," *Proceedings,* I (1926), 15-18.

HARTFORD SCIENTIFIC SOCIETY:
"Historical Sketch," *Bulletin,* I (1905), no. 3.

HISTORICAL AND PHILOSOPHICAL SOCIETY OF OHIO:
Venable, W. H. "Historical and Philosophical Society of Ohio," *Magazine of Western History,* III (1886), 499-506.

HISTORY OF SCIENCE SOCIETY:
Isis, XIX, no. 2 (1933). Notes on the history and work of this society are found on the inside front and back covers.

ILLINOIS NATURAL HISTORY SOCIETIES:
Forbes, S. A. "History of the Former State Natural History Societies of Illinois," *Science,* n. s., XXVI (1907), 892-898.
Hendrickson, Walter B. "Natural Science and Urban Culture in the Nineteenth Century Middle West," *Transactions of the Academy of Science of St. Louis,* XXXI, no. 9 (1958), 232-248.

ILLINOIS STATE ACADEMY OF SCIENCE:
Bailey, W. H. "The Beginning of the Illinois State Academy of Science," *Transactions,* XLIII (1950), 24-33.
Crook, A. R. "Organization Meeting of Illinois State Academy of Science," *Science,* n. s., XXVII (1908), 186-188.
——"The Relation of Academies of Science to the State," *Transactions,* III (1910), 32-43.
Leighton, M. M. "Twenty-five Years of the Illinois State Academy of Science," *Transactions,* III (1910), 32-43.

INDIANA ACADEMY OF SCIENCE:
Butler, A. W. "The Beginnings of the Indiana Academy of Science," *Proceedings,* XIX (1923), 14-18.
——"Early History of the Indiana Academy of Science," *Proceedings,* XIX (1923), 35-37.
Edington, W. E. "The Charter Members of the Indiana Academy of Science," *Proceedings,* LXI (1951), 261-263.
——"The First Five Years of the Indiana Academy of Science," *Proceedings,* LVI (1946), 221.
"Field Meetings," *Proceedings,* I (1891), 9-13.
"Greetings from Indiana Associations," *Proceedings,* XIX (1909), 39-47. 25th Anniversary.
"Indiana Academy of Science . . . ," *Proceedings,* V (1895), 7-11.

"Plans for the Indiana Academy of Science," *Proceedings,* XIX (1909), 48-71. 25th Anniversary.

Scovell, J. T. "The Indiana Academy of Science," *Proceedings,* XVIII (1908), 209-210.

Visher, S. S. Indiana Scientists, A Biographical Directory (Indianapolis, 1951).

IOWA ACADEMY OF SCIENCE:

Fairchild, D. S. "The Iowa Academy of Science," *Proceedings,* XXXI (1924), 69-77.

"Historical Note," I, no. 1 (1887-1889), 6-7.

Iowa Journal of History and Politics, II (1904), 305-308.

Martin, J. N. "The Iowa Academy of Science in Relation to Its Contributions to the Welfare of the Commonwealth of the State and Nation," *Proceedings,* LI (1944), 135-140.

Pammel, L. H. "Charter Members of the Iowa Academy of Science," *Proceedings,* XIX (1912), 27-41.

IOWA ANTHROPOLOGICAL ASSOCIATION:

Iowa Journal of History and Politics, II (1904), 143-146.

IOWA ENGINEERING SOCIETY:

Fitzpatrick, T. J., and S. Dean. "The Early History of Iowa Engineering Society," *Proceedings* (1905), 147-167.

KANSAS ACADEMY OF SCIENCE:

Dalton, Standlee. "Ninety-four Years of Progress," *Transactions,* LXVI (1963).

Harshbarger, W. A. "Symposium—Fifty Years of Scientific Development in Kansas. The Kansas Academy of Science," *Transactions,* XXIX (1920), 35-41.

"Historical Sketch of the Academy," *Transactions,* XIX (1904), 10.

Schoewe, W. H. "The Kansas Academy of Science—Past, Present, and Future," *Transactions,* XLI (1938), 399-416.

"The Seventy-fifth Anniversary of the Kansas Academy of Science," *Transactions,* XLVI (1943), 19-251.

Taft, Robert. "The Kansas Academy of Science," *Transactions,* XLVI (1943), 263-266.

KENTUCKY ACADEMY OF SCIENCE:

Buchner, G. D. "The Kentucky Academy of Science as a State Institution," *Transactions,* IV (1929-1930), 17-20.

LINNAEAN SOCIETY OF LANCASTER CITY AND COUNTY (Pa.):

Rathvon, S. S. An Essay on the Origin of the Linnaean Society . . . Its Objects and Progress . . . (Lancaster, Pa., 1866).

LINNAEAN SOCIETY OF NEW ENGLAND:
Bouvé, T. T. "Historical Sketch of the Boston Society of Natural History, with a Notice of the Linnaean Society Which Preceded It," Boston Society of Natural History, Anniversary Memoirs: 1830-1880 (1880), 3-14.

LOUISIANA ACADEMY OF SCIENCE:
"A Brief History of the Louisiana Academy of Science," *Proceedings,* VIII (1943), 115-116; XIX (1956), 60-62.
"Historical Sketch," *Proceedings,* I (1932), 4.

LOUISIANA STATE MEDICAL SOCIETY:
New Orleans Medical and Surgical Journal, LXXXII, no. 1 (1929).

MARYLAND ACADEMY OF SCIENCES:
Maryland Academy of Sciences, *Transactions,* n. s., I (1888-1890), 1-10.

MARYLAND STATE DENTAL ASSOCIATION:
Proceedings, Dental Centenary Celebration . . . (Baltimore, 1940).

MASSACHUSETTS AUDUBON SOCIETY:
Packard, Winthrop. "The Story of Audubon Society," Massachusetts Audubon Society, *Bulletin* (December, 1901).

MASSACHUSETTS HISTORICAL SOCIETY:
Handbook of the Massachusetts Historical Society, 1791-1948 (Boston, 1949).

MASSACHUSETTS MEDICAL SOCIETY:
Burrage, W. L. A History of the Massachusetts Medical Society (n.p., 1923).

MICHIGAN ACADEMY OF SCIENCE, ARTS AND LETTERS:
Michigan Academy of Science, Arts and Letters. *Annual Report,* I (1894), 5-10; II (1900?), 5-6.
Beal, W. J. "The Claims of the Michigan Academy of Science," *Report,* X (1908), 30-31.
McCartney, E. S. "The Beginnings and Growth of the Michigan Academy of Science, Arts, and Letters," *Papers,* XIX (1933), 1-19.

MINNESOTA ACADEMY OF SCIENCE:
"The Minnesota Academy of Science," *Proceedings,* XVIII (1950), 151-153.

MISSISSIPPI ACADEMY OF SCIENCES:
Nichols, R. J. "The State of the Academy," *Journal,* IV (1948-1950), 150-160.

MISSOURI HISTORICAL AND PHILOSOPHICAL SOCIETY:
Seever, W. J. "Missouri Historical and Philosophical Society: 1844-1851," *Missouri Historical Society Collections,* II (1900), 5-11.

MONTANA ACADEMY OF SCIENCES:

Castle, G. B. "The Montana Academy of Sciences—Past, Present and Future," *Proceedings,* I (1940), 8-12.

NATIONAL ACADEMY OF ENGINEERING:

Walsh, John. "NAE: Search for a Form Produces a National Academy of Engineering in a 'Partnership' with NAS," *Science,* CXLVI (1964), 1661-1662.

NATIONAL ACADEMY OF SCIENCES:

"Building of the National Academy of Sciences," *Scientific Monthly,* XIV (1922), 583-584.

Campbell, W. W. "The National Academy of Sciences: Address of the President . . . , Washington, D. C., April 25, 1933," *Science,* n. s., LXVII (1933), 549-552.

Compton, K. T. "The National Academy of Sciences: Address of Welcome," *Science,* n. s., LXXVIII (1933), 515-518. A discussion of the Science Advisory Board created by executive order of President Franklin D. Roosevelt, dated July 31, 1933.

Conduct of Scientific Work under United States Government. Message from the President of the United States (Washington, 1909).

Constitution and By-Laws of the National Academy of Sciences (Washington, 1864).

Hale, G. E. "National Academies of Science and the Progress of Research," *Science,* n. s., XXXVIII (1913), 681-698; XXXIX (1914), 189-200; XL (1914), 907-914; XLI (1915), 12-23 (Printed separately, Lancaster, Pa., 1915).

——"The Proceedings of the National Academy as a Medium of Publication," *Science,* n. s., XLI (1915), 815-817.

Livingston, B. E. "Relations of the American Association to the National Academy," *Science,* n. s., LXVI (1927), 493-495.

National Academy of Sciences. The Semi-Centennial Anniversary of the National Academy of Sciences, 1863-1913 (Washington, 1913).

True, F. W., edit. History of the First Half Century of the National Academy of Sciences, 1863-1913 (Washington, 1913).

Zwemer, R. L. "The National Academy of Sciences and the National Research Council," *Science,* CVIII (1948), 234-238.

NATIONAL ASSOCIATION FOR RESEARCH IN SCIENCE TEACHING:

Mallinson, G. G. "National Association for Research in Science Teaching," *Science,* n. s., CXXI (1955), 9A.

NATIONAL ASSOCIATION OF AUDUBON SOCIETIES FOR THE PROTECTION OF WILD BIRDS AND ANIMALS:

"History of the Audubon Movement," *Bird Lore,* VII (1905), 45-57.

NATIONAL COMMITTEE FOR MENTAL HYGIENE:
Beers, C. W. A Mind That Found Itself (Garden City, New York, 1923).

NATIONAL INSTITUTION:
Constitution and By-Laws of the National Institution for the Promotion of Science. Established at Washington, May, 1840.

Cutbush, E. An Address Delivered before the Columbian Institute, for the Promotion of Arts and Sciences at the City of Washington on the 11th January, 1817 (Gales and Seaton, 1817).

Du Ponceau, P. S. "Letter Respecting the Institution in General; Its Organization, Plan of Bulletin . . . ," *Proceedings of the National Institution for the Promotion of Science,* I (1841), 10-13.

Oliver, J. W. "America's First Attempt to Unite the Forces of Science and Government," *Scientific Monthly,* LIII (1941), 253-257.

Poinsett, J. R. Discourse on the Objects and Importance of the National Institution for the Promotion of Science (Washington, 1841).

Rathbun, Richard. The National Institute: 1840-1862 (Washington, 1924).

NATIONAL PEST CONTROL ASSOCIATION:
Davis, J. J. "The National Pest Control Association," *Scientific Monthly,* LX, no. 2 (1945), 151-152.

NATIONAL RESEARCH COUNCIL:
Angell, J. R. "Organization in Scientific Research," *Review,* II (1920), 251-253.

Hale, G. E. "A National Focus of Science and Research," *Scribner's Magazine,* CXXII (1922), 515-531.

——"National Research Council," *Science,* n. s., XLIV (1916), 264-266.

"History of the National Research Council, 1919-1933," *Science,* n. s., LXXVII (1933), 355-360, 500-503, 552-554, 618-620; LXXVIII (1933), 26-29, 93-95, 158-161, 203-206, 254-256. Also printed separately (National Research Council, Washington, D. C., 1933); very important for the study of the National Research Council.

Kellogg, Vernon. "National Research Council," *Educational Review,* LXII (1921), 365-373.

——"National Research Council," *International Conciliation,* CLIV (1920), 423-430.

——"National Research Council," *North American Review,* CCXII (1920), 754-765.

——"Organization and Work of the National Research Council," *Scientific American Monthly,* I (1920), 73-77.

——"Work of the National Research Council," *Science,* n. s., LVIII (1923), 337-341, 362-366.

National Research Council, *Bulletin,* I, pt. 1, no. 1 (Washington, D. C., 1919), 1-43.

"National Research Council, The: Organization of the National Research Council," *Science,* n. s., LI (1919), 458-462.

"Stimulation of Research," *Scientific American,* CXX (1919), 518-520.

NATIONAL RESEARCH COUNCIL, DIVISION OF ENGINEERING:

Clevenger, G. H. "The Division of Engineering: National Research Council," *Science,* n. s., L (1919), 58-60.

NATIONAL SCIENCE TEACHERS ASSOCIATION:

"National Science Teachers Association," *Science,* XCIX (1944), 316.

NATIONAL SPELEOLOGICAL SOCIETY:

Nicholas, G. "National Speleological Society," *Science,* CXX (1954), 5A.

NATIONAL TUBERCULOSIS ASSOCIATION:

Knopf, S. A. History of the National Tuberculosis Association (National Tuberculosis Association, New York, 1922).

NATURAL HISTORY SOCIETY OF HARTFORD:

Jarvis, S. F. "Address . . . in Behalf of the Objects of the Natural History Society," Hartford Natural History Society, *Transactions,* I (1836), 1-64.

DER NATURHISTORISCHE VEREIN VON WISCONSIN, MILWAUKEE:

Hendrickson, Walter B. "Natural Science and Urban Culture in the Nineteenth Century Middle West," *Transactions of the Academy of Science of St. Louis,* XXXI, no. 9 (1958), 232-248.

NEBRASKA ORNITHOLOGISTS' UNION:

"History of Ornithology in Nebraska, and of State Ornithological Societies in General," Nebraska Ornithologists' Union, *Proceedings* (1901).

NEW YORK ACADEMY OF SCIENCES:

Barnhart, J. H. "First Hundred Years of the New York Academy of Sciences," *Scientific Monthly,* V (1917), 463-475.

Fairchild, H. L. History of New York Academy of Sciences (New York, 1887).

Pregel, Boris. "Expansion of the Academy's Activities," *Transactions,* 2nd ser., XXI (1959), 204-206.

Taylor, H. F. "Learned Societies and the Educational Aspects of Future Science The New York Academy," *Transactions,* 2nd ser., X (1948), 112-116.

NEW ORLEANS ACADEMY OF SCIENCE:

Hendrickson, Walter B. "Natural Science and Urban Culture in the Nineteenth Century Middle West," *Transactions of the Academy of Science of St. Louis,* XXXI, no. 9 (1958), 232-248.

NORTH CAROLINA ACADEMY OF SCIENCE:
Brimley, C. S. "Twenty Years of the North Carolina Academy of Science," *Journal Elisha Mitchell Scientific Society,* XXXVIII (1922-1923), 46-50.

NUTTALL ORNITHOLOGICAL CLUB:
Allen, F. H. "Bird Clubs in America," *Bird Lore,* V (1903), 12-17.

OHIO STATE ACADEMY OF SCIENCE:
Alexander, W. H. "The Ohio Academy of Science," *Ohio Journal of Science,* XLI (1941), 288-311.
Ohio State Academy of Science. Constitution, By-Laws and Historical Sketch (1892).
Rice, E. L. "Ohio Academy of Science Quarter-Centennial Anniversary," *Ohio Journal of Science,* XVI (1915-1916), 109-112.

OKLAHOMA ACADEMY OF SCIENCE:
Decker, C. F. "The Oklahoma Academy of Science . . . ," *Proceedings,* XVI (1936), 9-13.
Lucas, E. L. "The Fiftieth Anniversary of the Oklahoma Academy of Science," *Proceedings,* XL (1959), 10-13.
Richards, A. "The First Two Decades of the Oklahoma Academy of Science," *Proceedings,* XXV (1945), 14-20.
Shannon, C. W. "Oklahoma Academy of Science," *Proceedings,* I (1910-1920), 8-12.

OPERATIONS RESEARCH SOCIETY OF AMERICA:
Morse, P. M. "The Operations Research Society of America," *Journal of the Operations Research Society of America,* I, no. 1 (1952), 1.

OREGON ACADEMY OF SCIENCE:
Gilfillan, F. A. "History of the Oregon Academy of Science," *Proceedings,* I (1943-1947), 1-3.

PEABODY ACADEMY OF SCIENCE:
Morse, S. E. A Brief Sketch of the Peabody Academy of Science (Salem, Mass., 1900).

PEARY ARCTIC CLUB:
Peary, R. E. "Fieldwork of the Peary Arctic Club," Geographical Society of Philadelphia, *Bulletin,* IV (1904), no. 1.

PENNSYLVANIA ACADEMY OF SCIENCE:
Derickson, S. H. "The Pennsylvania Academy of Science in Retrospect," *Proceedings,* IX (1935), 11-12.

Gress, E. M. "Organization of the Pennsylvania Academy of Science," *Proceedings,* XXIII (1949), 15-17.

Light, V. E. "High Lights of the Pennsylvania Academy's First Twenty-five Years," *Proceedings,* XXIII (1949), 17-22.

PHILADELPHIA ACADEMY:

Kingsley, J. S. "The Philadelphia Academy," *Popular Science Monthly,* XX (1882), 531-538.

PHILOSOPHICAL SOCIETY OF WASHINGTON:

Organization and Membership of the Philosophical Society of Washington (Washington, 1903).

PORTLAND SOCIETY OF NATURAL HISTORY, PORTLAND, MAINE:

Buffalo Society of Natural Science, *Bulletin,* X, no. 1 (1910), 99-100.

Portland Society of Natural History, *Proceedings,* I (1862-1869), 193-209.

Willis, William. History of Portland, Maine (1865), "The Portland Society of Natural History," 750.

PURDUE UNIVERSITY, ENGINEERING SOCIETIES:

"Engineering Societies and the University, The," *Purdue Engineering Review,* no. 2 (1906), 97-102.

RADIATION RESEARCH SOCIETY:

Failla, G. "The Radiation Research Society," *Science,* CXIV (1952), 27.

ROCHESTER ACADEMY OF SCIENCE:

Beckwith, Florence. "Early Botanists of Rochester and Vicinity and the Botanical Section," *Proceedings,* V (1910-1918), 39-58.

Fairchild, H. L. "History and Work of the Rochester Academy of Science," *Proceedings,* III, Brochure, 3 (1906), 32-339.

Suydam, G. B. "Early Botanists of Rochester and Vicinity and the Botanical Section, Part II," *Proceedings,* VIII (1941-1943), 124-149.

ROYAL SOCIETY OF LONDON:

Birch, Thomas. "The History of the Royal Society of London for Improving of Natural Knowledge, from Its First Rise" (London, 1756-1757).

Brasch, F. E. "The Royal Society of London and Its Influence upon Scientific Thought in the American Colonies," *Scientific Monthly,* XXXIII (1931), 337-355, 448-469.

Celebration of the Two Hundred and Fiftieth Anniversary of the Royal Society of London, The, July 15-19, 1912 (London, 1913).

Crowther, J. G. The Social Relations of Science (New York, 1941), ch. 62, "The Royal Society," 371-387.

De Beer, E. S. "The Earliest Fellows of the Royal Society," *Notes and Records of the Royal Society of London,* VII (1950), 172-192.

Denny, Margaret. "The Royal Society and American Scholars," *Scientific Monthly,* LXV (1947), 415-427.

Fäy, Bernard. "Learned Societies in Europe and America in the Eighteenth Century," *American Historical Review,* XXXVII (1932), 255-266.

Heindel, R. H. "Americans and the Royal Society," *Science,* n. s., LXXXVII (1938), 267-272.

Huggins, William. The Royal Society (London, 1906).

Kraus, Michael. "Scientific Relations between Europe and America in the Eighteenth Century," *Scientific Monthly,* LV (1942), 259-272.

Lyons, Sir H. The Royal Society, 1660-1940 (Cambridge, 1944).

Record of the Royal Society of London for the Promotion of Natural Knowledge, The (4th ed., London, 1940).

Sprat, Thomas. The History of the Royal Society of London, for the Improving of Natural Knowledge (London, 1667).

Stearns, R. P. "Colonial Fellows of the Royal Society of London, 1661-1788," *William and Mary Quarterly,* 3d ser., III, no. 1 (1946), 208-268; also in *Osiris,* VIII (1948), 73-121.

Stimson, Dorothy. Scientists and Amateurs; a History of the Royal Society (New York, 1948).

Thomson, Thomas. History of the Royal Society, from Its Institution to the End of the Eighteenth Century (London, 1912).

Weld, C. R. A History of the Royal Society with Memoirs of the Presidents, Compiled from Authentic Documents (2 vols., London, 1848).

Wheatley, H. B. Early History of the Royal Society (Hertford, 1905).

SAN DIEGO SOCIETY OF NATURAL HISTORY:

San Diego Society of Natural History. History of the San Diego Society of Natural History, 1874-1924.

SCIENTIFIC CONGRESS (FIRST NATIONAL):

Goode, G. B. "The First National Scientific Congress (Washington, April, 1844), and Its Connection with the Organization of the American Association," American Association for the Advancement of Science, *Proceedings,* XL (1891), 39-47; reproduced in Smithsonian Institution, Annual Report for 1897, pt. 2, 467-477.

SCIENTIFIC RESEARCH SOCIETY OF AMERICA:

Prentice, D. B. "Scientific Research Society of America," *Science,* CXXI (1955), 7A.

——"The Scientific Research Society of America," *Supplement to American Scientist,* XXXVII, no. 1 (1949).

SIGMA XI:

Benjamin, Marcus. "Sigma Xi," *Popular Science Monthly,* LXIX (1906), 281-282.

Jordan, D. S. "Comrades in Zeal," *Popular Science Monthly,* LXV (1904), 304-305.

McMahon, James. "Sigma Xi, a Sketch of the Origin and Growth of the Society: Its Aims and Ideals," *Cornell News,* II (1900), 207-208.

——"Sigma Xi, at the American Association and the Geological Society of America," *Science,* n. s., XII (1900), 196-197.

——"Sigma Xi, at the American Association for the Advancement of Science," *Science,* n. s., XI (1900), 965-966.

Marx, C. D. Preparatory Training for Scientific and Technical Studies (Syracuse, New York, 1888).

Ward, H. B. "The Part of Sigma Xi in Scientific Education," Society for the Promotion of Engineering Education, *Proceedings,* XV (1915), 285-294.

Ward, H. B., comp. Sigma Xi: Quarter Century Record and History, 1886-1911 (Chicago, 1913?).

SIOUX CITY ACADEMY OF SCIENCE AND LETTERS:

Iowa Journal of History and Politics, III (1905), 342-343. Sioux City Academy of Science and Letters, *Proceedings,* I (1903-1904), 32-36.

SMITHSONIAN INSTITUTION:

Abbot, C. G. "The Smithsonian Institution as an Illustration of Internationalism in Science," *Science,* n. s., XCV (1942), 639-641.

Blackford, C. M. "The Smithsonian Institution," *North American Review,* CLXXXIX (1909), 93-106.

Bolton, H. C. "The Smithsonian Institution: Its Origin, Growth, and Activities," *Popular Science Monthly,* XLVIII (1896), 449-464.

Goode, G. B., edit. The Smithsonian Institution, 1846-1896 (Washington, 1897).

Oehser, P. H. Sons of Science; the Story of the Smithsonian Institution and Its Leaders (New York, 1949).

Rhees, W. J., edit. The Smithsonian Institution: Documents Relative to Its Origin and History, 1835-1899 (Smithsonian Miscellaneous Collections, XVII, Washington, 1879).

——The Smithsonian Institution: Journals of the Board of Regents, Reports of Committees, Statistics, etc. (Smithsonian Miscellaneous Collections, XVIII, Washington, 1880).

Rhees, W. J., comp. The Smithsonian Institution: Documents Relative to Its Origin and History, 1835-1899 (Smithsonian Miscellaneous Collections, XLII, XLIII, Washington, 1901).

"Scientific Activities of the Smithsonian Institution," *Nature,* CI (1918), 176.

True, F. W. "An Account of the United States National Museum," Report of the United States National Museum for 1896, 287-374.

True, W. P. The First Hundred Years of the Smithsonian Institution, 1846-1946 (Washington, 1946).
——The Smithsonian, America's Treasure House (New York, 1950).
Wetmore, Alexander. "One Hundred Years After," *Science,* CIV (1946), 115-116.

SOCIETY FOR THE ADVANCEMENT OF NATURAL SCIENCE, LOUISVILLE:
Hendrickson, Walter B. "Natural Science and Urban Culture in the Nineteenth Century Middle West," *Transactions of the Academy of Science of St. Louis,* XXXI, no. 9 (1958), 232-248.

SOCIETY FOR EXPERIMENTAL STRESS ANALYSIS:
Holt, Marshall. "Society for Experimental Stress Analysis," *Science,* CXIX (1954), 3A.

SOCIETY FOR INDUSTRIAL AND APPLIED MATHEMATICS:
Society for Industrial and Applied Mathematics. "Recent Growth of the Society," *Newsletter,* III, no. 6 (1955), 3.

SOCIETY FOR SOCIAL RESPONSIBILITY IN SCIENCE:
"Society for Social Responsibility in Science," *Science,* CXVIII (1953), 3.

SOCIETY FOR THE ENCOURAGEMENT OF ARTS, MANUFACTURES, AND COMMERCE IN GREAT BRITAIN:
Wolfe, Abraham. A History of Science, Technology, and Philosophy in the Eighteenth Century (New York, 1939), 499-501.
Wood, H. T. A History of the Royal Society of Arts (London, 1913).

SOCIETY OF ECONOMIC PALEONTOLOGISTS AND MINERALOGISTS:
Journal of Paleontology, I, no. 1 (1927).

SOCIETY OF GENERAL PHYSIOLOGISTS:
Buck, John. "Society of General Physiologists," *Science,* CXVIII (1953), 3.

SOCIETY OF PROTOZOOLOGISTS:
Levine, N. D. "Society of Protozoologists," *Science,* CXXI (1955), 9A.

SOCIETY OF RHEOLOGY:
Journal of Rheology, I (1929), 93.

SOUTH CAROLINA ACADEMY OF SCIENCE:
Hoy, W. E., Jr. "History of the South Carolina Academy of Science," *Bulletin,* I (1935), 2-3.

SOUTH DAKOTA ACADEMY OF SCIENCE:

Churchill, E. P. "Janus View," *Proceedings,* XV (1935), 9-13.

Haines, A. L. "The First Third Century of the South Dakota Academy of Science," *Proceedings,* XXVII (1948), 20-25.

SOUTHERN CALIFORNIA ACADEMY OF SCIENCES:

Constitution, By-Laws, and List of Members 1894-1895 (Los Angeles, 1894), "Historical Sketch."

Collins, H. O. "A Few Words Relating to the Academy," *Bulletin,* VIII (1909), 6.

Comstock, J. A. "Historical Sketch of the Academy," *Bulletin,* XXXVII (1938), 146-153.

Davidson, Anstruther. "Early History of the Southern California Academy of Sciences," *Bulletin,* XXVII (1928), 61-62.

Howard, Hildegarde. The Southern California Academy of Sciences. A History Commemorating the Fiftieth Anniversary of Incorporation 1907-1957. (Los Angeles, 1957).

SOUTHERN SOCIETY FOR PHILOSOPHY AND PSYCHOLOGY:

Philosophical Review, XIII (1904), 390.

STATE HISTORICAL AND NATURAL HISTORY SOCIETY, COLORADO:

History, Constitution, and By-Laws of the Department of Natural History of the State Historical and Natural History Society (Denver, 1897).

STATEN ISLAND INSTITUTE OF ARTS AND SCIENCES:

Staten Island Association of Arts and Sciences. History, Act of Incorporation, etc. (New Brighton, New York, 1906).

TENNESSEE ACADEMY OF SCIENCE:

Ganier, A. F. "The Work of the Academy," *Journal,* I, no. 1 (1926), 18-19.

McGill, J. T. "The Tennessee Academy of Science," *Journal,* II, no. 4 (1927), 7-10.

TEXAS ACADEMY OF SCIENCE:

Leake, C. D. "Scientific Academies," *Texas Journal of Science,* I, no. 1 (1949). Abstract.

TORREY BOTANICAL CLUB:

Barnhart, J. H. "Historical Sketch of the Torrey Botanical Club," Torrey Botanical Club, *Memoirs,* XVII (1918), 12-21.

Burgess, E. S. "The Work of the Torrey Botanical Club," Torrey Botanical Club, *Bulletin,* XXVII, 552-558.

UNION OF AMERICAN BIOLOGICAL SOCIETIES:

Lewis, I. F. "The Union of American Biological Societies," *Science,* n. s., LVI (1922), 681-682.

——"The Union of American Biological Societies," *Science,* n. s., LVIII (1923), 256-257.

UTAH ACADEMY OF SCIENCES, ARTS AND LETTERS:

"Historical," *Transactions,* I (1908-1917), 3.

WASHINGTON ACADEMY OF SCIENCES, WASHINGTON, D. C.:

Gilbert, G. K. "First Annual Report of the Secretary," Washington Academy of Sciences, *Proceedings,* I (1889), 1-14.

WEST VIRGINIA ACADEMY OF SCIENCE:

Winter, J. E. "History of the West Virginia Academy of Science," *Proceedings,* XXI (1949), 15-17.

WESTERN ACADEMY OF NATURAL SCIENCES OF CINCINNATI:

Hendrickson, Walter B. "Natural Science and Urban Culture in the Nineteenth Century Middle West," *Transactions of the Academy of Science of St. Louis,* XXXI, no. 9 (1958), 232-248.

WESTERN ACADEMY OF NATURAL SCIENCES OF ST. LOUIS:

Hendrickson, Walter B. "Natural Science and Urban Culture in the Nineteenth Century Middle West," *Transactions of the Academy of Science of St. Louis,* XXXI, no. 9 (1958), 232-248.

Klem, M. J. "The History of Science in St. Louis," *Transactions of the Academy of Science of St. Louis,* XXIII, no. 2 (1914), 79-127.

WESTERN MUSEUM SOCIETY, CINCINNATI:

Hendrickson, Walter B. "Natural Science and Urban Culture in the Nineteenth Century Middle West," *Transactions of the Academy of Science of St. Louis,* XXXI, no. 9 (1958), 232-248.

WILD FLOWER PRESERVATION SOCIETY:

Plant World, V (1902), 76; VI (1903), 292.

Wild Flower, II, no. 5-6 (1925).

WILLIAMS COLLEGE LYCEUM OF NATURAL HISTORY:

Durfee, Calvin. A History of Williams College (Boston, 1860).

"Lyceum of Natural History, The, 1835-1885," Williams College, *The Fortnightly,* I (1885), 20.

"Semi-Centennial of the Lyceum of Natural History at Williams College," *Science, an Illustrated Journal Published Weekly,* V (1885), 385-386.

Smallwood, W. M. "The Williams Lyceum of Natural History, 1835-1885," *The New England Quarterly,* X (1937), 553-557.

Smallwood, W. M., and M. S. C. Smallwood. Natural History and the American Mind (New York, 1941), "Williams College Lyceum," 308-313.

Whipple, A. B., *et al.* Catalogue of the Lyceum of Natural History of Williams College, Instituted A. D., 1835 (Williamstown, 1852).

WISCONSIN ACADEMY OF SCIENCES, ARTS, AND LETTERS:

Birge, E. A. "The Medallion of the Academy, 1870-1920," *Transactions,* XX (1921), 711-716.

Chamberlin, T. C. "The Founding of the Wisconsin Academy of Sciences, Arts and Letters," *Transactions,* XX (1921), 693-701.

Hobbs, W. H. The Library of the Wisconsin Academy of Sciences, Arts, and Letters. State Historical Society of Wisconsin, Memorial Volume (1901).

Hoyt, J. W. "Origin, Organization, Plans and Necessities of the Academy," *Bulletin,* I, no. 1 (1870), 1-24.

WISCONSIN ARCHAEOLOGICAL SOCIETY:

Smith, H. I. "Recent Work of the Wisconsin Archaeological Society," *Science,* n. s., XXII (1905), 152-155.

WISCONSIN NATURAL HISTORY ASSOCIATION, MILWAUKEE:

Hendrickson, Walter B. "Natural Science and Urban Culture in the Nineteenth Century Middle West," *Transactions of the Academy of Science of St. Louis,* XXXI, no. 9 (1958), 232-248.

WISCONSIN NATURAL HISTORY SOCIETY:

"Geschichtlicher Ueberblick," Deutscher Naturhistorischer Verein von Wisconsin, Jahresbericht, 1879-1880, 6-7.

WORCESTER LYCEUM OF NATURAL HISTORY:

Jewett, C. F. History of Worcester County, Massachusetts (1879), 137-138.

Paine, Nathaniel. An Account of the Worcester Lyceum and Natural History Association (Worcester, 1876).

——Literary, Scientific and Historical Societies of Worcester (Worcester, 1898).

——The Worcester Lyceum and Natural History Association (Worcester, 1870).

YOUNG NATURALISTS' SOCIETY:

Henrick, C. J. "The Young Naturalists' Society," *Scientific Monthly,* LIV (1942), 251-258.

Selected References to the History of Science in America

Abelson, P. H. "Science and the 1964 Election," *Science,* CXLVI (1964), 17.

American Philosophical Society, *Proceedings,* LXXXVI, no. 1 (1942), and *ibid.,* LXXXVII, no. 1 (1943), "The Early History of Science and Learning in America."

Barber, Bernard. Science and the Social Order (Glencoe, Illinois, 1952).

Baxter, J. P. Scientists against Time (Boston, 1946).

Beard, C. A., edit. A Century of Progress (New York, 1933). See the chapters devoted to scientific progress in America.

Bedini, S. A. Early American Scientific Instruments and Their Makers. Smithsonian Institution Museum of History and Technology, *Bulletin* 231 (Washington, D. C., 1964).

Bell, E. T. The Development of Mathematics (New York, 1940).

Bell, W. J., Jr. Early American Science: Needs and Opportunities for Study (Williamsburg, Virginia, 1955).

Bidwell, P. W., and J. I. Falconer. History of Agriculture in the Northern United States 1620-1860. Contributions to American Economic History, no. 5 (1925).

Bridenbaugh, Carl. The Colonial Craftsman (New York, 1950).

Browne, C. A., edit. A Half Century of Chemistry in America, 1876-1926. Jubilee Number, American Chemical Society, *Journal,* XLVII, no. 8A (1926).

Cajori, Florian. The Early Mathematical Sciences in North and South America (Boston, 1928).

Century of Science in America, A, with Special Reference to the American Journal of Science, 1818-1918 (New Haven, 1918).

Chamberlin, T. C. "Seventy-five Years of American Geology," *Science,* n. s., LIX (1924), 127-135.

Chittenden, R. H. The Development of Physiological Chemistry in the United States (American Chemical Society, Monograph Series, New York, 1930).

Cohen, I. B. Some Early Tools of American Science (Cambridge, Mass., 1950).

Coolidge, W. D. "Seventy Years of Physical Science," *Popular Science,* CXL, no. 5 (1942), 52-57, 198-202.

Cox, D. W. America's New Policy Makers. The Scientists Rise to Power (Philadelphia, 1964).

Dana, E. S., and others. A Century of Science in America (New Haven, 1918). This refers especially to the history of the *American Journal of Science.*

Dupré, J. S., and Lakoff, S. A. Science and the Nation; Policy and Politics (Englewood Cliffs, N.J., 1962).

Dupree, A. H., edit. Science and the Emergence of Modern America, 1865-1916 (Chicago, 1963).

Edwards, E. E. "American Agriculture—the First 300 Years," in Yearbook of Agriculture for 1940, 171-276.

——"A Bibliography of the History of Agriculture in the United States," United States Department of Agriculture, Miscellaneous Publication, no. 84 (Washington, 1930).

Fairchild, H. L. "The Development of Geologic Science," *Scientific Monthly*, XIX, no. 1 (1924), 77-101.

Fenton, C. L., and M. A. Fenton. Giants of Geology (Garden City, New York, 1952).

First Century of the Republic, The (New York, 1876). See F. A. P. Bernard, "The Exact Sciences," and T. Gill, "Natural Science."

Fuller, Edmund. Tinkers and Genius; the Story of Yankee Invention (New York, 1955).

Gilpin, Robert. American Scientists and Nuclear Weapons Policy (Princeton, N.J., 1962).

Goode, G. B. "The Beginnings of Natural History in America," Smithsonian Institution, Annual Report, 1897, U. S. National Museum, pt. II, 355-406; and "The Beginnings of American Science," Smithsonian Institution, Annual Report, 1897, U. S. National Museum, pt. II, 409-466.

Gray, L. C. History of Agriculture in the Southern United States to 1860 (2 vols., Washington, 1933).

Greenberg, D. S. "Venture into Politics; Science and Engineers in the Election Campaign" (I) *Science*, CXLVI (1964), 1440-1444; and (II), 1461-1463.

Havemeyer, Loomis, edit. Conservation of Our Natural Resources (New York, 1936).

Hindle, Brooke. The Pursuit of Science in Revolutionary America, 1735-1789 (University of North Carolina, 1956).

Howard, L. O. "A History of Applied Entomology," Smithsonian Miscellaneous Collections, Vol. LXXXIV (Washington, 1930).

Hull, T. G. "A Century of Progress in Medicine," *Hygeia*, XI (1933), 1109-1112, 1144-1146.

Kilgour, F. G. "The Rise of Scientific Activity in Colonial New England," *Yale Journal of Biology and Medicine*, XXII (1949), 123-138.

Kremers, Edward. History of Pharmacy (Philadelphia, 1940), pt. III, "Pharmacy in the United States," 125-324, esp. ch. 12, "The Growth of Associations," 173-196.

Lamb, A. B. "A Century of Progress in Chemistry," *Science*, n. s., LXXVIII (1898), 307-320.

Langer, Elinor. "Science Goes to Lunch," *Science*, CXLVI (1964), 1145-1149. Describes the impact of the Cosmos Club on government-sponsored science.

McAdie, Alexander. The Principles of Aërography (Chicago and New York, 1917), ch. I, "A Brief History of Meteorology," 1-24.

McGee, W. J. "Fifty Years of American Science," *Atlantic Monthly,* LXXXII (1898), 307-320.

Merrill, G. P. The First One Hundred Years of American Geology (New Haven, 1924); "Contributions to the History of American Geology" in U. S. National Museum, Annual Report . . . for 1904 (Washington, 1906), 189-233; edit. and comp., Contributions to a History of American State and Natural History Surveys (Washington, 1920).

Miner, W. H. "Some Notes on the Beginning of American Science," *Magazine of History,* X (1909), 282-286; XI (1910), 27-32, 129-132; XII (1910), 261-272; XIII (1911), 87-94.

Morison, S. E. The Intellectual Life of Colonial New England (New York, 1956), ch. X, "Scientific Strivings," 241-275.

Nye, R. B. The Cultural Life of the New Nation, 1776-1830. The New American Nation Series (New York, 1960).

Oliver, J. W. History of American Technology (New York, 1956).

Packard, F. R. History of Medicine in the United States (2 vols., New York, 1931).

Poore, B. P. "History of the Agriculture of the United States," Annual Report for 1866 of the U. S. Commissioner of Agriculture (Washington, 1867), 498-527.

Rafinesque, C. S. "Survey of the Progress and Actual State of Natural Sciences in the United States of America, from the Beginning of this Century to the Present Time," *American Monthly Magazine and Critical Review,* II, no. 2 (1817), 81-88.

Raisz, Erwin. General Cartography (New York, 1938), ch. IV, "American Cartography," 57-70.

Rand, Christopher. Cambridge, U.S.A.; Hub of a New World (New York, 1964).

Rice, W. N. "The Contributions of America to Geology," *Science,* XXV (1907), 161-175.

Savelle, Max. The Colonial Origins of American Thought (Princeton, N.J., 1964).

——Seeds of Liberty; the Genesis of the American Mind (New York, 1948).

Schlesinger, A. M., Jr., and Morton White, edits. Paths of American Thought (Boston, 1963).

Schuchert, Charles. "A Century of Geology—the Progress of Historical Geology in North America," *American Journal of Science,* ser. 4, XLVI (1918), 45-103.

Sedgwick, W. T., and H. W. Tyler. A Short History of Science (New York, 1939). See especially chs. XV, XVI, and XVII.

Silliman, Benjamin. ". . . Progress of Science in America," *American Journal of Science,* L (1847), preface, 3-9.

Smith, D. E. History of Mathematics (2 vols., Boston, 1923).

Smith, D. E., and Jekuthiel Ginsburg. A History of Mathematics in America before 1900. Carus Mathematical Monographs, published by the American Mathematical Association, no. 5 (Chicago, 1934).

Smith, E. F. Chemistry in America (New York, 1914).

Struik, Dirk. Yankee Science in the Making (Boston, 1948).

True, A. C. A History of Agricultural Education in the United States (Washington, 1929).

——A History of Agricultural Experimentation and Research in the United States, 1607-1925; Including a History of the United States Department of Agriculture. United States Department of Agriculture, Miscellaneous Publication, 251 (Washington, 1937).

——A History of Agricultural Extension Work in the United States, 1785-1923. United States Department of Agriculture, Miscellaneous Publication, 15 (Washington, 1928).

Waterfield, R. L. A Hundred Years of Astronomy (London, 1938).

Williams, H. S. The Story of Nineteenth Century Science (New York and London, 1900). This work gives a brief survey of the progress made in many fields of science during the nineteenth century.

Woodruff, L. L., edit. The Development of the Sciences (New Haven, 1923).

Wright, Louis. The Cultural Life of the American Colonies. The New American Nation Series (New York, 1957).

Young, R. T. Biology in America (Boston, 1923).

Biographical Guides to American Men of Science

American Men of Science. Edit. by Jacques Cattell, 9th ed., Vol. I: Physical Sciences (Lancaster, Pa., 1955, and New York, 1955); Vol. II: Biological Sciences (Lancaster, Pa., 1955, and New York, 1955); Vol. III: The Social and Behavioral Sciences (New York, 1956).

Appleton's Cyclopedia of American Biography (6 vols., New York, 1887-1889).

Bell, W. J., Jr. Early American Science Needs and Opportunities for Study (Williamsburg, Virginia, 1955). Contains biographical and bibliographical material on "Fifty Early American Scientists," 45-80.

Bibliography of American Natural History. Comp. by Max Meisel (3 vols., New York, 1924-1929). "Selected Bibliography of Biographies . . . of the Principal American Naturalists of the Pioneer Century and of Colonial Times," I, 156-244.

Carnegie Library of Pittsburgh. Men of Science and Industry: A Guide to the Biographies of Scientists, Engineers, Inventors and Physicians in the Carnegie Library of Pittsburgh (Pittsburgh, 1915).

Darrow, F. L. Masters of Science and Invention (1923).

Dictionary of American Biography. Edit. by Allen Johnson and Dumas Malone (Centenary ed., 20 vols., New York, 1946).

Harvard Guide to American History. Edit. by Oscar Handlin and others (Cambridge, Mass., 1954).

Hubert, P. G. American Inventors Past and Present (1896).

Jaffe, Bernard. Men of Science; the Role of Science in the Growth of Our Country (New York, 1944).

Jordan, D. S., edit. Leading American Men of Science (New York, 1910).

National Cyclopedia of American Biography, Vols. I-XXI (New York, 1892).

Preston, Wheeler. American Biographies (New York, 1940).

Who's Who in America, I— (Chicago, 1899—).

Youmans, W. J. Pioneers of Science in America (New York, 1895).

Selected References to American Scientific Agencies Other Than Scientific Societies

Allen, E. W. "Coöperation with the Federal Government in Scientific Work," National Research Council, *Bulletin,* no. 26 (Washington, 1922).

American Association of Museums. Handbook of American Museums (Washington, 1932).

American Foundations News Service, 1949— (New York, 1949—).

Angell, J. R. "Development of Research in the United States," National Research Council, *Reprint and Circular Series,* no. 6 (Washington, 1919).

Bartlett, H. R. "The Development of Industrial Research in the United States," Research—A National Resource, Section 2, 19-77, *Report of the National Research Council to the National Resources Planning Board* (April, 1941).

Bragg, L. M. "The Birth of the Museum Idea in America," *Charleston Museum Quarterly,* I, no. 1 (1923), 3-13.

Bush, Vannevar. "OSRD and Joint Research and Development Board," *Science,* CV (1947), 89-90.

Carmichael, Leonard. "The National Roster of Scientific and Specialized Personnel," Progress Reports, *Science,* n. s., XCII (1940), 135-137, XCIII (1941), 217-219, and XCV (1942), 86-89.

——"The National Roster of Scientific and Specialized Personnel," *Scientific Monthly,* LVIII (1944), 141-152.

Clifford, C. W., comp. Bibliography of Museums and Museology (New York, 1933).

Coleman, L. V. The Museum in America (3 vols., Washington, 1939).

Consolazio, W. V., and M. C. Green. "Federal Support of Research in the Life Sciences," *Science,* CXXIV (1956), 522-526.

Cox, D. W. America's New Policy Makers: The Scientists' Rise to Power (Philadelphia, 1964).

Dupree, A. Hunter. Science in the Federal Government; a History of Policies and Activities to 1940 (Cambridge, Massachusetts, 1957).

Farrington, O. C. "The Rise of Natural History Museums," American Association of Museums, *Proceedings,* IX (1915), 36-53.

Goode, G. B. "Museum History and Museums of History," American Historical Association, *Papers,* III (1889), 497-519.

Hale, G. E. "A National Focus of Science and Research," National Research Council, *Reprint and Circular Series*, no. 39 (Washington, 1932).

Handbook of Scientific and Technical Awards in the United States and Canada, M. A. Firth, edit. (New York, 1956).

Hayes, C. B. "The American Lyceum: Its History and Contribution to Education," United States Department of the Interior, Office of Education, *Bulletin*, no. 12 (Washington, 1932).

Historical Societies in the United States and Canada: A Handbook (Washington, 1944). Earlier editions, 1926 and 1936.

Hull, Callie, comp. Industrial Research Laboratories of the United States, 7th ed. National Research Council, *Bulletin*, no. 104 (Washington, 1940).

Hull, Callie, and C. J. West, comps. Funds Available in the United States for the Support and Encouragement of Research in Science and Its Technologies. National Research Council, *Bulletin*, no. 66 (Washington, 1928).

Jameson, J. F. History of Historical Societies (Savannah, 1914).

Jewett, F. B. "The Mobilization of Science in National Defense," *Science*, n. s., XCV (1942), 235-241.

Kellogg, Vernon. "The University and Research," National Research Council, *Reprint and Circular Series*, no. 19 (Washington, 1921).

Klopsteg, P. E. "Role of Government in Basic Research," *Science*, n. s., CXXI (1955), 781-784.

Means, J. H. Doctors, People and Government (Boston, 1953).

Medical Research: A Midcentury Survey, Vol. I: American Medical Research in Principle and Practice (New York, 1955).

Mees, C. E. K. "Organization of Industrial Scientific Research," *Nature*, XCVII (1916), 411-413, 431-434. (Published separately, New York, 1920.)

Merrill, F. J. H. Natural History Museums of the United States (Albany, 1903).

"Museums of the United States (a Directory)," *Museum Work*, V (1923), 103-120; VIII (1926), 121-155.

"National Institute of Health, The," *Science*, CVII (1948), 615.

National Roster of Scientific and Specialized Personnel. Report of the National Roster of Scientific and Specialized Personnel to the National Resources Planning Board (Washington, 1943).

National Science Foundation. Employment Profile of Scientists in the National Register of Scientific and Technical Personnel, 1954-1955 (Washington, 1956).

——Federal Funds for Science IV. The Federal Research and Development Budget, Fiscal Years 1954, 1955, and 1956 (Washington, 1955).

——Federal Grants and Contracts for Unclassified Research in the Life Sciences, Fiscal Year 1952 (Washington, 1954).

——Federal Grants and Contracts for Unclassified Research in the Life Sciences, Fiscal Year 1954 (Washington, 1955).

——Federal Support for Science Students in Higher Education (Washington, 1956).

——List of International and Foreign Scientific and Technical Meetings, 1955-1958 (Washington, 1955).

——Organization of the Federal Government for Scientific Activities (Washington, 1956).

"Natural History Museums of the United States and Canada," New York State Museum, *Bulletin*, LXII (1903), 105-110.

Perry, E. D. "The American University," N. M. Butler, edit., Education in the United States, I (1900), 253-318, 297-300; lists publications issued at leading universities and shows the intimate relation between the universities and the societies.

Rea, P. M. Directory of American Museums of Art, History and Science. Buffalo Society of Natural Sciences, *Bulletin*, X, no. 1 (1910).

Reverdin, Henri. Principal American Foundations. League of Nations Committee on Intellectual Cooperation, Brochure, no. 11. This contains an account of the Smithsonian Institution and its branches (Smithsonian Miscellaneous Collections, LXXVI, no. 4, Washington, 1923). This report gives a detailed account of the aid which the Smithsonian Institution gave to various projects.

Rich, W. S. American Foundations and Their Fields (7th ed., New York, 1955).

"Science and Loyalty," *Science*, CXXIII (1956), 651.

"Scientific Manpower Commission," *Science*, CXXII (1955), 1213.

Stewart, Irvin. Organizing Scientific Research for War: The Administrative History of the Office of Scientific Research and Development (Boston, 1948).

Voris, Le Roy. "Agricultural Research Institute," *Science*, CXVI (1952), 3.

West, C. J., and E. L. Risher, comps. Industrial Research Laboratories of the United States, Including Consulting Research Laboratories. National Research Council, *Bulletin*, no. 60 (Washington, 1927).

White, L. D. "Scientific Research and State Government," National Research Council, *Reprint and Circular Series*, no. 61 (Washington, 1925).

Wolfle, Dael. America's Resources of Specialized Talent; a Current Appraisal and a Look Ahead. The Report of the Commission on Human Resources and Advanced Training (New York, 1954).

INDEX

Abbot, C. G., 130, 140, 153
Abilene Geological Society, 205
Abstracting Board of the International Council of Scientific Unions, 221
Academia Naturae Curiosum, 1
Academia Secretorum Naturae, 1
Académie des États-Unis de l'Amérique, 11, 12, 13, 14, 27
Académie des Sciences, 1, 78
Académie Française, 11
Académie Internationale d'Histoire des Sciences, 165, 166
Academy of Arts and Sciences, 25
Academy of Dentistry for the Handicapped, 232
Academy of Medicine of Philadelphia, 19, 65
Academy of Natural Sciences of Philadelphia, 43, 44, 68, 176
Academy of Psychosomatic Medicine, 213
Academy of Science and Art of Pittsburgh, 120
Academy of Time, 200
Academy Society (Maryland), 15
Accademia dei Lincei, 1
Accademia del Cimento, 1
Accademia delle Scienze, 1
Acoustical Society of America, 146
Adams, C. F., 123
Adams, J. Q., 46, 52, 69
Adams, John, 9, 10, 11, 58, 77
Adams, Samuel, 11
Adams Papers, 196
Addams, Jane, 107
Aerial Phenomena Research Organization, 229
Aero Club, 106
Aero Medical Association of the United States, 151
Aeronautical Chamber of Commerce of America, 148
Aeronautical societies, 148
Affiliated Chiropodists-Podists of America, 232
Agassiz, Alexander, 90, 122, 123
Agassiz, Louis, 33, 37, 48, 69, 73, 75, 76, 77, 79, 80, 81, 88, 90, 108

Agassiz Association, 110
Agricultural Aircraft Association, 231
Agricultural History Society, 150
Agricultural Relations Council, 232
Agricultural Research Institute, 209, 231
Agricultural societies, 57-63, 114-115, 150
Agricultural Society of South Carolina, 22
Aigster, F., 53
Air Conditioning, Heating, and Ventilating Engineers of Baltimore, 148
Air Force and Space Digest, 226
Air Force Association, 226
Air Pollution Control Association, 204
Airy, Sir George Biddell, 82
Aitken, R. G., 139
Alabama Academy of Science, 152
Alabama Anthropological Association, 112
Alabama Industrial and Scientific Society, 120
Alabama Pharmaceutical Association, 118
Alabama Psychological Association, 215
Alabama Society of Engineers, 106
Alamogordo, N. M., 224
Albany Institute, 23
Albany Institute of History and Art, 42
Albany Lyceum of Natural History, 23, 42
Alembert, Jean le Rond d', 30
Alexander Dallas Bache Fund, 130
Allegheny Astronomical Society (Pittsburgh), 52
Alliance of Universities and Learned Societies, 122
Allport, Floyd, 141
Alpha Delta Epsilon Scientific Fraternity, 133
Aluminum Research Institute, 149
Amaryllis Society, 150
Amateur Astronomers Association, 142
Amateur Astronomers League, 143, 198-199
Amateur Astronomers of Roanoke, 199
Amateur Astronomical Society (New Orleans), 142